# INDUSTRIAL USES OF
# BIOMASS ENERGY

# INDUSTRIAL USES OF BIOMASS ENERGY

## The example of Brazil

Edited by

### *Frank Rosillo-Calle*

King's College London, University of London, UK

### *Sergio V. Bajay*

State University of Campinas – UNICAMP, Brazil

### *Harry Rothman*

University of the West of England, UK

Routledge
Taylor & Francis Group

LONDON AND NEW YORK

First published 2000 by Taylor & Francis

2 Park Square, Milton Park, Abingdon, Oxfordshire OX14 4RN
52 Vanderbilt Avenue, New York, NY 10017

*Routledge is an imprint of the Taylor & Francis Group, an informa business*

First issued in paperback 2019

Copyright © 2000 Taylor & Francis

Typeset in Sabon by
Keystroke, Jacaranda Lodge, Wolverhampton

*British Library Cataloguing in Publication Data*
A catalogue record for this book is available from the British Library

*Library of Congress Cataloging in Publication Data*
A catalogue record for this book has been requested

ISBN 978-0-7484-0884-9 (hbk)
ISBN 978-0-367-39886-6 (pbk)

# IN MEMORIAM

David Oakley Hall

It is with great sadness that we have to report that a leading contributor to this book, Professor David Oakley Hall, passed away on 22nd August 1999. David Hall was born in South Africa in 1935 and was Professor of Biology at King's College London from 1974.

He was a world leading authority in biomass energy to which he dedicated most of his professional life. He worked tirelessly and published over 400 articles in scientific journals, books, etc., on all areas related to biomass energy, bioproductivity of tropical grasslands, photosynthesis and stress physiology, etc. He lectured in more than 45 different countries and also presented his work on radio and TV and in films, magazines and newspapers.

Those of us who had the privilege of working with him know his endless enthusiasm, driving force, dedication, and his enormous capacity for work. It is in recognition of such enormous contribution to the field of biomass energy, and his personal qualities, that we proudly dedicate this book to him.

The Editors

# CONTENTS

CONTENTS

# FIGURES

# TABLES

# CONTRIBUTORS

**Richard L. Bain,** Senior Program Manager II, National Renewable Energy Laboratory (NREL), USA

**Sergio Valdir Bajay,** Professor of Energy Systems, State University of Campinas (UNICAMP), Brazil

**Ausilio Bauen,** Researcher, King's College, London, UK

**Mauro Donizeti Berni,** Research fellow, Interdisciplinary Centre for Energy Planning (NIPE), State University of Campinas (UNICAMP), Brazil

**Guilherme Bezzon,** State University of Campinas (UNICAMP), Brazil

**Oscar A. Braunbeck,** Faculdade de Engenharia Agrícola (School of Agricultural Engineering), State University of Campinas (UNICAMP), Brazil

**Eliane Bezerra de Carvalho,** Brazilian Regulatory Agency, Electric Power Sector, Brazil

**Luís A. B. Cortez,** Faculdade de Engenharia Agrícola (School of Agricultural Engineering), State University of Campinas (UNICAMP), Brazil

**Kevin R. Craig,** Technology Manager, Biomass Power Program, National Renewable Energy Laboratory (NREL), USA

**André Faaij,** Assistant Professor, Department of Science, Technology and Society, Utrecht University, Netherlands

**André Luis Ferreira,** Research fellow, Interdisciplinary Centre for Energy Planning (NIPE), State University of Campinas (UNICAMP), Brazil

**André Furtado,** Assistant Professor, Science and Technology Policy Department, State University of Campinas (UNICAMP), Brazil

**David O. Hall,** Former Professor of Biology, King's College, London, UK

**Jo I. House,** Research assistant, King's College, London, UK

**Carlos Roberto de Lima,** Lecturer, Department of Forestry Engineering, Federal University of Paraiba, Brazil

**Isaías de Carvalho Macedo,** Director of Industrial Technology, Copersucar Technology Centre, Brazil

**José Roberto Moreira,** Biomass Users Network (BUN) and Biomass Reference Center, São Paulo, Brazil

**Ralph P. Overend,** Principal Scientist, National Renewable Energy Laboratory (NREL), USA

**José Dilcio Rocha,** Visiting Research Fellow, National Renewable Energy Laboratory (NREL), USA

**Frank Rosillo-Calle,** Research Fellow, King's College, London, UK

**Harry Rothman,** Professor of Science and Technology Policy and co-Director of the Science and Technology Policy Unit, University of the West of England, Bristol, UK

**Ivan Scrase,** Consultant in development, energy and the environment

**Arnaldo Walter,** Assistant Professor, Department of Energy, State University of Campinas (UNICAMP), Brazil

# PREFACE

The use of biomass has long been considered synonymous with poverty and under-development. Since little biomass is used in industrialised countries and a lot in developing countries, the message one gets is that advanced countries continue to use 'commercial' fuel (oil, gas, coal, hydro and nuclear), while poor countries continue to use – as they have for decades – 'non-commercial' fuels – a sure sign of under-development.

This book, edited by Rosillo-Calle, Bajay and Rothman, dispels this erroneous view. We know that biomass is the 'oil of the poor' and represents a large proportion of the energy used in many developing countries, but that the primitive ways of using it suffered from the lack of advanced technologies and were therefore characterised by low efficiency. The case of industrial uses of biomass in Brazil, as the book shows, is that such perceptions can and have been dramatically changed.

Biomass is used in Brazil in a highly efficient way to produce large quantities of ethanol (an excellent automotive fuel!). The alcohol programme in Brazil is not a laboratory experiment, but a fully-fledged industrial activity, involving sales of approximately US$5 billion a year, which has replaced half of the gasoline that would otherwise be used in that country.

The other uses of biomass, as wood, in the modern pulp and paper industry, as charcoal, and in the steel industry, as well as electricity generation, are also discussed and the authors consistently prove that modern uses of biomass have come to age.

This book makes clear that the biomass contribution to the world energy matrix is mainly constrained by the question of 'externalities'. The inclusion of social and environmental costs in the final price of energy will strengthen biomass energy uses.

Jose Goldemberg
Chairman of the World Energy Assessment
and former Secretary of State for Education in Brazil
February 1999

# ACKNOWLEDGEMENTS

The editors gratefully acknowledge and thank all contributors for their contribution efforts, and thank the many other people, too numerous to name, who in one way or another have made this book possible.

In particular we thank the British Council and CAPES (Coordenadoria de Aperfeçoamento de Pessoal de Nivel Superior) for their financial contribution to the project.

# NOTE

Please note that throughout this book 'billion' denotes a thousand million.

# ABBREVIATIONS

| | |
|---|---|
| ABRACAVE | Brazilian Association of Charcoal Producers, Belo Horizonte, MG. |
| ACESITA | Company of Special Steels-Itabira, Brazil |
| ANEEL | Brazilian National Agency for Electricity Regulation |
| ANT | Actor Network Theory |
| ARBRE | arable biomass renewable energy |
| ARE | alternative renewable energies |
| ARENA | a software trademark |
| BCL | Battelle Columbus Laboratory |
| BDP | Brazilian demonstration project |
| BDT | bonne-dry tonne |
| BIG-CC | biomass integrated gasification combined cycle |
| BIG-GT | biomass integrated gasification-gas turbine |
| BNDES | National Bank for Economic and Social Development, Brazil |
| BOD | biological oxygen demand |
| boe | barrel oil equivalent |
| BRACELPA | Brazilian Association of Pulp and Paper Producers |
| CCGT | combined cycle gas turbine |
| CEC | Commission of the European Community |
| CEMIG | Electricity Utility of Minas Gerais, Brazil |
| CENBIO | National Reference Center for Biomass, Brazil |
| CEST | condensing extraction steam turbine |
| CETESB | Technical Environmental Company, S.P., Brazil. |
| CFB/ST | circulating fluidised bed and steam turbine |
| CHP | combined heat and power |
| CNPq | Brazilian National Research Council for Technological Development |
| COALBRA | Coke and Alcohol from Wood Company |
| COD | chemical oxygen demand |
| CONAMA | Brazilian Council for the Environment |
| COPERSUCAR | Co-operative of Sugar and Ethanol Producers of the State of Sao Paulo |
| CTA | Constructive Technology Assessment |
| CTC | Centre of Technology, COPERSUCAR, Piracicaba, Brazil |
| DFA | damage-function approach |
| DMB | trademark |
| DMC | direct microbial conversion |

| | |
|---|---|
| DNAEE | National Department of Water and Electricity, Brazil |
| DOE | Department of Energy of the USA |
| ECES | Environmentally Compatible Energy Scenario |
| EGT | European Gas Turbine |
| EJ | exajoules |
| Eletrobras | Brazilian Electricity Utility |
| ENEA | Italian National Agency for New Technology, Energy and the Environment |
| EPRI | Electric Power Research Institute |
| ERF | exposure-response function |
| FAST | Forecasting and Assessment in Science and Technology, EU. |
| FERCO | Future Energy Resources Corporation |
| FFES | Fossil-Free Energy Scenario |
| FINEP | Federal Research Projects Funding Agency, Brazil |
| GIS | Geographical Information Systems |
| GJ | gigajoule |
| GtC | gigatonnes carbon |
| Gtoe | gigatonnes of oil equivalent |
| ha | hectare |
| HECs | herbaceous energy crops |
| HHV | high heating value |
| HNEI | Hawaii Natural Energy Institute |
| HPO | high pressure operation |
| IAA | Alcohol and Sugar Institute, Brazil |
| IAC | Agricultural Research Institute, Campinas, SP, Brazil |
| IBAMA | Brazilian Institute for the Environment and Natural Resources |
| IBDF | Brazilian Institute for Forestry Development |
| IEA | International Energy Agency, Paris |
| IGCC | integrated gasification combined cycle |
| IGT | Institute of Gas Technology, USA |
| IIASA | International Institute for Applied Systems Analysis, Austria |
| IIT | Indian Institute of Technology |
| IPCC | Intergovernmental Panel on Climate Change |
| IPP(s) | independent power producers |
| IRR | internal rate of return |
| ISTIG | intercooled steam injection gas turbine |
| KTP | Cuban trademark |
| kWh | kilowatt hour |
| LCV | low calorific value |
| LHV | lower heating value |
| LPO | low pressure operation |
| MAFLA | Mannesmann Florestal, Brazil |
| MCFC | molten carbonate fuel cell |
| MG | State of Minas Gerais, Brazil |
| Mha | million hectares |
| MICT | Brazilian Ministry of Industry, Trade and Tourism |
| MME | Brazilian Ministry of Mines and Energy |
| MNVAP | Minnesota Valley Alfafa Producers, USA |

| | |
|---|---|
| MSW | municipal solid waste |
| Mtoe | million tonnes oil equivalent |
| NEPA | National Environmental Policy Act, USA |
| NOVEM | The Netherlands Agency for Energy and Environment |
| NREL | National Renewable Energy Laboratory, USA |
| OTA | Office of Technology Assessment of the USA Congress |
| PAH | poliaromatics |
| PICHTR | Pacific International Center for High Technology Research, USA |
| PLANALSUCAR | Brazil's National Plan for Sugarcane |
| PNA | National Alcohol Program, Brazil |
| PORCEL | Brazilian Program for the Reduction for Electric Energy Losses |
| ppb | parts per billion |
| ppm | parts per million |
| PRM | Producer Rice Mill Systems |
| PROALCOOL | Brazilian Alcohol Program. Also, PNA (used interchangeably) |
| PSA | pressure swing adsorption |
| PURPA | Public Utility Regulatory Policies Act, USA |
| PyNE | a pyrolysis database |
| RE | renewable energy |
| RIGES | Renewables-Intensive Global Energy Scenario |
| SETI | Science, Engineering, Technology and Innovation, EU |
| SHF | separate hydrolysis and fermentation |
| SINDIFER | Sindicato da Industria do Ferro (Metal Union), Belo Horizonte, Brazil. |
| SOFC | solid oxide fuel cell |
| SRWCs | short rotation woody crops |
| SSF | simultaneous saccharification and fermentation |
| STIG | steam injected gas turbine |
| stwb | steres with bark |
| t | tonne |
| TCUF | thermochemical users facility |
| toe | tonne oil equivalent |
| tpd | tonnes per day |
| TPS | Thermiska Processor, Sweden |
| TSC | tonne of crashed sugarcane |
| UASB | upflow anaerobic sludge – Blanket Reactor |
| UNCED | United Nations Conference on Environment and Development |
| VTT | Technical Research Center of Finland |
| wb | wet basis |
| WEC | World Energy Council |
| WTA | willingness to accept |
| WTP | willingness to pay |

# INTRODUCTION

Biomass provides a renewable source of energy on a scale sufficient to play a significant role in the development of sustainable energy programmes, so vital if we are to create a greener, more ecologically sensitive society. Despite its ancient image, its full potential remains to be tapped. After a long period of neglect, biomass energy is undergoing a revival of interest and new technical advances are showing that it is capable of becoming more efficient and competitive. Brazil has been at the forefront of the revival of biomass energy systems.

This book is the outcome of a joint project, which took place from 1996 to 1999, between the Faculty of Mechanical Engineering's Energy Department and the Interdisciplinary Centre for Energy Planning, both at the State University of Campinas, São Paulo, Brazil, and the Division of Life Sciences, King's College London, University of London. The project was partially funded jointly by the British Council and its Brazilian counterpart CAPES (Coordenadoria de Aperfeicoamento de Pessoal a Nivel Superior).

The project's main aims were to carry out a detailed study of the three largest industrial sectors that use biomass as a major energy source:

1   production of ethanol fuel from sugar-cane;
2   implication of biomass energy in the pulp and paper industry; and
3   use of charcoal as a thermal and reductor agent in the pig-iron and steel industry.

This book brings together the work and experience of many people who have been involved in this project in field work, workshops, seminars, and so on. Whenever possible, we have tried to involve industrialists directly. The Brazilian experience in the modern industrial applications of biomass energy is unique in many ways. This is a rich experience which we would like to share with a wider public.

However, the book goes further and includes wider aspects of biomass energy. It also presents a global overview of current and potential future use of biomass energy, the potential role of Technology Assessment (TA) in creating a new energy paradigm, the implications of the internalisation of the external costs of energy, together with the potential contribution of new technologies to energy supply and consumption. Biomass energy is currently the main source of energy in many developing countries and is also increasing its role in many industrial countries, albeit for quite different reasons, e.g. environmental and socio-economic. In most global future energy scenarios biomass energy, particularly in its modern forms, plays an increasingly significant role in the energy supply matrix.

We are entering a new millennium with a big question mark against our current unsustainable and highly polluting technological paradigms. We need to change to less polluting energy sources and to sustainable energy technologies, and to a sustainable economic system capable of serving our current and future wants, while ensuring a sound and sustainable environmental and ecological future.

Biomass energy technologies can play a significant role in moving toward such alternative energy paradigms. For instance, significant breakthroughs are being made in genetic engineering and processes which may radically change the viability of ethanol as a transportation fuel. The large-scale production and use of biomass as a modern energy carrier could have very significant positive socio-economic and environmental impacts. It could, for example, democratise the world's energy market and better redistribution of resources by allowing many millions of farmers to contribute to energy supply. This could bring substantial social and economic benefits to many rural areas, creating greater energy self-sufficiency and stability. At regional and global levels the environmental benefits of biomass energy could also be very significant.

Thus, it is clear that biomass energy is not going away, as was often predicted two decades or so ago. Biomass energy is not necessarily the 'poor man's fuel', its role is rapidly changing for a combination of environmental, energy, climatic, social and economic reasons. It is increasingly becoming the fuel of the environmentally-conscious, rich society.

The use of biomass energy has many pros and cons, which are presented in this book. One of the major barriers confronting renewable energy is that the conventional fuels do not take into account the external costs of energy, such as environmental costs. It is important to create a new situation in which all sources of energy are put on a more 'equal footing'. For biomass energy, which has little or no environmental costs, the internalisation of the cost of energy could be a major determinant for its large-scale implementation. This, together with agricultural productivity and technological advances, could be a key determinant in ensuring greater competitiveness with fossil fuels.

In Chapter 1 Hall *et al.* present an excellent global analysis of current and future potential uses of biomass energy. The authors also explore the wider social, economic, agricultural and environmental implications of biomass energy.

Chapter 2 presents an overview of energy in Brazil, with particular emphasis on biomass energy in the sugar-cane, pulp and paper and the charcoal-based industrial sectors. It includes an analysis of the recent political and institutional changes of the energy sector in the country.

Chapter 3 explores the role of Technology Assessment in helping to create a new energy technology paradigm, with emphasis on Brazil through various case studies.

Externalities evaluation has received considerable attention during recent years and this is likely to increase even further in the future. The externality issue will not be going away since it can have major implications for the future supply of energy. These issues are discussed in Chapter 4.

Brazil is the world's largest producer of sugar-cane and the implications of this culture, current practices and recent technological advances in the Brazilian sugar and alcohol industry are taken up by Cortez and Braunbeck in Chapter 5. Chapter 6 examines in detail the industrial processes of the sugar and ethanol plants, together with past and prospective future technological developments.

Pulp and paper making is addressed in Chapter 7. An extensive set of issues is discussed under three main headings: sustainable forestry, energy production and use and the environmental impacts, and finally paper recycling.

Chapter 8 deals with industrial uses of charcoal. Brazil is the world's largest producer and consumer of industrial charcoal. This has major social, economic and environmental implications. This industry, however, faces an uncertain future unless it modernises to respond effectively to changing economic and environmental circumstances.

Chapter 9 presents a detailed overview of technologies for converting biomass to modern energy carriers, under four major technological categories: gasification, electricity generation, ethanol from cellulosic material, and pyrolysis.

Finally Chapter 10 summarises the main conclusions of the book, and indicates future directions.

The editors

# 1

# OVERVIEW OF BIOMASS ENERGY

*David O. Hall, Jo I. House and Ivan Scrase*

## Summary

Biomass currently supplies about one-third of the developing countries' energy – varying from about 90 per cent in countries like Uganda, Rwanda and Tanzania, to 45 per cent in India, 30 per cent in China and Brazil, and 10–15 per cent in Mexico and South Africa. These percentages are changing only slightly even as countries use more commercial/fossil fuels. The crucial questions are whether the two billion or more people who are now dependent on biomass for energy will actually decrease in numbers in the next century, and what are the continuing consequences to development and environment (local and global) from this dependence 'forever' on biomass. The World Bank in 1996 recognised that 'energy policies will need to be as concerned about the supply and use of biofuels as they are about modern fuels'.

A number of developed countries derive a significant amount of their primary energy from biomass: USA 4 per cent, Austria 14 per cent, Sweden 18 per cent and Finland 20 per cent. Presently biomass energy supplies at least 2 EJ/year in Western Europe, which is about 4 per cent of primary energy (54 EJ). Estimates show a likely potential in Europe in 2050 of 9.0–13.5 EJ depending on land areas, yields, and recoverable residues. This biomass contribution represents 17–30 per cent of projected total energy requirements up to 2050.

There is considerable potential for the modernisation of biomass fuels to produce convenient energy carriers, such as electricity, gases and transportation fuels, whilst continuing to provide for traditional uses of biomass; this modernisation of biomass and the industrial investment is already happening in many countries. When produced in an efficient and sustainable manner, biomass energy has numerous environmental and social benefits compared with fossil fuels. These include improved land management, job creation, use of surplus agricultural land in industrialised countries, provision of modern energy carriers to rural communities of developing countries, a reduction of $CO_2$ levels, waste control, and nutrient recycling. Greater environmental and net energy benefits can be derived from perennial and woody energy cropping than from annual arable crops which are short-term alternative feedstocks for fuels. Agroforestry systems can play an important role in providing multiple benefits to growers and the community, besides energy. In order to ameriolate $CO_2$ emissions, using biomass as a substitute for fossil fuels (complete replacement, co-firing, etc.) is more beneficial from a socio-economic perspective than sequestering the carbon in forests.

Case studies are presented for several countries and the constraints involved in modernising biomass energy and the potential for turning them into entrepreneurial opportunities are discussed. It is concluded that the long-term impacts of biomass programmes and projects depend mainly on ensuring income generation, environmental sustainability, flexibility and replicability while taking account of local conditions and providing multiple benefits which is an important attribute of agroforestry systems. Biomass for energy must be environmentally acceptable in order to ensure its widespread adoption as a modern energy source. Implementation of biomass projects requires governmental policy initiatives that will internalise the external economic, social and environmental costs of conventional fuel sources so that biomass fuels can become competitive on a 'level playing field'.

## 1.1 Introduction

The world derives one-fifth of its energy from renewable resources – 13–14 per cent from biomass and 6 per cent from hydro (Figure 1.1). In the case of biomass this represents about 25 million barrels of oil per day (55 EJ/year). In developing countries it is the most important source of energy (33 per cent of total) for the three-quarters of the world's population who live in them. Because biomass survey data are not available for many countries, and because existing databases are often not maintained and updated, accurate values for the global use of biomass are not available. In some developing countries, biomass provides 90 per cent or more of total energy. Biomass is also used for energy in some industrialised countries such as the United States (4 per cent, equivalent in energy content to 1.9 million barrels of oil per day), Austria (14 per cent), and Sweden (18 per cent).

Biomass is generally and wrongly regarded as a low status fuel, and rarely finds its way into the energy statistics when, in fact, it should be considered as a renewable equivalent to fossil fuels. It offers flexibility of fuel supply, due to the range and diversity of fuels which can be produced. Biomass can be burnt directly to produce electricity or heat, or it can be converted into solid, gaseous and liquid fuels using conversion technologies such as fermentation to produce alcohols, bacterial digestion to produce biogas, and gasification to produce a natural gas substitute. Industrial, agricultural and forest residues can be used as a biomass source, or energy crops, such as trees and sugar-cane, can be grown specifically for conversion to energy.

Biomass use through the course of history has varied considerably, greatly influenced by two main factors: population size and resource availability. Since the annual photosynthetic production of biomass is about eight times the world's total energy use, and this energy can be produced and used in an environmentally sustainable manner, while emitting no net $CO_2$, there can be little doubt that this potential source of stored energy must be carefully considered in any discussion of present and future energy supplies. The fact that nearly 90 per cent of the world's population will reside in developing countries by about 2050 probably implies that biomass energy will be with us into the foreseeable future, unless there are drastic changes in the world energy trading patterns.

Information on biomass production and use patterns is grossly inadequate even as a basis for informed guesses, let alone the making of policy and the implementation of plans. In the 1980s there were extensive debates as to whether a number of countries

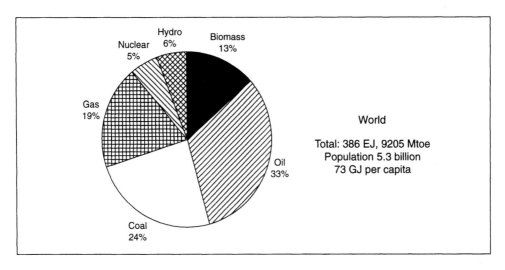

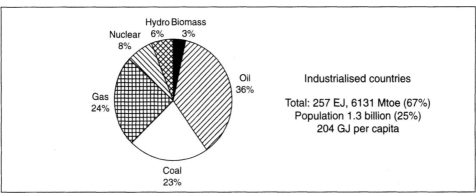

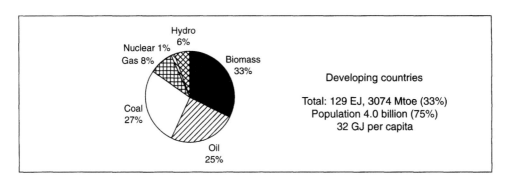

*Figure 1.1* World energy consumption
Source: Hall and House (1995)

**Industrialised countries:** BP Statistical Review of World Energy 1990, British Petroleum PLC, UK;
(industrialised countries = all OECD countries plus E. Europe and ex-USSR).
**Developing countries:** 1990 UN Energy Statistics Yearbook, United Nations, New York, 1992.
Biomass: 1987 data from Biomass Users Network (BUN) (Rosillo-Calle and Hall, 1991). Nuclear and
hydro converted on basis of oil equivalent to produce same amount of electricity.
Mtoe = million tonnes of oil equivalent

3

had a 'woodfuel gap' between fuelwood supplies from forests and woodlands and the use of fuelwood. Would the last tree disappear in Rwanda or Tanzania, to name two examples, by 1987 or 1990, as was calculated by various protagonists? Of course, there are still many trees standing in these countries! We have little idea, however, why the original calculations of biomass supply and use were wrong. Nor do we know how data can be gathered more efficiently, when increasing concern about biomass energy is not matched by the funding available, compared to the support on offer for analysing commercial fuel production and use.

One of the problems in concentrating on 'fuelwood' or 'firewood' as the biomass sources is that it ignores charcoal and the extensive and seasonal use of crop residues, forest residues and dung in many countries. Such residues can amount to as much as 30–40 per cent of total biomass supply. Also, fuelwood is not necessarily cut trees and branches, but very often consists mainly of twigs and small branches, harvested without serious damage to trees. Thus biomass supply and use statistics must include all forms of biomass monitored over the seasons on an annual basis. These data should be available on a time-series basis for use in discerning trends and planning future action. Naturally, biomass use data must be combined with accurate biomass supply information so that national and regional balances can be monitored to ascertain whether resource depletion is occurring, or whether a sustainable biomass supply is plausible. How to make people adapt to limiting biomass supplies so as to ensure sustainability is also an important question.

It is also often not recognised, or documented, that biomass is used as an energy source not only for cooking in households, many institutions and service industries, but also for agro-industry processing and in the manufacture of bricks, tiles, cement, fertilisers, etc. These non-cooking uses can be numerous, especially in and around towns and cities. Also, rural-based village and small-scale industries are frequently biomass energy driven and play an important role in rural and national economies and are increasingly a focus of industrial development policies. Biomass energy is likely to remain the major source of heat energy for these industries for many years, since it is, at present, the only medium-to-large-scale heat-energy resource which is economically viable and potentially sustainable. In India, for example, these industries account for as much as 50 per cent of the manufacturing sector and provide a large part of national employment, second only to the agricultural sector.

Biomass energy use, particularly in its traditional forms, is difficult to quantify, thus creating additional problems. There are two major reasons for this:

1  biomass is generally a lowly regarded fuel and thus rarely finds its way into official statistics, and when it does it tends to be downgraded. Traditional uses of bioenergy, e.g. fuelwood, charcoal, animal dung, and crop residues, are inaccurately associated with problems of deforestation and desertification. For example, in Central Zambia, the main charcoal producing area of the country, there was no evidence of land degradation due to deforestation caused by woodfuel harvesting either for firewood or for charcoal;
2  difficulties with measuring, quantifying, and handling biomass since it is a dispersed energy source, together with its inefficiency of use, results in little final energy being obtained.

Charcoal, for example, is a very important fuel in many developing countries but charcoal yields are notoriously low, e.g. about 12 per cent in Zambia, and 8–10 per cent in Rwanda on a dry weight basis. There is great potential for increasing charcoal production efficiency, e.g. in Brazil the best kilns have an efficiency of about 35 per cent.

The relatively poor image of biomass is now changing for three main reasons:

1   considerable efforts over recent years to present a more balanced and realistic picture of the existing use and potential of biomass through new studies, demonstrations, and pilot plants;
2   increasing utilisation of biomass as a modern energy carrier, particularly in industrialised countries;
3   increasing recognition of the local and global environmental advantages of biomass and of the measures necessary to control net $CO_2$ and sulphur emissions.

Contrary to the general view, biomass utilisation worldwide remains steady or is growing for two reasons: 1) population growth, and 2) urbanisation and improvement in living standards. As living standards improve, many people in rural and urban areas in developing countries convert to different uses of biomass, e.g. charcoal and wood instead of residues and twigs, building materials, cottage industries, etc. Thus, urbanisation does not necessarily lead to a complete shift to fossil-based fuels.

Although present biomass energy-use is predominantly domestic use in the rural areas of developing countries, it is increasingly realised that it also provides an important fuel source for the urban poor and for many small-to-medium-scale industries. We should thus address the issues of, first, equity for the poor, and especially for women who are important factors in biomass energy provision and use through their cooking activities; second, environmentally sustainable land-use, which requires that biomass in all its forms for food, fuel, etc., be sustainably produced; and, third, development and increased living standards which require increased energy provision. It is therefore imperative that we focus on the efficient production and use of biomass energy so that it can provide modern fuels such as electricity and liquid fuels (as is occurring in a number of more industrialised countries around the world) in addition to the more traditional, and very large role, as a heat supplier.

We can expect demand for biomass to rise considerably in the future, first, because as emphasised already, population growth will be high in developing countries; second, because of greater use in the industrial countries, due partly to environmental considerations; and, third, technologies presently being developed will allow either the production of new or improved biomass fuels, or the improved conversion of biofuels into more efficient energy carriers and thus stimulate demand for feed-stock.

When considering large-scale bioenergy programmes, whether on a global or local scale, the following questions need to be addressed: land availability (short and long term); productivities, species and mixtures; environmental sustainability; social factors; economic feasibility; ancillary benefits; disadvantages and perceived problems.

Perceived problems concern nutrient cycling, fertiliser and pesticide requirements, energy input/output ratios, effects on biodiversity and the landscape, possible contributions to erosion, possible conflicts with food production on high-productivity land, and the level of subsidies required. Many of these problems diminish if biomass energy is seen providing multiple local and global benefits and as a long-term entrepreneurial opportunity for improved land management, based on optimal productivities using minimum inputs of resources while providing environmental and social benefits.

The remainder of this chapter is divided into seven main sections. In Section 1.2 the potential role of biomass as a source of energy in the world in the next century is examined, highlighting seven recent scenario studies. The question of the availability of land is addressed. It is emphasised that biomass must be modernised to provide modern energy services.

Section 1.3 examines the costs of modern forms of biomass energy and the potential markets for such types of bioenergy. The barriers to successful commercialisation of bioenergy systems and the possible effects of carbon taxes are discussed.

Examples of the use of biomass for energy are given in Section 1.4. The examples encompass electricity, combined heat and power, district and commercial heat, biogas, and liquid biofuels using various types of feedstocks.

The question as to the use of forests to sequester $CO_2$ or to use biomass as a substitute for fossil fuels in order to decrease atmospheric $CO_2$ emissions is examined in Section 1.5. The options are complex since they depend on existing forest and land use patterns, but ultimately global environmental health requires substitution for $CO_2$-emitting fuels.

Section 1.6 discusses various environmental issues which need to be addressed when implementing biomass energy projects. These concerns involve energy and carbon balances, nutrient recycling, species selection, biodiversity, soil health, and landscaping. The drawing up of guidelines acceptable to the growers, planners and the public is discussed.

A brief concluding section stresses that certain types of biomass energy are already competitive and fully economic in some markets. In the future a 'level playing field' with respect to other types of fuels will favour biomass energy because of environmental and socio-economic concerns, coupled with the increasing interest of major energy companies and governments concerned with climate change and sustainability issues.

Please note that this chapter has been written on the basis of six articles recently published from our biomass energy group – see Bibliography. All the detailed references are given in these six papers.

## 1.2 Future global scenarios for biomass energy

In the 1990s a number of global energy scenarios have been published which include substantial roles for energy efficiency and renewable energies, while some have studied biomass and incorporated large roles for bioenergy. Here we consider seven such studies.

*The Renewables – Intensive Global Energy Scenario (RIGES)* prepared as part of the UNCED Rio de Janeiro Conference in 1992 proposes a significant role for biomass in the next century. They propose that by 2050 renewable sources of energy could account for three-fifths of the world's electricity market and two-fifths of the market for fuels used directly, and that global $CO_2$ emissions would be reduced to 75 per cent of their 1985 levels and such benefits could be achieved at no additional cost. Within this scenario, biomass should provide about 38 per cent of the direct fuel and 17 per cent of the electricity use in the world. Detailed regional analysis shows how Latin America and Africa might become large exporters of biofuels.

*The Environmentally Compatible Energy Scenario (ECES)* for 2020 assumes that past trends of technological and economic structural change will continue to prevail in the future and thereby serve, to some extent, economic and environmental objectives at the same time. Primary energy supply is predicted to be 12.7 Gtoe (533 EJ) of which biomass energy would contribute 11.6 per cent (62 EJ) derived from wastes and residues, energy plantations and crops, and forests – this excludes traditional uses of non-commercial biomass energy for fuelwood in developing countries.

*A Fossil-Free Energy Scenario (FFES)* was developed as part of Greenpeace International's study of global energy warming. Greenpeace forecast that in 2030 biomass could supply 24 per cent (= 91 EJ) of primary energy (total = 384 EJ) compared to their low estimate of only 7 per cent today (= 22 EJ). The biomass supply could be derived equally from developing and industrialised countries.

*The World Energy Council (WEC)* examined four 'Cases' for global energy supply to 2020 spanning energy demand from a 'low' (ecologically driven) case of 475 EJ to a 'very high' case of 722 EJ, with a 'reference' case total of 563 EJ. In the ecologically-driven case traditional biomass could contribute about 9 per cent of total supply, while modern biomass would supply 5 per cent of the total, equal to 24 EJ or 561 Mtoe. New renewables combined (modern biomass, solar and wind, etc.) could supply 12 per cent of the total. In the high growth case these contributions could be 8 per cent and 5 per cent, respectively, of a higher total supply.

*Shell International Petroleum Company* carried out a scenario analysis of what might be the major new sources of energy after 2020, when renewable energies have progressed along their learning curves and become competitive with fossil fuels. After 2020, in their business-as-usual scenario, the renewables biomass, wind, solar and geothermal become the major new suppliers of energy. In their conservation scenario, where less new energy is needed, biomass becomes the major supplier, with smaller roles for the other renewables.

In both Shell scenarios fossil fuel use expands its market share initially, but after 2020 renewables provide for all further growth. In the business-as-usual (Sustained Growth) scenario total global energy use in 2060 amounts to over 1,500 EJ (compared to 400 EJ today). Only one-third of this is from fossil fuels or nuclear. Biomass provides 221 EJ (14 per cent of the total), with 179 EJ coming from plantations rather than traditional non-traded sources. Solar and wind would provide 260 and 173 EJ, respectively. In the conservation (Dematerialisation) scenario total energy use in 2060 amounts to under 940 EJ, with fossil fuels and nuclear providing 41 per cent of the total. Biomass provides 207 EJ (22 per cent of the total), with 157 EJ from dedicated bioenergy sources. Solar and wind provide 36 and 144 EJ, respectively. Shell have recently set up Shell International Renewables as part of their core business, and will invest US$0.5 billion in biomass and solar energy over the next five years.

*The Intergovernmental Panel on Climate Change (IPCC)* has considered a range of options for mitigating climate change, and increased use of biomass for energy features in all of its scenarios. In their five scenarios biomass takes an increasing share of total energy over the next century, rising to 25–46 per cent in 2100. In the biomass intensive energy scenario, with biomass providing for 46 per cent of total energy in 2100, the target of stabilising $CO_2$ in the atmosphere at present-day levels is approached. Annual

$CO_2$ emissions fall from 6.2 GtC in 1990 to 5.9 GtC in 2025 and to 1.8 GtC in 2100: this results in cumulative emissions of 448 GtC between 1990 and 2100, compared to 1300 GtC in their business-as-usual case.

*The IIASA/WEC (1998) study 'Global Energy Perspectives'* examined three main scenarios – Case A: High Growth; Case B: Middle Course; Case C: Ecologically Driven. Within Case A, three sub-scenarios were envisaged: Case A1 with oil and gas, Case A2 with coal, and Case A3 with biomass and nuclear. In scenario A3 biomass could contribute nearly 17 per cent (316 EJ) of total energy by 2100 and in Scenario C1 (renewables and no nuclear) biomass provides nearly 30 per cent (245 EJ) of a lower total energy (878 EJ for C1 and 1,855 EJ for A3).

*The IEA (1998) study 'World Energy Lookout'* addressed for the first time the current role of biomass energy and its future potential. It is estimated that by 2020 biomass will be contributing 60 EJ (compared to their estimate of 44 EJ today = 11 per cent of total energy) thereby providing 9.5 per cent of total energy supply. The period 1995–2020 will show a 1.2 per cent annual growth rate in biomass provision compared to a 2.0 per cent rate for 'conventional' energy.

Table 1.1 summarises the estimates for future biomass use from these seven studies. The figures should be compared with current global energy use of approximately 400 EJ, about 55 EJ of which are derived from biomass mostly used as a traditional energy source.

The large role biomass is expected to play in future energy supply can be explained by several considerations. First, biomass fuels can substitute more or less directly for fossil fuels in the existing energy supply infrastructure. Intermittent renewables such as

*Table 1.1* The role of modernised biomass in future global energy use (EJ) from Hall and Scrase (1998). Present biomass energy use is about 55 EJ/year

| Scenario | Year of scenario | | |
|---|---|---|---|
| | *2025* | *2050* | *2100* |
| IEA (1998) | 60[a] | – | – |
| IIASA/WEC (1998) | 82[b] | 153[b] | 316[b] |
| | 59[c] | 97[c] | 245[c] |
| Shell (1996) | 85 | 200–20 | – |
| IPCC (1996) | 72 | 280 | 320 |
| Greenpeace (1993) | 114 | 181 | – |
| Johansson *et al.* (1993) | 145 | 206 | – |
| WEC (1993) | 59 | 94–157 | 132–215 |
| Dessus *et al.* (1992) | 135 | – | – |
| Lashof and Tirpak (1991) | 130 | 215 | – |

*Sources:* IEA (1988) and IIASA/WEC (1998), Hall and Scrase (1998)

*Notes*
[a] 2020 (Total primary energy supply)
[b] Scenario A3 (High growth – biomass and nuclear)
[c] Scenario C1 (Ecologically driven – large renewables, no nuclear)

wind and solar energy are more challenging to the ways we distribute and consume energy. Secondly, the potential resource is large (see Section 1.2.2) since land is available which is not needed for food production and as agricultural food yields continue to rise in excess of the rate of population growth. Thirdly, in developing countries demand for energy is rising rapidly, due to population increase, urbanisation and rising living standards. While some fuel switching occurs in this process, the total demand for biomass also tends to increase. Evidence from Myanmar, Madagascar, Zambia and Rwanda, for example, has shown that urbanisation raises the demand for biomass, particularly charcoal for household and industrial use. Even in East Asia and the Pacific, where there has been considerable economic growth and increase in the use of fossil fuels, biomass still accounts for 33 per cent of energy supplies. As noted earlier, the World Bank recently concluded that 'energy policies will need to be as concerned about the supply and use of biofuels as they are about modern fuels . . . [and] . . . they must support ways to use biofuels more efficiently and sustainably'. This is an important change of perspective for the World Bank, though the quote perpetuates the misconception that biofuels cannot necessarily be 'modern'.

### 1.2.1 Biomass as a modernised fuel

There is an enormous untapped biomass potential, particularly in improved utilisation of existing forest and other land resources, higher plant productivity and efficient conversion processes using advanced technologies. Much more useful energy could be extracted from biomass than at present. It can then form part of a matrix of fuel sources offering increased flexibility of fuel supply and energy security. Bioenergy can be used on small and large scales in a decentralised manner, bringing substantial benefits both to rural and urban economies. Job creation is also of importance in industrialised countries, and growing biomass will provide an economically viable use for the land being taken out of production in 'set-aside' schemes aimed at decreasing production of surplus agricultural commodities in Europe, North America, and elsewhere.

The development of large-scale energy production from biomass will probably rely in the future on specifically-grown energy crops, such as tree plantations, sugar-cane, perennial herbaceous plants and grasses, and rape seed, for example. For this to become successful, biomass productivities must be improved since they are generally low; for example, woody species have productivities much less than 5 t (dry weight)/ha/yr without good management. It is now possible with good management, continuous research and breeding, and planting of selected species and clones on appropriate soils to obtain 10–15 t/ha/yr in temperate areas and 15–25 t/ha/yr in tropical countries. Record yields of 40 t/ha/yr have been obtained with eucalyptus in Brazil and Ethiopia. High yields are also feasible with herbaceous (non-woody) crops; for example, in Brazil, the average annual yield of sugar-cane has risen from 47 to 65 t (harvested weight)/ha over the last 15 years while over 100 t/ha/yr are common in a number of areas such as Hawaii, South Africa, Zambia, and Queensland. Current and projected productivities and production cost of plantation biomass in the US have also been calculated along with energy balances. Net yields range from 9 t (oven dry)/ha/yr for switchgrass in 1990 and 14.4 t/ha/yr in 2010 to 18.5 t/ha/yr for energy cane in 1990 and 29.3 t/ha/yr in 2010. Energy balances (ratio of energy output to energy inputs during production) rise from 10.3 and 15.3 in 1990 for switchgrass and hybrid poplar, respectively, to 11.7

and 18.5 in 2010, while costs fall from US$2.7–3.9/GJ in 1990 to US$1.9–2.7/GJ in 2010.

### 1.2.2 Land availability

Land availability is perceived as a constraint to large-scale production of biomass but there are considerable areas potentially available. In tropical countries there are large areas of deforested and degraded lands that would benefit from the establishment of bioenergy plantations. For example, from a fairly detailed analysis of 117 tropical countries it was found that eleven countries were suitable for expansion of the forest area up to 553 Mha. The World Resources Institute found that within fifty tropical countries, 67 Mha could be realistically converted to plantations over the next sixty years, more than 200 Mha could be regenerated and a further 63 Mha are available for agroforestry. Yet another study found that 265 Mha were available for reforestation and 85 Mha were available for agroforestry.

The UN's Food and Agriculture Organisation (FAO) study *Agriculture Towards 2010* assessed the potential cropland resources in over 90 developing countries. It is possible to calculate the likely future need for cropland and subtract this from the estimated total potential agricultural land resource to estimate the theoretical remaining productive land in 2025. The potential energy production on this 'remaining' land was then calculated assuming a biomass productivity of 10 t/ha/yr (150 GJ/ha/yr). As a whole, the developing countries will be using only 40 per cent of their potential cropland in 2025, but there are wide regional variations. Asia (minus China, for which figures were unavailable) will have a deficit of 47 Mha, but yields of most food crops are low and there is great potential for improvement using better genetic strains and management. Agricultural technologies that boost yields, such as improved varieties and management techniques, etc., have not yet reached many rural farmers and could increase yields by 50 per cent world-wide. Africa currently only uses one-fifth of its potential cropland, and would still have 75 per cent remaining in 2025, on which it could theoretically produce nearly ten times its present energy consumption. Latin America, presently using only 15 per cent of its potential cropland, would still have 77 per cent left in 2025, capable of producing nearly eight times its present energy consumption.

Large areas of surplus agricultural land in North America and Europe could become significant biomass-producing areas. In the USA, farmers are paid not to farm about 10 per cent of their land, and in the EU, up to 15 per cent of arable farmland can be 'set-aside', although the percentage has varied over the last five years. Apart from over 30 Mha of cropland already set-aside in the USA to reduce production or conserve land, another 43 Mha of croplands have high erosion rates and a further 43 Mha have 'wetness' problems, which could be eased with a shift to various perennial energy crops. The United States Department of Agriculture estimates that a further 60 Mha may be idled over the next 25 years.

In the European Union (twelve countries), at least 15–20 Mha of good agricultural land (an area the size of England and Wales) is expected to be taken out of production by the years 2000/2010 (and this could reach over 50 Mha in the next century). If all this land were used to plant trees, it would represent an annual sink of 90–120 Mt C for the near future (assuming that oven-dry biomass is about 50 per cent carbon and the EU Biomass for Energy Programme productivity target of 12 t/ha is met). Alternatively, this area of land could provide 3.6–4.8 EJ/yr of biomass energy, displacing 90–120 Mt of carbon emissions from coal, 72–96 Mt of carbon from oil, or 50–67 Mt of carbon from natural gas (7–17 per cent of total EU carbon emissions in 1991).

How much land would theoretically be needed worldwide to meet all of the present energy consumption with biomass (a very unlikely outcome) can be estimated. Assuming a productivity of 12 t/ha/yr is reached in the next century, an energy content of oven-dry biomass of 20 GJ/t, and that recoverable residues are also used, this shows that 950 Mha would be needed to grow biomass energy crops to substitute for fossil fuels in industrialised areas, yet developing countries will 'only' require 305 Mha. Therefore, on a global scale, there is enough land available to allow biomass to make a significant impact on atmospheric carbon levels and energy production. Developing countries could also take economic advantage of their land by producing biomass fuels for export to industrialised regions, or by allowing the richer countries to pay for plantation establishment to offset their carbon emissions, as is already happening in trial schemes. However, this may not be practical in the long run and is probably not equitable unless local benefits arise during the growth of the biomass and the biomass is used to offset fossil fuel.

## 1.3 Costs and markets of modern bioenergy

As pointed out in the *Economist*, 'greenery needs to be profitable as well as virtuous'. The costs of bioenergy are already competitive in certain instances, and as the technologies become more developed they will continue to decline further.

However, most biomass energy technologies have not yet reached a stage where market forces alone can make the adoption of these technologies possible. One of the principal barriers to the commercialisation of all renewable energy technologies is that current energy markets mostly ignore the social and environmental costs and risks associated with fossil fuel use. Conventional energy technologies are able to impose upon society various external costs (such as environmental degradation and health-care expenditures) which are difficult to quantify. Meanwhile, renewable energy sources, which produce few or no external costs and may even cause positive external effects (such as decreased $SO_2$ and $CO_2$ emissions, job creation, rural regeneration and savings in foreign exchange), are systematically put at a disadvantage. Furthermore, conventional energy sources tend to receive large subsidies. Internalising external costs and benefits and re-allocating subsidies in a more equitable manner must become a priority for all renewables to be in a better ('level playing field') position to compete with fossil fuels.

Pursuit of all policy options needs stimulus from appropriate policy instruments, such as sulphur credits and carbon taxes. In Sweden carbon taxes were introduced in 1991 – and were roughly equivalent to US$150/tC, although they have now been reduced. Norway's carbon tax is about US$120/tC. Cost estimates for reducing $CO_2$ emissions vary quite considerably – often by a factor of two or three. An OECD working paper estimated that to cut the output of $CO_2$ by 20 per cent between 1990 and 2010, and to stabilise it thereafter, would need a tax averaging US$210/tC for the world as a whole, which is equivalent to US$36 per barrel of oil (petrol taxes in some European countries are presently equivalent to a carbon tax of US$200/tC). The EU proposed a carbon tax of US$3/barrel of oil in 1993, rising to US$10/barrel of oil in 2000 (which will be equivalent to US$0.05/litre of gasoline or US$0.015/kWh for electricity).

As productivities and conversion technologies are improving, costs are falling, and they will continue to do so as this industry becomes more developed. For example, capital costs of building a plant fall dramatically as more plants are built. The cost of building one prototype BIG-GT gasifier (see below) is in the region of US$3,000/kW, yet the cost of building ten identical plants falls to US$1,300/kW. The Electric Power Research Institute (EPRI) in the US has calculated that woody biomass production costs in 1990 were US$25/t, and this could be reduced to US$15/t in 2010. The cost of electricity generation using higher efficiency conversion technologies and dedicated biomass crops will be around 6 c/kWh.

## 1.4 Examples of bioenergy use and projects

### Finland

Finland obtains over 20 per cent of its primary energy from biomass, with a total biomass consumption of 6.1 Mtoe (million tonnes oil equivalent) in 1995. About 70 per cent is produced from derived fuels (mainly waste liquors from pulping but also firewood and wood waste), 30 per cent from peat. The pulp and paper industries use waste wood and liquor for 60 per cent of their fuel requirements, with modern pulp mills capable of providing all their own energy requirements as well as significant amounts of excess electricity and liquid fuels. District heating has been used since 1952 and supplies more than 40 per cent of the country's heat demand. Over half of the large district heating plants use biofuels, as do many smaller stations, and peat-fired combined heat and power plants are being introduced in some larger towns.

One of the reasons for the success of the bioenergy industry in Finland is significant government support. There are enough biomass resources available to allow for a doubling of their use for energy. Finland's considerable forest resources would allow for a large increase in this amount, as will land released as set-aside. The government set a target to increase the use of bioenergy by 1.5 Mtoe by 2005 thereby reducing $CO_2$ by a further 4.2 Mt corresponding to 6.7 per cent of 1996 levels.

### Sweden

Sweden obtains 18 per cent of its energy (87 TWh or 315 PJ per annum) from biofuels. The use of these fuels can be split into three different sectors:

1   The forest products industry has traditionally converted its by-products into heat and electricity for its own use. In 1996, it obtained 36 TWh from pulp digester liquors, 7 TWh from cellulose waste and 9 TWh from sawmill wastes.
2   Individual households used about 12 TWh of woodfuels in 1996 mostly in the form of logs for heating.
3   The use of biofuels for district heating is growing fast and accounted for 23 TWh in 1996. Of this, 12.4 TWh were obtained from woodfuels (mostly unprocessed), 4.5 TWh were obtained from refuse and 3.5 TWh came from peat. Energy crops

such as trees, straw and grass were also used, but as yet contribute only a very small amount of energy.

There is much more potential to produce energy from indigenous biomass fuels, particularly from agroindustrial wastes and energy crops grown on marginal and other land. Currently over 18,000 ha of short rotation willow is being grown under bioenergy schemes. Sweden has also imported a small amount of biomass fuels indicating the potential for the future development of an international trade in biofuels.

### Austria

Austria obtains over 14 per cent of its primary energy from wood. Bioenergy use has increased by 5 per cent in total over the last decade and is expected to rise further, maybe accounting for 20 per cent or more of total energy consumption in the near future. This is important for a country like Austria which is dependent on imported fuel for about 90 per cent of their oil and 80 per cent of their gas and coal.

Biomass is mostly used for space heating. Over 50 per cent of farmhouses use fuelwood for this purpose, and district heating has been introduced as an efficient and convenient way of heating houses in towns and villages. Austria now has thousands of differently sized biomass heating systems based on woodchips with a combined capacity in excess of 1,500 MW (equivalent to a large coal-fired power station). Over 300 district heating schemes have already been installed (25 per cent annual growth rate 1987/ 97) at an average investment cost of US$1,000/kW, and these may be developed to incorporate combined heat and power (CHP) systems in the future. This has been achieved by political commitments creating and encouraging an institutional framework that includes favourable legislation, capital grants, cheap finance and education. Farmers' groups developing woodfuel-fired district heating schemes can receive grants covering up to 50 per cent of the capital costs.

### Denmark

Denmark is a leading example and its success reflects the country's commitment to renewable energy as a whole, e.g. currently about 8 per cent of the energy consumption comes from renewables (half from biomass) and is expected to reach 35 per cent in 2035 under the latest *Energy 21* (1996) energy plan. Considerable efforts are made to use the annual straw surplus of 3 Mt with a proposed 1.2 Mt straw and 0.2 Mt woodchips to be used by the electric power industries by 2001. The Danish Government programme for CHP plants utilising biomass or natural gas has a target of 450 MWe. More than 50 straw-fired district heating plants totalling 170 MW (thermal) produce energy for district heating schemes at a cost of US$12/GJ, competitive with coal- and oil-fired heating plant. Four co-generation (CHP) plants fired by straw are in operation, ranging in size from 7 MW to 28 MW (thermal), and two more rated at 60 MW (thermal) are under construction. Co-firing of straw and coal is now proceeding and looks environmentally very interesting because of the greatly reduced sulphur emissions.

Over the past ten years the interest has concentrated on large centralised biogas plants because of technological advances and other advantages, e.g. these plants provide facilities for disposing of surplus manure which is a serious environmental problem in some areas of Denmark. As with other renewables, the government provides fiscal incentives for biogas production, e.g. about 20 per cent on funding investments for centralised biogas and 30 per cent for individual plants. In June 1996 there were 18 centralised biogas plants in operation, and others at the planning stage, with an annual input of 1.2 M tonnes of biomass (75 per cent animal manure and 25 per cent organic waste), and biogas production of 40–5 million $m^3$, equivalent to about 25 M $m^3$ of natural gas, about 2 PJ; the objective is 4 PJ in 2000 and 6 PJ in 2005. With the ambitious 'Energy 21' plan it is expected that $CO_2$ emissions will be reduced by 50 per cent by 2030.

## USA

The USA currently obtains over 4 per cent of its energy from biomass (a total of 3.9 EJ). Of this the forest industry self-generates 1.7 EJ, electricity generation amounts to 0.9 EJ, residential use 0.7 EJ, wastes 0.5 EJ, corn ethanol 0.1 EJ. The UN Renewable Intensive Energy Scenario estimated that the contribution of biofuels in the US could rise to 22 EJ by 2025, and then fall to 19 EJ in 2050 as increased energy efficiency reduced energy consumption by about a quarter. At present (1998) there are about 350 grid-connected biomass power plants with a total capacity of about 7,000 MW (down from about 8,500 MW in the 1980s when the industry was created by PURPA and generated 66,000 jobs with an investment of US$15 billion). In 1994 the Electric Power Research Institute (EPRI) and the Department of Energy put out a request for host utilities to propose biomass energy case studies that they will cost share in establishment and production. The reason for this, according to EPRI, is:

> Projections of emerging renewable energy markets in the first part of the 21st century consistently suggest that biomass energy systems will play a major role. . . . The projections are based on the following facts: 1) vast availability of suitable land (tens of millions of acres) for growing energy crops; 2) environmental benefits ($CO_2$ neutral, low $SO_2$, and recyclable ash); and 3) favourable economics compared to coal-derived energy in the 1990s. Further, biomass plantations can help the economy in rural America.

## Brazil

The use of ethanol fuel dates from the early 1900s, but it was not until the 'energy crisis' of the 1970s that a large-scale development of fuel alcohol production from sugar-cane was promoted. In the 1997/8 sugar-cane season about 15 billion litres of ethanol were produced for vehicle use (equivalent to 220,000 barrels of oil a day) and a further 1.6 billion litres for non-energy use. Anhydrous fuel ethanol production was 5.3 billion litres and the hydrated form at 9.6 billion litres. In addition 14.2 million tonnes of sugar were produced. 'Ethanol fuel is important for its economic, social and environmental aspects; savings in oil imports favouring trade balance, over one million job opportunities, and zero net $CO_2$ emissions' (Andrade *et al.* 1998).

It is also significant that Brazil, the eighth-largest industrial country in the world, obtains 21 per cent of its total energy from biomass in the form of ethanol, charcoal and fuelwood – used in the transport, industrial, commercial and residential sectors. Such biomass use combined with hydroelectricity generation means that Brazil obtains over half its energy from non-$CO_2$ emitting fuel sources; car pollution levels have also been reduced by 20 per cent.

It has been estimated that total investment in the ProAlcool industry has been about US$12 billion since 1976 and that this has saved the country about US$29 billion in foreign exchange. The investment has resulted in more than doubling the ethanol yields per hectare and a reduction in costs averaging 3.5 per cent per year over the last 20 years. It is calculated that a further reduction of 23 per cent is possible from combined savings in the agricultural and industrial sectors (Rosillo-Calle and Cortez 1998).

Future developments in efficient combined heat and power generation using surplus bagasse (with direct sales into the grid and local industries) are stimulating investment in Brazil – and elsewhere in the world – as it is realised that sugar-cane production can give multiple energy (and other) products with economical, environmental and social advantages when appropriately managed.

### 1.4.1 Biomass energy in Western Europe to 2050

Presently biomass energy supplies at least 2 EJ/year (47 Mtoe) in the EU which is about 3.6 per cent of total primary energy consumption (54.1 EJ). Estimates of the potential for bioenergy in the next century range from 2 to 20 EJ/year. We have estimated a potential of 9.0–13.5 EJ in 2050, which represents 17–30 per cent of projected total energy requirements. This depends on assumptions of available land areas (10 per cent of useable land, 33 Mha), achievable yields, and the amount of recoverable residues utilised. Within OECD Europe about 76 per cent (333 Mha) of the total land area is classified here as 'useable', i.e. it can potentially be used for growing biomass for energy. How much of this 'useable' land could be put to energy cropping will depend on policies regarding food surpluses, energy self-sufficiency and rural economics. The role of agricultural subsidies in particular will be crucial in future land use policies. Total subsidies for OECD Europe were about US$167 billion in 1990, out of a total for all OECD countries of about US$320 billion – they have not changed significantly since then. Whether this level of subsidies will be maintained in the future (as it was in the 1990s), and whether they could be shifted away from food production towards bioenergy and other non-food crops are crucial questions. Just a small fraction of the US$167 billion would have a significant effect on stimulating the large-scale development of bioenergies.

An example of near-term relevance of biomass energy is in the EU, where a recent (1997) White Paper on Renewable Energy proposed that Europe could double its renewable energy contribution from 6 per cent today to 12 per cent by the year 2010. This would substantially help meet the Kyoto Protocol targets. It was proposed that biomass energy in total could contribute an additional 90 Mtoe per year compared to today's annual contribution of about 47 Mtoe. Of this additional energy, 'energy crops' (trees, woody grasses, and so on) are proposed to provide 45 Mtoe/year which could be grown on about 13 million hectares of land (4 per cent of total land at a yield of 10 t/ha and assuming a conversion efficiency of 75 per cent). This extra 45 Mtoe/year of renewable, $CO_2$-neutral biomass energy would reduce $CO_2$ emissions by 50 million tonnes (Mt) carbon/year compared to the present EU total $CO_2$ emissions of 890 Mt carbon/year. The contribution of all forms of biomass (137 Mtoe) to reducing $CO_2$ emissions would total about 150 Mt carbon by the year 2010, that is, a reduction of 17 per cent, which is twice the EU's obligation under the Kyoto Protocol.

Greater environmental and net energy benefits can be derived from perennial and woody energy crops compared to annual arable crops as alternative feedstocks for fossil fuels. In order to ameliorate $CO_2$ emissions, using biomass as a substitute for fossil fuels is more beneficial (from a socio-economic perspective) than sequestering the carbon in forests. The relative contribution of biofuels in the future will ultimately depend on markets and incentives, on R&D progress, and on environmental requirements. Land constraints are not considered significant because of the predicted surpluses in land and food and the near balance in wood production and wood requirements.

## 1.5 $CO_2$ sequestration versus fossil fuel substitution

Since it was first proposed in 1977, there have been numerous analyses of the potential for forests to mitigate the global $CO_2$-induced greenhouse effect by sequestering carbon in their standing biomass. Global estimates of the potential range from 1–3 Pg carbon sequestered per annum into the middle of the next century, depending on land afforested, enhanced forest management, and longevity of the stored carbon.

The strategies proposed for increasing the role of trees in stabilising atmospheric $CO_2$ levels and ameliorating global warming are varied. They include a rapid reduction of deforestation, and major global reforestation in the forms of afforestation and rehabilitation of degraded lands, agroforestry, and reforestation/plantation establishment. Carbon is absorbed by trees as they grow to maturity over a long period, e.g. 20–60 years, but then it ceases. The trees can be harvested and the woody biomass (which is about 50 per cent carbon) can theoretically be put into permanent storage ('pickled') or can be used as long-lived forest products. However, the storage option is quite costly, while the demand for long-lived wood products may be relatively too small to have a major impact on the global carbon cycle. Alternatively, using biomass produced on available land as a source of energy to substitute for fossil fuels would be more effective in decreasing atmospheric $CO_2$ than simply using the biomass as a carbon store, and would also provide a wider range of benefits. This would allow the land to be used indefinitely for carbon removal, and, since biomass energy crops have relatively high growth rates and higher yields compared to conventional tree cover, they can make a larger and more rapid contribution.

Bioenergy can be sold and is competitive in many circumstances, therefore the net cost of offsetting $CO_2$ emissions by substitution could be near zero or negative. Growing trees as a long-term carbon store will be important where the creation of new forest reserves is deemed desirable for environmental or economic reasons and on low productivity land. In general, however, using biomass to substitute for fossil fuels is more advantageous and appropriate. A detailed analysis in 1990 by Hall, Mynick and Williams outlined the economic advantages of the fossil fuel substitution route compared to the carbon sequestration alternatives. With advanced biomass electric conversion systems, profits could be made (depending on feedstock costs) while decreasing $CO_2$ emissions to the atmosphere compared to coal electric systems.

Sequestering carbon into biomass would incur costs between US$8 and 60/t carbon in the US in order to offset 25 per cent of current $CO_2$ emissions. Recent estimates (1995) for a global afforestation programme over the next century on 335 Mha (total sequestration = 104 GtC) gave 'integrated costs of carbon sequestration' at US$6.01/tC for the USA, US$13.86/tC for Europe and US$8.20/tC for the FSU (CIS). Tropical Asia, Africa and Latin America have costs of only US$1.66–2.41/tC.

A 1994 global programme estimated that OECD Europe could plant 7.8 Mha at 0.23 Mha/year to sequester 2.18 GtC over the next century. An examination of the fluxes of carbon through forests in seventeen West European countries and the potential for ameliorating net $CO_2$ emissions was made and it was concluded that a combination of afforestation, forest enhancement and management of removed biomass (e.g. long-lasting products, recycling, bioenergy) could result in a sustainable potential sequestration of 150–250 Tg C annually (0.15–0.25 GtC/yr) in the long term. This is a twofold to fourfold increase over the present accumulation of 39–58 Tg C/year, mainly due to the build up of biomass in existing middle-aged and mature stands. Accumulation of biomass can continue temporarily but is unsustainable in the long term. Therefore, this sink is expected to disappear. Afforestation and controlling the C flux associated with removal could substitute this temporary sink and have a more long-term impact. In 1995 it was concluded that if sequestration and substitution are linked, fossil fuel substitution by biomass power could have a greater impact on the carbon cycle than carbon storing forests in the medium and long term. Furthermore, commercial reality would favour energy forestry and there would be long-term social benefits.

The most effective strategies for using forest land to ameliorate $CO_2$ increases in the atmosphere have been modelled by Marland and Marland from Oak Ridge National Laboratory, USA. The options depend on the current status of the land, the productivity that can be expected, the efficiency with which the forest harvest is used to substitute for fossil fuels, and the time perspective of the analyses. Forests with large-standing biomass and low productivity should be protected as carbon stores. Deforested land with low productivity should be reforested and managed for forest growth and carbon storage. Productive land could be forested and managed for a harvestable crop which is used with optimum efficiency for long-lived products or as a substitute for fossil fuels. The longer the time perspective, the more likely that harvesting and replanting will result in net carbon benefits. It has also been concluded that short rotations do not achieve high carbon storage and that conifers and broad-leaf plantations will give different carbon storage benefits at 50 and 100 year time horizons.

In 1990 Hall, Mynick and Williams concluded that displacing fossil fuel with biomass grown sustainably and converted into useful energy with modern conversion tech-nologies would be more effective in decreasing atmospheric $CO_2$ than sequestering carbon in trees. The extent to which biomass energy would decrease OECD $CO_2$ emissions depends on the ability of, say, wood to displace coal; this is the more probable short- to medium-term option compared to biomass-derived liquid fuels offsetting coal and gas-derived liquid fuels. The greater reactivity and lower sulphur content of wood compared to coal gives considerable advantages in advanced conversion tech-nologies. Thus, if biomass is considered primarily as a substitute for coal using modern conversion technologies for producing either electricity or liquid synfuels, the effect on atmospheric $CO_2$ would be comparable to what could be achieved with C sequestration, per tonne of biomass produced.

A clear point for policy-makers is that trees and other forms of biomass can act as carbon sinks but at maturity, or at their optimum growth rate, there must be plans to use the biomass as a source of fuel to offset fossil energies (or as very long-lived timber products). Otherwise, the many years of paying to sequester and protect the carbon in trees will simply be lost as they decay and/or burn uncontrollably.

The evolution of land uses for forestry and biofuel can be considered as an ongoing process with joint fuel and timber/fibre outputs. This has been modelled dynamically (by Read) to show that the rapid accumulation of a 'buffer stock' of carbon in forest biomass yields a number of benefits. The buffer stock concept involves policy-driven land use change with an initial long rotation of plantation afforestation – the product being sold partly as biofuel and partly as conventional timber/fuel product – and with the land released by felling used partly for intensive food production and partly for intensive short rotation biofuel production.

Biomass has many advantages for an environmentally-friendly future. To obtain maximum benefit, trees, other than in primary forests, should be used as an energy source (or long-lived product) at the end of their growing life. It is probably preferable in most circumstances (except mature and primary forests) to use the biomass on a continuous basis as a substitute for present and future fossil fuel use.

The Kyoto Protocol offers good opportunities for joint implementation and clean development mechanism (CDM) projects which should involve biomass energy in both developed and developing countries to offset fossil fuels and to sequester carbon in biomass and soil sinks. It is important that people involved in the Kyoto Protocol recognise the significant opportunities for using biomass energy, which also complements the Biodiversity and Desertification Conventions, by providing opportunities for enhancing biodiversity and reforestation specially in degraded lands.

As the IEA so clearly stated in 1998: long-term and continuous reduction of $CO_2$ emissions through land-based activities will to a large extent have to come from the use of wood for bioenergy and products. The provision of the Kyoto Protocol with respect to sinks can be seen as a valuable incentive to protect and enhance carbon sinks now, while at the same time providing the resources needed for the substitution of fossil fuels in the future.

## 1.6 Environmental concerns

With the realisation that biomass energy could become part of the modern energy economy on a large scale there have been increasing concerns as to the short- and long-term environmental effects of such a strategy. Fortunately a number of environmental and biomass energy groups acknowledged some time ago that if biomass was to play an important role in future energy policy then its production, conversion and use must be environmentally acceptable and also accepted by the public. These latter two factors are likely to be the most crucial constraints on future biomass developments and must be addressed in some detail in order for scenario builders to appreciate the opportunities and problems associated with biomass energy production and use. These environmental aspects should also have the highest priority in future R&D policy as the authors firmly believe that ensuring environmental sustainability will be the single most important determining factor in future biomass for energy development.

### 1.6.1 The energy cost of growing biomass

If biomass energy is to be used as a fossil-fuel substitute, the energy provided should clearly be greater than the fossil-fuel energy needed to produce it. Determining the overall energy balance requires taking into account both 1) the energy required to produce the biomass; and 2) the energy required to convert the biomass feedstock into the energy carrier that will be marketed (e.g. electricity or a liquid fuel).

This relationship has been examined for eleven situations involving biomass from trees, ranging from natural forests to conifer plantations managed at average intensity to intensively-managed short-rotation hardwood plantations. A strong positive energy balance relationship (in which the net energy yield was twelve times the energy input) was found. Estimates of the energy costs of plantation biomass (including both woody and herbaceous crops) grown under US conditions have been made. With near-term expected biomass yields (net of harvesting and storage losses) in the range 9–13 dry tonnes per hectare per year, net energy yields have been estimated to be in the range of 10–15 times the energy inputs for these crops. With projected higher future yields, these ratios would be somewhat higher. A modelling study in the UK has shown that energy output to input ratios of 30:1 could be achieved depending on yield and management efficiency.

Thus the energy content of biomass grown on modern plantations will generally be much greater than the fossil-fuel inputs required for its production.

### 1.6.2 Energy and $CO_2$

Studies with relevance to Europe showed that utilising starch, sugar and oil crops for liquid fuels can give net energy and $CO_2$ benefits, provided that the agricultural phase is well managed and that the by-products of grain production, e.g. straw, are used in the energy conversion processes (which also should be efficient).

Comparing electricity from wood with liquid fuel from biomass options highlights the fact that the solid fuel (from perennial plants) to electricity options are the most favourable in terms of energy ratios, fuel substitution and $CO_2$ abatement costs. As will be discussed below, woody crops (trees and herbaceous plants such as Miscanthus) generally have more environmental advantages (especially in the production phase) than annual, arable crops. The use of conventional agricultural practices compared to those in forestry will have to be carefully weighed as arable land is taken out of food production.

### 1.6.3 Environmental impacts

If plantations were to replace natural forests, there would be destructive environmental effects and negative effects on carbon inventories; this should be avoided. On the other hand, forestry projects established on degraded land or abandoned cropland can have numerous positive environmental effects. Negative environmental effects might include soil compaction and loss of fertility, increased run-off, water pollution from agrochemicals, lowering of water tables in arid areas, loss of biodiversity, exotic species becoming weeds or introducing new pests and diseases, uncontrolled forest fires, and creation of roads, buildings and traffic. With careful planning and management these problems can be avoided.

The possible positive effects far outweigh these negative effects. Reforestation/ afforestation of degraded or agricultural land can improve soil structure, soil organic matter and nutrients, reduce run-off and increase soil water storage, therefore reducing flooding downstream, increase local rainfall and modify local temperatures, increase biodiversity, reduce pressure on natural forests, create windbreaks, and, of course, store carbon. In this way, afforestation can protect fragile regions by conserving watersheds and aquifers, reducing erosion, and preserving wildlife habitats. Furthermore, if the biomass is used to replace fossil fuel, it can offset many other environmental impacts.

The following sections discuss these factors in detail. They are taken from papers by Hall and by Williams (see Bibliography) who discussed the role of biomass energy in sustainable development.

### 1.6.3.1 Sustainable bioenergy plantations

If biomass plantations are likely to play major roles in the global energy economy, strategies are needed for achieving and sustaining high yields over large areas and long periods (Table 1.2). There is little long-term experience with high-yielding crops that can be drawn upon for guidance in formulating such strategies. Although the experience of sustaining high sugar-cane yields for centuries in the Caribbean and elsewhere suggests that high sustained yields are feasible, research is needed to ascertain the optimal strategies for achieving such biomass energy yields under a wide range of conditions. The practical issues that must be dealt with include site establishment, species selection, soil fertility, pests and diseases, erosion, water pollution, and the biological diversity of the plantation and its environs.

*Table 1.2* Estimates of tropical plantation woody biomass productivity potential under different management situations (oven dry t/ha/yr)

| Situation | Semi-arid | | Sub-humid | |
|---|---|---|---|---|
| | low | high | low | high |
| 1  No genetic improvement, no fertiliser added and no water added | 2 | 5 | 5 | 10 |
| 2  Genetic improvement, no fertiliser added and no water added | 4 | 10 | 10 | 22 |
| 3  Genetic improvement, fertiliser added but no water added | 6 | 12 | 12 | 30 |
| 4  Genetic improvement, water added but no fertiliser added | 8 | 18 | 11 | 25 |
| 5  No genetic improvement and no water added, but fertiliser added | 3 | 7 | 8 | 15 |
| 6  Genetic improvement, fertiliser and water added | 20 | 30 | 20 | 35 |

*Source:* Ravindranath and Hall (1996)

### 1.6.3.2 Site establishment

With many energy crops, planting is far less frequent than for food crops. For short-rotation forestry (SRF) crops, replanting takes place every fifteen to twenty years, if there are three cuts (one from a planted rotation and two from coppice rotations) per planting; for perennial grasses, replanting occurs perhaps once a decade. With such energy crops, herbicide applications would be much less frequent than for agricultural crops.

### 1.6.3.3 Species selection

Species selected for the plantations should be fast-growing and well-matched to the plantation site. This applies, for example, in respect to water requirements and its seasonality, drought resistance, soil pH, nutrition, tolerance of saline soils, as well as susceptibility to herbivores, fire, disease, and pests.

### 1.6.3.4 Soil fertility

Attention must be given to long-term soil fertility, which also involves soil management and enhancing micro- and macrofauna in the soil. Nutrients removed during harvesting must either be generated naturally or restored by adding fertilisers. Fertilisers are usually required when establishing plantations on degraded lands and may also be needed to realise high yields even on good sites. Despite the favourable net energy yield for intensively managed biomass plantations, it will often be desirable to reduce fertiliser inputs. For SRF, it will usually be desirable to leave leaves and twigs in the field, as nutrients tend to concentrate in these parts of the plant. Also, the mineral nutrients recovered as ash at the energy conversion facilities should be returned to the site.

Another strategy for reducing fertiliser requirements is to select species that are especially efficient in their use of nutrients. There is a wide range of nutrient-use efficiency among plants; in addition, either selecting a nitrogen-fixing species or intercropping the primary crop with a nitrogen-fixing species can make the plantation self-sufficient in nitrogen. In the future it may be feasible to reduce fertiliser requirements through the use of techniques being developed for matching nutrient applications more precisely to the plant's time-varying need for nutrients.

### 1.6.3.5 Pests and diseases

Plantations of only one or two species with plants of comparable age can be more vulnerable to attack by pests and pathogens than are natural forests or grasslands. Care in species selection, varying ages and good plantation design can be helpful in controlling pests and diseases, rendering the use of chemical pesticides unnecessary in all but special circumstances. A good plantation design, for example, will include: 1) areas set aside for native flora and fauna to harbour natural predators for plantation pest control; and perhaps 2) blocks characterised by different clones and/or species. If a pest attack breaks out in a block of one clone, a now common practice in well-managed plantations is to let the attack run its course and to let predators from the set-aside area help halt the pest outbreak.

### 1.6.3.6 Erosion

The potential for erosion control will be an important criterion in selecting the plantation species wherever erosion is a problem. Erosion tends to be significant during the year following planting. Thus, annual herbaceous crops like sorghum are no better in controlling erosion than annual agricultural row crops like maize, whereas either SRF tree or perennial grass crops, for which planting is infrequent, can provide good erosion control. The effectiveness of perennial grasses and trees in controlling erosion is indicated by recent experience with the Conservation Reserve Program in the USA.

### 1.6.3.7 Water pollution

Nutrient leaching from plantations can contaminate ground water and runoff, thereby degrading drinking water supplies and promoting algal blooms. Fortunately, the various strategies for limiting fertiliser input also reduce the potential for water pollution associated with nutrient leaching. Reduced soil disturbance in perennial systems also decreases chances of pollution.

### 1.6.3.8 Biological diversity

Biomass plantations are often criticised because the range of biological species they support is much narrower than for natural forests. While true, the criticism is not always relevant. It would be relevant if a biomass plantation replaced a virgin forest. However, it would not be relevant if a plantation and associated natural reserves were established on degraded lands; in this instance, the restored lands would be able to support much greater biological diversity than was possible before restoration. If biomass energy crops were to replace monocultural food crops, the effect on the local ecosystem would depend on the plantation crop species chosen, but in most cases the shift would be to a more biologically diverse landscape.

Achieving sustainable biomass production while maintaining biological diversity may ultimately require a shift to polycultural strategies (e.g. mixed species in various planting configurations). Polycultural establishment and management techniques therefore warrant high priority in research and development programmes for energy crops.

Establishing and maintaining natural reserves within and around plantations can help control crop pests while enhancing the local ecosystem. This has been the experience in Brazil over the last decade or longer with their eucalyptus plantations, where about one-third of the land is not planted but left for indigenous regrowth. Preserving biodiversity on a regional basis will require land-use planning in which natural forest patches are connected via a network of undisturbed corridors (riparian buffer zones, shelterbelts, and hedgerows between fields), thus enabling species to migrate from one habitat to another.

Major expansions in research are needed to provide a sound analytical and empirical basis for achieving and sustaining high biomass yields in environmentally acceptable ways. There is, however, time for such research as well as for extensive field trials, because major bioenergy industries can be launched with residues from the agricultural and forest products industries.

### 1.6.4 Landscape

Countryside landscapes are highly valued by many people. Energy forestry represents a possible major change to well-loved landscapes. Thus it is essential for biomass energy developers to recognise local and national perceptions from the outset of any large-scale developments. Fortunately the relatively poor image of commercial plantation forestry is now widely recognized so that guidelines put out by forestry organisations are more acceptable to developers and the public alike.

It should be noted that 'horizon-to-horizon' monocultures are not necessary (or acceptable) since only a relatively small fraction of the landscape needs to be planted to the biomass plantation/crop. This results from the smaller scale which favours bioenergy systems such as 20–50 MW power plants which can be modular. In the case of a 20 MW plant with 45 per cent efficiency (80 per cent load), using biomass produced at 10 odt/ha would require 7,000 ha but this represents only 10 per cent of the land within a 14 km radius of the generation plant. For a 50 MW plant with plantation yields of 15 t/ha and 11,000 ha requirements with a 15 per cent land use, the radius of the planted area would be 16 km.

If biomass energy systems are well managed, they can form part of a matrix of energy supply which is environmentally sound and therefore contributes to sustainable development. When compared, for example, to conventional fossil fuels, overall the impacts of bioenergy systems may be less damaging to the environment (since they produce many, but local and relatively small, impacts on the surrounding environment), than the fewer, but larger and more distributed, impacts for fossil fuels. It is these qualities which may make the environmental impacts of bioenergy systems more controllable, more reversible and, consequently, more benign.

### 1.6.5 Guidelines

These should address the main concerns which environmentalists and the public have while ensuring that the biomass growers receive both monetary and social benefits. Such a balancing act needs careful consideration and planning if long-term benefits are to accrue to both groups.

Considerable effort has gone into the development of good practice guidelines for the use of biomass for energy for example in the US, the UK and Sweden, and work is under way to develop Europe-wide guidelines. These guidelines have been published and distributed widely and recognise the central importance of site-specific factors, and the breadth of social and environmental issues which should be taken into consideration. All concentrate on short rotation coppice, while the US guidelines also consider perennial grasses and some residues. In general residues, biogas and liquid fuels have not been subject to such detailed recommendations. This bias reflects the anticipated environmental benefits of woody and perennial energy crops for solid fuels compared to the other biomass options discussed above. However, given that residues may remain more widely used than energy crops for some time, guidelines are needed on when it is appropriate to use residues for energy, and how best to maximise the environmental advantages this may offer.

A key message of these guidelines is that site selection and crop selection must be made carefully, and the crop must be managed sensitively. Land should be primarily agricultural, and energy crops should never displace land uses of high ecological value. Consideration needs to be given to the landscape and visibility, soil type, water use, vehicle access, nature conservation, archaeology, pests and diseases, and public access. Planning permission is not usually needed before energy crops can be grown, but something akin to an environmental assessment ought to be made nonetheless. The guidelines also stress the importance of consulting local people at the planning stage, and ongoing community involvement in the development stages. Issues such as changes to the landscape, increased traffic movements or new employment opportunities in rural areas may prove to be very significant to local people.

Careful site selection can bring positive environmental benefits. If SRC is planted in riparian (riverside) zones it can absorb lateral flows of nutrients from adjacent agriculture, thus preventing water pollution. Energy crops are generally more environmentally acceptable than intensive agriculture, since chemical inputs are lower and the soil undergoes less disturbance and compaction. In terms of wildlife energy crops are sometimes considered similar to agricultural monocultures, though different species will be favoured. Given good site selection, field layout, species mixing and sensitive management practices SRC can be beneficial for birds, wild plants, soil organisms and other species.

There is no single best way to use biomass for energy, and the environmental acceptability will depend on sensitive and well-informed approaches to new developments in each location. It is clear that biomass for energy *can* be environmentally friendly, and steps *must* be taken to ensure that it is, if biomass is to be accepted as an important fuel of the future.

## 1.7  Conclusions

There are good prospects that biomass-derived electricity, heat and liquid fuels can be produced competitively with fossil fuels in the future, in addition to the niche markets which are already economic in a number of countries around the world. The site specificity of biomass requires considerable local inputs, trained personnel and an ongoing commitment to ensure continued high productivities. Local and regional involvement in decision-making are essential for efficient long-term sustainability of biomass energy production in an environmentally and socially acceptable manner. Technical and financial support will need to be made available to developing regions.

Enhanced biomass availability on a sustainable basis requires support and development of new biomass systems in which production, conversion and utilisation are performed efficiently in an environmentally sustainable manner. Efforts to modernise biomass energy should concentrate on those applications for which there are favourable prospects for rapid market development, e.g. the generation of electricity (and heat) from residues and biomass plantations using advanced gas turbines fired by gasified biomass, and the production of alcohol fuels from sugar-cane. Although biomass conversion systems are at present typically less efficient than those for fossil fuels, development of modern energy conversion technologies is improving this situation. The most likely technology to be used to convert biomass to electricity at modest scales in the near-term is biomass integrated gasifier/gas turbine cycles (BIG-GT), which will be more efficient than conventional coal–steam–electric power generation and coal gasification, and will have lower capital costs.

Bioenergy industries have already been launched in numerous developed and developing countries. Nevertheless the techniques and technologies for growing biomass and converting it into modern energy carriers must be more fully developed. If bioenergy R&D is given high priority, and if policies are adopted to nurture the development of bioenergy industries, these industries will be able to innovate and diversify as they grow and mature. Suitable policy options include a reallocation of agricultural subsidies and a review of energy subsidies to create a 'more level playing field', internalising external negative costs of fossil fuels by methods such as carbon taxes, and the use of these taxes to support bioenergy and other alternative carbon mitigation options.

Biomass is an important, environmentally-friendly source of energy which can also provide sound energy alternatives to ameliorate global warming. It is potentially the world's largest source of energy and it should play a key role in the future energy mix. It must also be recognised that biomass is already a major source of energy in both its traditional and modern forms.

The Kyoto Protocol, when finally implemented, will have a significant impact on the way we use and produce energy; it represents an important step forward to a more environmentally sustainable energy future. However, it suffers from shortcomings as far as biomass energy is concerned. For example, the Protocol does not seem to pay sufficient attention to the implications of the 'carbon substitution strategy', as it appears to advocate mostly the planting of trees (afforestation and reforestation) to create carbon sinks. The difficulties with this strategy have been outlined in this chapter. In our view using a biomass energy-based 'carbon sequestration and substitution strategy' merits greater attention since it will bring far greater benefits than a 'carbon sequestration strategy', e.g. income is generated continuously thereby creating local jobs and other benefits. It is important to strike a balance between these two strategies and that all pros and cons are fully assessed before planning and investments are made – the long-term implications of the different strategies are crucial.

## 1.8 Bibliography

*Note*: Many relevant reviews and articles can be found in the journal *Biomass and Bioenergy* and in the proceedings of the biennial Biomass Conferences in the USA and Europe (obtainable from National Renewable Energy Laboratory, Golden, CO, and Elsevier Publications, Oxford, respectively).

Andrade, A. M., Andrade, C. M. and Bodinaut, J. A. (1998) 'Biomass energy use in Latin America: Focus on Brazil', *Proceed. Biomass Energy: Data Analysis and Trends*, OECD/IEA, Paris, 87–95.

Hall, D. O. (1997a) 'Biomass energy in industrialised countries – a view of the future', *Forest Ecology and Management*, **91**, 17–45.

Hall, D. O. (1997b) in: 'Biomass energy: key issues and priority needs', OECD/IEA, Paris.

Hall, D. O. and Rosillo-Calle, F. (1991) 'Why biomass matters: Energy and the environment', *Biomass Users Network*, SLLN (Special Issue).

Hall, D. O. and House, J. (1995) 'Biomass energy in Western Europe to 2050', *Land Use Policy*, **12**, 37–48.

Hall, D. O. and Rosillo-Calle, F. (1999) 'Biomass as an environmentally acceptable fuel and implications for the Kyoto Protocol', *Ecological Engineering* (forthcoming).

Hall, D. O. and Scrase, I. (1998) 'Will biomass be the environmentally-friendly fuel of the future?' *Biomass and Bioenergy*, **15**, 357–67.

Hall, D. O., Mynick, H. E. and Williams, R. H. (eds) (1990) *Carbon Sequestration vs. Fossil Fuel Substitutions: Alternative Roles for Biomass in Coping with Greenhouse Warming*, Report No. 255, Centre for Energy and Environmental Studies, Princeton University.

—— (1991) 'Cooling the greenhouse with bioenergy', *Nature*, 353, 11–12.

—— (1991) 'Alternative roles for biomass in coping with greenhouse warming', *Science and Global Security*, 2, 113–51.

Hall, D. O., Rosillo-Calle, F. R., Williams, R. H. and Woods, J. (1993) 'Biomass for energy: supply prospects' in Johansson, T. B. *et al.* (eds) *Renewable Energy: Sources for Fuels and Electricity*, Washington DC: Island Press, 593–652.

IEA (1998) 'World Energy Outlook', International Energy Agency, Paris.

IIASA/WEC (1998) 'Global Energy Perspectives', Nakicenovic, N. *et al.* (eds), Cambridge University Press.

Marland, G. and Marland, S. (1992) 'Should we store carbon in trees', *Water Air and Soil Pollution*, vol. 64 (1–2): 181–95, Oak Ridge Natl Lab, Div. Environmental Sciences, Oak Ridge, TN.

Ravandrith, N. H. and Hall, D. O. (1996) 'Estimates of feasibility production of short rotation tropical forestry', *Energy for Sustainable Development*, 2, 14–20.

Read, P. (1998) 'Dynamic interaction of short rotations and conventional forestry in meeting demand for bioenergy in the least cost mitigation strategy', *Biomass and Bioenergy*, vol. 15(1): 7–15.

Rosillo-Calle, F. and Cortez, A. B. (1998) 'Towards PROALCOOL II – A Review of the Brazilian Bioethanol Programme', *Biomass and Bioenergy* 14:(R): 115–24.

# 2

# ENERGY FROM BIOMASS IN BRAZIL

*Sergio Valdir Bajay, Eliane Bezerra de Carvalho*
*and André Luis Ferreira*

## 2.1 Introduction

This chapter presents a general overview of biomass energy in Brazil, including possible future outcomes, depending on adopted public policies and regulatory actions. Recent data are presented and discussed, and current policies and practices are also critically evaluated. We deal with general issues concerning biomass energy, with particular reference to the sugar and alcohol, paper and pulp, and iron and steel industrial sectors. We highlight the most important issues in each of these three industrial branches, which are highly dependent on biomass resources as raw material and energy (these individual sectors are discussed in more detail in later chapters).

The chapter is divided into six complementary sections. Sections 2.2 and 2.3 present an updated summary of energy demand in Brazil, including production, conversion and consumption of the main biomass fuels in the country. Section 2.4 describes briefly the use of biomass energy in the industrial sectors, the object of analysis in this book, together with obstacles and possible solutions. Section 2.5 assesses new opportunities for electricity generation in alcohol distilleries and paper and pulp mills, which have arisen in the light of the new institutional setting of the Brazilian power sector. Regulatory obstacles for such opportunities are pointed out, together with the actions required to overcome them. Possible financial incentives to co-generation and power generation with renewable sources of energy are also discussed. Finally, Section 2.6 examines some general issues concerning biomass and the environment, discusses the main environmental effects of extracting energy from biomass in the three industrial branches evaluated in this book, and shows the limits and potentialities of environmental policy tools to provide incentives to the use of biomass.

## 2.2 Energy consumption in the Brazilian economy and its main sectors

The contribution of renewables to primary energy consumption in Brazil has increased by about 46 per cent between 1983 and 1997, compared to a 61 per cent increase in non-renewable primary energy sources in the same period. Hydroelectricity increased its share of renewable sources from about 49 per cent in 1983 to 61.5 per cent in 1997.

27

Biomass energy makes up the rest of this group. Despite the fact that the share of renewable energy in Brazil's total primary energy consumption decreased, overall there has been an increased consumption of about 10 per cent in absolute terms. Alcohol and sugar-cane were responsible for this increase, balancing out the fall in consumption of firewood.

Figures 2.1–2.4 show the evolution of the total secondary energy consumption and its main components – oil products, coal, natural gas, electricity, fuels from biomass and 'other fuels' – in the economy as a whole and in the industrial, transportation, household and service sectors, respectively, during the period 1983–97.

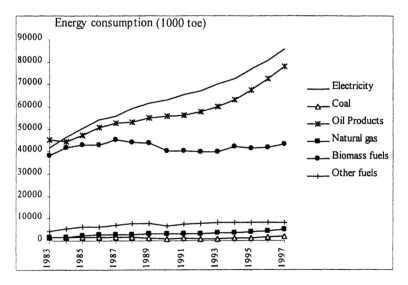

*Figure 2.1* Evolution of energy consumption in the Brazilian economy

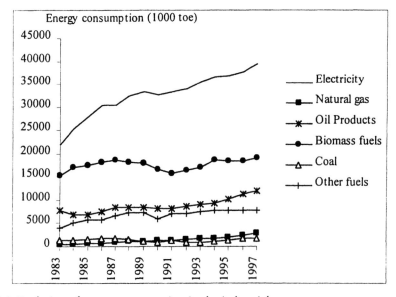

*Figure 2.2* Evolution of energy consumption in the industrial sector

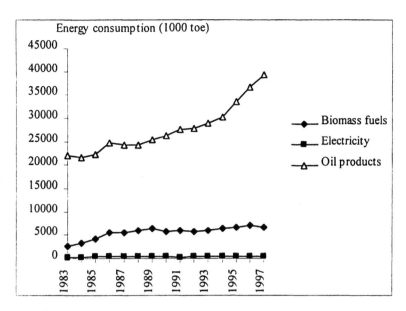

*Figure 2.3* Evolution of energy consumption in the transportation sector

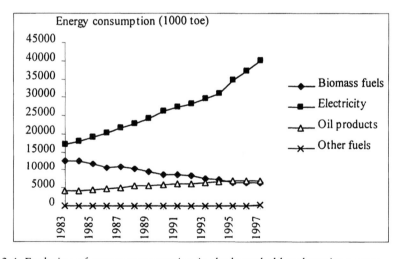

*Figure 2.4* Evolution of energy consumption in the household and service sectors

The total secondary energy consumption and its components in the industrial, transportation, and household and service sectors increased by about 67, 64, 86 and 58 per cent, respectively, for the above period. In 1983, the three components were about 38, 19 and 26 per cent, respectively, of total consumption. In 1997 the market shares were about 37, 21 and 24 per cent, respectively, with the industrial sector still remaining the most important throughout. Biomass energy lost market share to competitors in this period because of the large fall in the relative consumption of firewood, from about 15 per cent of the total secondary energy consumption in 1983 to 6 per cent in 1997. In the industrial sector, the consumption of oil products increased by about 54 per cent between 1983 and 1997. This figure is less than the total secondary energy consumption in the sector because some substitution of electricity, biomass energy, natural gas and coal for oil products took place during that period. There was a considerable increase of about 157 per cent in the consumption of alcohol in transportation and a sharp decline of about 48 per cent in firewood demand in the household and service sectors between 1983 and 1997.

## 2.3 Production, conversion and final consumption of firewood, charcoal, alcohol, sugar-cane bagasse and black liquor

Tables 2.1–2.6 present the Brazilian production, stock variations and losses, conversion and consumption of firewood, charcoal, anhydrous and hydrated alcohol, sugar-cane bagasse and black liquor, in the period 1983–97.

Despite the sharp fall in consumption of firewood during the period in question, the household sector still remains the largest consumer, followed by industry, e.g. the food and drink, ceramics, and pulp and paper industries. The conversion to charcoal requires large amounts of firewood, with a peak consumption registered in 1989. The industrial sector, particularly the pig-iron and steel branches, is the largest consumer of charcoal in Brazil (see Table 2.2), followed by the household sector; consumption of industrial charcoal has been falling since the 1990s (see Chapter 8).

Table 2.3 shows the rather erratic production patterns of anhydrous alcohol in Brazil during the 1983–97 period. The production of hydrated alcohol for the same period is shown in Table 2.4. As can be observed, there was a strong upward trend until 1987, followed by stagnation. The overwhelming use of both types of alcohol has been as a fuel in the transportation sector. The exports of alcohol during the 1980s, mostly to the USA, were soon halted because of pressure from American alcohol producers, who argued that Brazil was subsidising ethanol. The American government responded by placing prohibitive excise duties on ethanol imports from Brazil. Some exports of hydrated alcohol were initiated again in 1993. Imports of alcohol from 1991 onwards were a consequence of alcohol and sugar-cane prices being kept artificially low by the government in its vain efforts to control inflation in the early 1990s. As prices for ethanol fuel were kept artificially low, there was a shift to a larger sugar production as it fetched a higher price. This coincided with high sugar prices in the international market. Local production of both hydrated and anhydrous alcohol decreased and the government was forced to import to meet domestic demand. More recently, higher prices for anhydrous alcohol have increased production, eliminating the need to import. The same cannot be said for hydrated ethanol or methanol imports because of the continual fall in fuel prices throughout the 1990s (Goldemberg 1996).

Sugar-cane bagasse production in Brazil, 1983–97 is shown in Table 2.5. From 1988 to 1994 production stagnated because of the overall lower sugar-cane production, while the years following witnessed a substantial recovery in production. The bagasse is consumed almost exclusively as a fuel in sugar and alcohol production, as well as in some other branches of the food and drink industry and, on a far smaller scale, in other industrial sectors.

The production of black liquor grew continually during the period 1983–97, as shown in Table 2.6. Black liquor is consumed by the paper and pulp mills, where a significant proportion of it is used to generate electricity in the co-generation units.

Table 2.1 Production, conversion to charcoal and consumption of firewood in Brazil, in $10^3$ tonnes, 1983–97

| | 1983 | 1985 | 1987 | 1989 | 1991 | 1993 | 1995 | 1997 |
|---|---|---|---|---|---|---|---|---|
| Production | 97,566 | 106,252 | 105,774 | 106,343 | 86,167 | 80,043 | 75,066 | 71,597 |
| Conversion to charcoal | 32,350 | 41,963 | 43,196 | 50,879 | 36,551 | 35,504 | 32,971 | 29,834 |
| Consumption | 65,216 | 64,289 | 62,578 | 55,464 | 49,616 | 44,539 | 42,098 | 41,765 |
| Households | 37,387 | 34,735 | 32,850 | 28,735 | 25,583 | 22,402 | 19,710 | 19,562 |
| Services | 578 | 533 | 500 | 378 | 356 | 324 | 291 | 294 |
| Agriculture | 10,166 | 8,500 | 8,400 | 7,600 | 6,800 | 6,050 | 6,081 | 5,973 |
| Industry | 17,075 | 20,511 | 20,820 | 18,743 | 16,872 | 15,763 | 16,016 | 15,936 |
| Ceramics | 5,440 | 5,744 | 6,270 | 5,888 | 4,783 | 4,715 | 4,533 | 5,136 |
| Chemistry | 860 | 970 | 1,060 | 810 | 680 | 568 | 469 | 338 |
| Food and drink | 5,963 | 7,031 | 7,120 | 6,540 | 6,467 | 5,776 | 5,694 | 5,692 |
| Paper and pulp | 2,475 | 3,167 | 2,505 | 2,369 | 2,173 | 2,464 | 2,932 | 2,475 |
| Textile | 450 | 750 | 730 | 570 | 475 | 326 | 333 | 323 |
| Other branches | 1,887 | 2,849 | 3,135 | 2,566 | 2,294 | 1,914 | 2,055 | 1,972 |

Source: MME (1998)

Table 2.2 Production, changes in stocks/losses and consumption of charcoal in Brazil, in $10^3$ tonnes, 1983–97

| | 1983 | 1985 | 1987 | 1989 | 1991 | 1993 | 1995 | 1997 |
|---|---|---|---|---|---|---|---|---|
| Production | 7,782 | 10,075 | 10,346 | 12,268 | 8,784 | 8,489 | 7,909 | 7,426 |
| Losses and stock changes | -466 | -502 | -517 | -613 | -418 | -334 | -304 | -251 |
| Consumption | 7,316 | 9,573 | 9,829 | 11,655 | 8,366 | 8,139 | 7,611 | 7,181 |
| Households | 1,354 | 1,328 | 1,210 | 1,092 | 950 | 823 | 672 | 612 |
| Services | 117 | 115 | 98 | 90 | 90 | 94 | 90 | 95 |
| Agriculture | 15 | 14 | 20 | 20 | 20 | 12 | 11 | 10 |
| Industry | 5,830 | 8,116 | 8,501 | 10,453 | 7,306 | 7,210 | 6,838 | 6,464 |
| Cement | 635 | 1,126 | 795 | 634 | 387 | 353 | 438 | 520 |
| Iron and steel | 4,374 | 5,915 | 6,575 | 8,249 | 5,700 | 5,825 | 5,517 | 5,012 |
| Steel alloys | 463 | 653 | 697 | 1,027 | 755 | 775 | 590 | 865 |
| Other metals | 158 | 190 | 266 | 398 | 316 | 175 | 226 | 28 |
| Other branches | 200 | 232 | 168 | 145 | 148 | 82 | 67 | 39 |

Source: MME (1998)

Table 2.3 Production, imports, exports, changes in stocks/losses and consumption (as a fuel and other uses) of anhydrous alcohol in Brazil, in 10³ m³, 1983–97

| | 1983 | 1985 | 1987 | 1989 | 1991 | 1993 | 1995 | 1997 |
|---|---|---|---|---|---|---|---|---|
| Production | 2,556 | 3,144 | 2,155 | 1,494 | 2,044 | 2,526 | 3,003 | 5,686 |
| Imports | 0 | 0 | 0 | 0 | 191 | 0 | 487 | 0 |
| Exports | –299 | –207 | –5 | 0 | 0 | 0 | 0 | 0 |
| Losses and stock changes | –21 | –697 | 63 | 208 | –588 | 24 | 1 | –1,177 |
| Consumption | 2,236 | 2,240 | 2,213 | 1,702 | 1,647 | 2,550 | 3,491 | 4,509 |
| As a fuel | 2,197 | 2,121 | 2,136 | 1,622 | 1,647 | 2,430 | 3,372 | 4,374 |
| Other uses | 39 | 119 | 77 | 80 | 0 | 120 | 119 | 135 |

Source: MME (1998)

Table 2.4 Production, imports, exports, changes in stocks/losses and consumption (as a fuel and other uses) of hydrated alchohol in Brazil, in 10³ m³, 1983–97

| | 1983 | 1985 | 1987 | 1989 | 1991 | 1993 | 1995 | 1997 |
|---|---|---|---|---|---|---|---|---|
| Production | 5,395 | 8,419 | 10,185 | 10,315 | 10,818 | 8,869 | 9,742 | 10,547 |
| Imports | 0 | 0 | 0 | 0 | 893 | 1,456 | 1,938 | 882 |
| Exports | –135 | –217 | –35 | 0 | 0 | –170 | –403 | –430 |
| Losses and stock changes | –1,590 | –1,423 | –604 | 1,409 | –772 | 290 | –256 | –1,174 |
| Consumption | 3,670 | 6,779 | 9,546 | 11,724 | 10,939 | 10,445 | 11,021 | 9,825 |
| As a fuel | 2,950 | 6,088 | 8,919 | 11,068 | 10,251 | 9,675 | 9,946 | 8,934 |
| Other uses | 720 | 691 | 627 | 656 | 688 | 770 | 1,075 | 891 |

Source: MME (1998)

Table 2.5 Production, conversion to electricity and consumption of sugar-cane bagasse in Brazil, in $10^3$ tonnes, 1983–97

| | 1983 | 1985 | 1987 | 1989 | 1991 | 1993 | 1995 | 1997 |
|---|---|---|---|---|---|---|---|---|
| Production | 48,887 | 56,867 | 62,653 | 55,330 | 58,801 | 60,564 | 69,847 | 82,032 |
| Conversion to electricity | 1,809 | 1,790 | 1,963 | 1,856 | 1,995 | 1,929 | 2,463 | 3,809 |
| Consumption | 47,078 | 55,077 | 60,690 | 53,474 | 56,806 | 58,635 | 67,384 | 78,223 |
| In fuel alcohol distilleries | 20,929 | 31,517 | 34,373 | 33,331 | 35,347 | 32,155 | 33,680 | 40,397 |
| In the chemical industry | 63 | 172 | 190 | 210 | 185 | 210 | 220 | 0 |
| In food and drink | 26,081 | 23,218 | 25,793 | 19,558 | 20,965 | 26,096 | 33,420 | 37,713 |
| In paper and pulp | 0 | 161 | 320 | 350 | 283 | 142 | 22 | 113 |
| In other branches | 5 | 9 | 14 | 25 | 26 | 32 | 42 | 0 |

Source: MME (1998)

Table 2.6 Production, conversion to electricity and consumption of black liquor in Brazil, in $10^3$ tonnes, 1983–97

| | 1983 | 1985 | 1987 | 1989 | 1991 | 1993 | 1995 | 1997 |
|---|---|---|---|---|---|---|---|---|
| Production | 3,184 | 3,876 | 4,231 | 4,221 | 5,192 | 7,029 | 7,375 | 8,285 |
| Conversion to electricity | 446 | 497 | 592 | 556 | 998 | 1,161 | 1,505 | 1,646 |
| Consumption | 2,738 | 3,379 | 3,639 | 3,665 | 4,194 | 5,868 | 5,870 | 6,639 |

Source: MME (1998)

## 2.4 Biomass-based energy-intensive industrial sectors in Brazil

The energy-intensive industrial sectors in Brazil that depend substantially on biomass, both as raw material and for energy supply, are the sugar-cane-fed ethanol distilleries, the charcoal-based pig-iron and steel plants, and the paper and pulp mills.

Bagasse and black liquor are industrial by-products of the sugar-cane and the pulp and paper industries, respectively, consumed on-site to produce steam and electricity.

The main features of each of these biomass-dependent energy-intensive industrial branches are discussed in this section, including their problems and possible solutions. Detailed assessment and other relevant issues are dealt with in later chapters.

### 2.4.1 Fuel ethanol distilleries

From 1975, with the creation of the National Alcohol Programme (PROALCOOL or PNA) by Federal Government Decree no. 76.593, Brazil has produced anhydrous alcohol from sugar-cane to blend it with gasoline, in Otto cycle car engines, in the proportion of up to 22 per cent. With the second phase of the PNA, which started in 1979 (Federal Government Decree no. 83.700), hydrated alcohol has also been produced to be used in Otto cycle engines modified to run on 100 per cent ethanol, or neat alcohol.

Currently Brazil is the world's largest producer of sugar-cane, with over 300 million tonnes crushed in the 1997/8 harvesting season. Prior to PROALCOOL, the Brazilian share was less than 15 per cent of world production. During the 1970s many ethanol distilleries were installed in the country, either new plants or distilleries annexed to existing sugar mills. In 1997 there were 339 distilleries in the country, 132 in the state of São Paulo. The main alcohol-producing states are São Paulo (about 68 per cent), Rio de Janeiro, Alagoas and Pernambuco.

The production of alcohol in the country increased 119 per cent between 1983 and 1992, reaching a maximum of $12.75 \times 10^6$ m$^3$ at the end of that period. In 1993–6 production practically stagnated at figures below the 1992 peak, but in 1997 a new peak was reached with $14.23 \times 10^6$ m$^3$. The production rate has been changing according to the prices fixed by the government per tonne of sugar-cane and per litre of alcohol, which are the main factors affecting production, besides climatic ones. The recurrent government policy of cutting the prices of electricity, oil products, alcohol and public services to control inflation has created serious financial difficulties for many entrepreneurs in this industry.

The stated objective of rapid growth, together with the facility to obtain subsidies to increase alcohol production capacity in the 1970s and early 1980s, facilitated the building of a large alcohol industry which still has considerable scope for increasing energy efficiency and reducing production costs. Government policies for this industry have so far failed to address the important issue of its cost-effectiveness.

Restructuring PROALCOOL is necessary because the continuing low oil prices on the international market urgently require that alcohol fuel be more competitive with gasoline. The share of alcohol-fuelled cars in total sales of new cars dropped from 96 per cent in 1985 to 0.07 per cent in 1997. Currently (1998) the PNA is kept afloat only by the compulsory addition of anhydrous ethanol to gasoline up to 24 per cent in volume, and by an ageing neat alcohol-fuelled car fleet of about 3.8 million. New policies are being considered by the government to boost PROALCOOL once more. The most important measures being envisaged are the compulsory addition of ethanol to diesel oil, in buses and lorries, and the setting up of government 'green fleets', to run on neat ethanol. One of the problems is that there are many conflicting interests ranging from 'free market apologists', agricultural, industrial, and energy interests, to all sorts of pressure groups. All are unable to agree on what could be politically and economically feasible to implement.

In 1997, about 54 per cent of the sugar-cane bagasse available was converted to electricity in co-generation systems at sugar and alcohol mills, while the remainder was used to generate process steam in the sugar and alcohol industry, and in other sectors, e.g. food and drink (particularly fruit juice producers), located near to the sugar-cane growing areas. Only about 5 per cent of the bagasse produced in 1997 was used to generate surplus electricity. There are already some firms in Brazil which specialise in upgrading and trading sugar-cane bagasse, whose volume has increased by 68 per cent between 1983 and 1997 (MME 1998).

There have been significant improvements in the productivity of both the sugar-cane culture and the ethanol-based industrial sector. These gains have been due to a combination of factors, including:

1   introduction of new and improved sugar-cane varieties;
2   economies of scale from larger and more efficient new plants; and
3   technological improvements and energy conservation measures in old plants.

As a result, Borges (1994) reported that anhydrous alcohol production costs have been decreasing on average at an annual rate of 3 per cent. The average cost of ethanol in the state of São Paulo fell from around US$75 per barrel of oil equivalent (boe) at the beginning of PROALCOOL to US$45 boe in 1993 (Borges 1994). The most efficient plants in Brazil can produce ethanol at around US$35/boe (Bajay and Carvalho 1994). However, there is still room for a further cost reduction of around 22 per cent, achievable with minor technological improvements already developed by the COPERSUCAR Technology Centre.

Although there are good reasons to justify the existence of a programme like PROALCOOL on environmental, social, technological and strategic grounds (Borges 1994), a concerted effort of government, alcohol producers, car manufacturers and R&D institutions should be started as soon as possible to make the PNA more competitive and to prepare the ground for a new growth cycle. This 'Third Phase' would be characterised by high levels of productivity and low production costs, together with further gains in environmental and social issues. The expected increase in oil prices in the near future, combined with the tightening legislation on environmental protection, should foster PROALCOOL. Nevertheless, a broad, decentralised and properly co-ordinated planning exercise should be undertaken to avoid the types of problem that faced the first and second phases of the programme when, for example, increasing production was the overwhelming goal.

Finding better uses for the sugar and alcohol by-products, like sugar-cane bagasse and vinasse, is certainly an excellent route to improving the economic performance of these plants. Walter *et al.* (1994) carried out a thorough technical and economic assessment of efficient co-generation technologies that could be used to burn or to gasify sugar-cane bagasse. This showed that substantial amounts of surplus electricity (about 6,000 MW) could be generated, which could be sold to the public supply grid. Currently a typical sugar and alcohol plant in Brazil generates (in a low-efficiency co-generation unit) about 14 kWh per tonne of crushed sugar-cane (tsc), which could be increased to 120–250 kWh/tsc with conventional but high-efficiency steam turbine cycles, and to about 500 kWh/tsc with the Biomass Integrated Gasifier-Steam Injected Gas Turbine (BIG-STIG) system, although it still needs to be commercially proven. The institutional, behavioural and financial impediments to the adoption of such technologies in Brazil were also detected in this study. The authors calculated that if all the benefits of surplus electricity sales of a BIG-STIG system were transferred to the hydrated alcohol cost, the latter would drop from about US\$0.182/l to US\$0.126/l (Goldemberg and Macedo 1994).

A wise and balanced government intervention is essential for the success of such a 'Third Phase' of PROALCOOL. This intervention should include the following measures:

1   to give up the vicious old practice of artificially keeping down fuel prices;
2   to set alcohol prices that appropriately reward only the most efficient producers, forcing inefficient ones either to modernise their installations and managerial practices or to withdraw from the fuel alcohol business;
3   to provide credit facilities and tax relief to upgrade old plants and to set up new, efficient units and;
4   to negotiate with the alcohol producers, trade unions and environmental groups a balanced share of the expected benefits among the capital owners, employees, and society as a whole, in terms of better environmental conditions.

### 2.4.2 *Iron and steel plants*

Coke, charcoal, coke furnace gas, natural gas and fuel oil are the main fuels currently employed in the Brazilian iron and steel industry (as seen in Table 2.7). Some of the plants that produce pig-iron and steel in Brazil, particularly in the state of Minas Gerais, use charcoal as a thermal and reducing agent in the industrial process, instead of coke, with consequent advantages in terms of product quality and emission of pollutants.

During the 1970s, the Brazilian government set up the National Programme for the Development of Steel, which was to transform Brazil into a large steel exporter, with the help of reforestation and charcoal. The programme can be considered a success in the sense that the country did become a large steel exporter, thanks to several large state-owned mills. Unfortunately the role of charcoal production has been disappointing. Together with the expansion of agricultural and pasturable lands, it shares responsibility for deforestation.

Environmental concerns about the destruction of native forests in Minas Gerais (e.g. by thousands of small charcoal producers), and the gradual exhaustion of such forests around the pig-iron producing areas, led both the state and federal governments to conclude that charcoal production from native forests was not the answer to the energy

*Table 2.7* Energy inputs (in $10^3$ toe) employed in the Brazilian iron and steel industries, 1983–97

| | 1983 | 1985 | 1987 | 1989 | 1991 | 1993 | 1995 | 1997 |
|---|---|---|---|---|---|---|---|---|
| Charcoal | 2,755 | 3,727 | 4,141 | 5,197 | 3,591 | 3,670 | 3,475 | 3,157 |
| Coke | 3,284 | 4,804 | 5,401 | 5,918 | 5,828 | 6,150 | 6,474 | 6,450 |
| Coke furnace gas | 545 | 756 | 1,020 | 1,068 | 899 | 907 | 1,016 | 976 |
| Electricity | 2,699 | 3,581 | 3,921 | 4,338 | 3,744 | 4,185 | 4,164 | 4,284 |
| Fuel oil | 506 | 471 | 383 | 411 | 371 | 453 | 380 | 316 |
| Natural gas | 103 | 192 | 351 | 348 | 351 | 373 | 585 | 690 |
| Other fuels | 143 | 159 | 272 | 336 | 231 | 453 | 512 | 1,057 |

*Source:* MME (1998)

supply problems facing the industry. This, together with the privatisation drive of the federal government, the high cost of charcoal (e.g. high transportation costs) and low cost of imported coke, caused a strong fall in consumption after 1989 (see Table 2.7). The companies are now concentrating on producing charcoal mostly from reforestation projects.

At present a large part of firewood supply for the industrial sector and for charcoal already comes from reforestation, e.g. 70 per cent in the case of charcoal. However, much more needs to be done to ensure that in future wood comes from environmentally sustainable forests, either plantations or natural forests.

The revitalisation of the charcoal-based industrial sectors in Brazil, particularly the pig-iron and steel plants, requires, necessarily, the use of highly productive forest plantations and large-scale efficient furnaces to produce cheap and good quality charcoal with the minimum environmental costs. The charcoal-based pig-iron and steel industry is learning from the paper and pulp industrial experience, which shows many economic benefits from industrialists owning their own plantations.

### 2.4.3 Paper and pulp mills

In 1974, the Brazilian government put forward the National Programme for Paper and Pulp, which aimed to attain self-sufficiency in paper and pulp and to export 2 million tonnes in 1980 and 20 million tonnes in 2000. Other objectives of this programme and the National Programme for the Development of Steel were to diversify rural economies, to create jobs, to attract foreign investment and to preserve native forests. To create a job in forestry requires only about 8 per cent of the investment required to create a job in industry.

The Brazilian pulp industry uses wood derived exclusively from forest plantations. In 1997, replanting and reforestation programmes totalled 102,000 hectares, increasing to 1.42 million hectares the land owned by companies in this sector. Wood consumption in 1997 was estimated at 48.6 million $m^3$, of which 89 per cent went for the production of pulp and paper and 11 per cent was used for energy generation (BRACELPA 1997a).

Reforestation has been a success in Brazil due to the following factors:

1  high productivity in certain areas;
2  availability of cheap land;
3  availability of cheap labour in rural areas; and
4  tax incentives, introduced during the 1960s – 25 per cent of all costs excluding land.

Nevertheless, despite the success, some of these plantations are badly managed, with low yields. The productivity of eucalyptus and pine plantations varies from approximately 3.12 t/ha/year (wb) in dry and poor soils to 27.30 t/ha/year in experimental plots; the Brazilian average is in the range of 7.80 to 11.70 t/ha/year. There is the potential to achieve 15.60–21.50 t/ha/year in many parts of the country with appropriate management methods. In the pulp industry, in 1997, average wood productivity was about 17.20 t/ha/year for eucalyptus and 14.40 t/ha/year for pine (BRACELPA 1997a).

In 1997, 73 per cent of the pulp produced in Brazil used eucalyptus as the raw material (hardwood pulp) and the rest was mainly pine, with some small amounts of sugar-cane bagasse, bamboo, jute, etc. for softwood pulp. Brazil is the world's largest exporter of bleached pulp from eucalyptus. Approximately 30 per cent of the Brazilian production of paper in 1997 was for printing and writing, 45 per cent was for packaging and wrapping, 10 per cent was for board, 9 per cent was for sanitary and household, 4 per cent was for newsprint, and the remainder was for industrial and special purpose papers (BRACELPA 1997a). At this time Brazil was the world's seventh-largest producer of pulp and the twelfth-largest producer of paper. Yet Brazil's per capita consumption of paper is just 38.6 kg/inhabitant/year – low in comparison with developed nations or other countries at a similar stage of economic development (BRACELPA 1998).

The paper and pulp mills can be classified into four main categories:

1   plants producing only pulp via chemical, mechanical or mixed processes;
2   plants producing paper only;
3   plants specialising in paper used for tissues; and
4   integrated plants, producing both paper and pulp.

Table 2.8 shows the recent evolution of the energy inputs used in the Brazilian pulp and paper industry. As can be observed, the largest inputs are electricity, black liquor, firewood and fuel oil. The first three substituted fuel oil, to some extent, in the early 1980s as a result of energy conservation and substitution programmes, with government participation, aimed at reducing the consumption of fuel oil. The initial targets of these programmes were fully achieved, while the medium- to long-term ones were, to a large extent, dropped, mainly because the government withdrew the financial and planning support to these programmes as soon as the oil prices fell (Boas 1990).

Black liquor is a by-product of pulp production by the chemical Kraft process. It is consumed as a fuel in the steam generators of pulp plants and provides a substantial part of the steam requirements of such plants, often integrated with paper mills. This steam often also generates electricity in co-generation units. The production of black liquor increased over 142 per cent in 1983–97 (MME 1998).

In 1996, the integrated plants and those producing just pulp generated respectively 54.4 per cent and 61.6 per cent of their electricity needs. The corresponding share for the plants manufacturing just paper was 6.5 per cent. A figure of 10.7 per cent of this generation, for the sector as a whole, was achieved with hydroelectric plants and the rest with co-generation units, fuelled by black liquor (53.2 per cent), firewood (21.3 per cent), fuel oil (9.5 per cent), coal (5.3 per cent), and natural gas (0.1 per cent) (BRACELPA 1997b).

The electricity consumption figures in Tables 2.7 and 2.8 correspond to the input fuel equivalent of the electricity consumed, assuming the latter is generated in a conventional steam cycle power station, requiring 300 g of fuel oil per kWh generated. According to this assumption, adopted in the Brazilian national energy balances, 1 toe = 0.29 MWh.

To increase the biomass fuels demand in the Brazilian paper and pulp industry, efforts and investments should be directed mainly to improve management of the planted forests, to energy efficiency enhancement programmes in the mills, and to the modernisation of the old co-generation units and the installation of new, more efficient ones that burn firewood or black liquor.

Table 2.8 Energy inputs (in $10^3$ toe) employed in the Brazilian pulp and paper industry, 1983–97

| | 1983 | 1985 | 1987 | 1989 | 1991 | 1993 | 1995 | 1997 |
|---|---|---|---|---|---|---|---|---|
| Black liquor | 770 | 949 | 1,023 | 1,030 | 1,179 | 1,649 | 1,649 | 1,865 |
| Coal | 122 | 131 | 152 | 149 | 129 | 111 | 93 | 85 |
| Electricity | 1,654 | 1,922 | 1,976 | 2,262 | 2,461 | 2,793 | 2,842 | 2,982 |
| Firewood | 757 | 969 | 766 | 725 | 665 | 754 | 897 | 757 |
| Fuel oil | 563 | 352 | 507 | 533 | 600 | 657 | 743 | 902 |
| Natural gas | 9 | 14 | 30 | 30 | 63 | 86 | 122 | 143 |
| Other fuels | 1,352 | 1,317 | 1,549 | 1,584 | 1,802 | 2,332 | 2,426 | 2,804 |
| Other residues | 0 | 84 | 194 | 341 | 315 | 378 | 373 | 348 |
| Sugar-cane bagasse | 0 | 34 | 67 | 73 | 59 | 30 | 5 | 24 |

Source: MME (1998)

## 2.5 New opportunities for electricity generation in the alcohol distilleries and paper and pulp mills

### 2.5.1 Institutional changes in the Brazilian power sector

In 1995 the Brazilian government took the decision to privatise the state-owned electricity utilities, which were responsible for supplying the majority of the electricity market. The British and the Argentinean privatisation models inspired the Brazilian authorities. The main motivation was the relief provided to the ever-growing public debt. However, as a sort of by-product of the privatisation process, a more efficient and autonomous regulatory system is emerging, with a new goal of promoting competition wherever possible in the power supply chain (Bajay and Carvalho 1996).

The institutional change in the Brazilian electric power sector started in 1995 with Law No. 8,987, that set new rules for the concession of public services. Also in 1995, Law No. 9,074 introduced two new agents into the sector: the independent power producers and the 'free' consumers, who can choose their suppliers. In order to make feasible the very existence of these new agents, 'open access' was granted by the same law to both transmission and distribution grids.

The process continued in 1996 with Law No. 9,427, which created the National Agency for Electricity Regulation (ANEEL) to replace the former National Department of Water and Electricity (DNAEE). ANEEL has much wider powers than DNAEE did, and also some autonomy from the federal government, including a separate budget. In 1998, Law No. 9,648 instituted a 'wholesale market' for electricity in the country and a National Operator of a nation-wide interconnected power system. The latter operates the centrally-despatched power stations and the so-called 'basic grid', that connects those power stations with the main load centres in the country. Law No. 9,648 also separated the activities of distribution and supply of electricity, creating, in this way, another new agent in the market: the 'supplier', who should not necessarily own a distribution grid. By the end of 1998 several of these new institutions were still not fully operational, e.g. the wholesale market and associated spot market.

The Brazilian government's urgent wish to privatise its power supply companies has been pushing forward the necessary regulatory changes at a steady pace. However, the other two areas of government intervention in a modern power sector – energy policies and an indicative form of planning – are lagging well behind schedule (Bajay and Carvalho 1998).

### 2.5.2 Regulation of the sales of surplus electricity from self-producers to the public grid

Small, independent electricity producers in Brazil were not allowed to sell surplus electricity to the national grid until 1981. After this time, Federal Government Decree No. 1,872 allowed such trade, but restricted it to isolated electricity systems (DNAEE Order No. 984). Sales to interconnected systems were authorised only in 1989 for long-term contracts of at least ten years (DNAEE Order No. 094) and for short-term ones of about one to three years (DNAEE Order No. 095). In terms of prices, both orders set upper limits, or caps, based on the interconnected grids' long- and short-term generation marginal costs, respectively. The potential avoided costs in the transmission and distribution grids have to be proved by the interested utilities in order to be added to the long-term contracts' price cap.

DNAEE Order No. 220, issued in 1991, allows the power supply utilities to transport electricity generated by a self-producer through their grids, to be consumed elsewhere, provided the consumer premises belong to the same company owned by the self-producer.

Federal Government Decree No. 2,003, of 1996, that details Law No. 9,074, ensures that self-producers and independent power producers (IPPs) have open access to electricity utilities' transmission and distribution grids, after paying due connection and transport fees. These fees were published for the first time by DNAEE, in November 1997 (Order No. 459). There has been widespread criticism that some of the figures are too high and that the distribution costs do not properly reflect the marginal costs.

### 2.5.3 Improved opportunities for biomass for electricity generation

The new institutional setting of the Brazilian power sector will provide price signals that should reflect better the opportunity costs. Industrial and service sector entrepreneurs will be better able to assess the likely benefits of building new plants for self-production, using such plants for load scheduling, selling surplus power to the public grid, and even becoming independent power producers themselves.

The spot market will be signalling continuously the power system's short-term marginal costs, which should constitute the basic reference not just for short-term contracts, but also for the new long-term bilateral contracts and trading in the market of futures and derivatives.

The sugar and alcohol and the paper and pulp sectors are the Brazilian industrial branches that rely most heavily on self-production to meet their electricity needs. As mentioned in Section 2.4, they use mostly co-generation plants burning industrial residues from biomass, e.g. sugar-cane bagasse, firewood and black liquor. The 'new rules of the game' tend to encourage great use of such industrial residues for process steam and power generation in co-generation units, together with forestry residues and sugar-cane trash (barbojo), that recent developments on harvesting machinery and new collection practices are making more cost-effective for power generation, particularly if gasification is involved. All biomass residues will be competing with natural gas and, to a lesser extent, with fuel oil in such applications. However, there is some scope for complementary schemes, as is explained elsewhere (see particularly Chapters 5 and 7).

### 2.5.4 Regulatory barriers to the diffusion of electricity self-sufficiency and independent power production

All self-sufficient producers and some IPPs need a back-up supply of electricity to meet their demand when the generating units suffer forced outages or during maintenance. When the self-producer or IPP is connected to a public grid, it usually takes this supply from the local utility. The utilities often refuse supply or set prohibitively high tariffs as a way of discouraging independent production. The best way to remedy this problem is through firm action and fair and transparent regulation. This behaviour is a formidable barrier to diffusing these forms of power generation and, unfortunately, it has been very much the reality in the Brazilian power sector. DNAEE Order No. 283, of 1985, tried to regulate this issue. It left it up to the utility whether to accept the supply and stated that the demand component of the tariff should be charged all the time, even, as is often the case, when the supply is seldom requested. It set no upper limit for tariffs nor provided a clear methodology to calculate one. As a result, the electricity utilities have been able to impose high tariffs for this type of service, causing many complaints by self-producers and, more recently, from IPPs.

In a comparative study of similar tariffs in France, Bajay (1998) found that the back-up tariffs practised in Brazil are two to four times higher, depending on the tariff group considered. He proposes a balanced method to calculate this type of tariff in Brazil, for small- and medium-sized self-producers and IPPs. The demand component should be proportional to the actual use of the supply. While the new electricity wholesale market is not fully operational, the energy component should correspond to the currently used time of day and seasonal tariff structure applicable to each tariff group. The supply should be provided by the local utility. When the wholesale market becomes operational, the self-producer or IPP should seek its energy supply contract like any other free consumer: in a bilateral arrangement or in the spot market. Initially, the demand charge at peak times (in the time of day and seasonal tariff structure referred to before above) can be used as the capacity component of the back-up tariff. When more realistic figures are available for the use of the transmission and distribution grids, they should supersede the demand charges at peak times. ANEEL is studying these proposals so as to issue an updated order about back-up tariffs as soon as possible.

Another important barrier to the diffusion of electricity self-production and IPP generation is found when, despite the potential benefits of open access to the transmission and distribution grids, they do not materialise because of unrealistically high or badly distributed fees for the use of such grids. ANEEL is well aware of such pitfalls in DNAEE Order No. 459 and is trying to improve the methodology and database employed in the calculation of such fees in order to have more realistic figures available soon.

### 2.5.5 Financial incentives to co-generation and power generation units employing renewable sources of energy

Co-generation plants have received financial incentives in some countries because they have a high efficiency of use of their fuel sources, much higher than any other type of thermal power plant. Such incentives have also been granted to power generation units employing renewable sources of energy. During the 1970s and part of the 1980s the major reason was that they represented indigenous sources of energy, which lowered the dependence on foreign sources. More recently, with globalisation and the formation of economic blocs of countries, this argument has lost much of its early appeal. However, the potential of these generating units to create environmental benefits has started to be rewarded (Bajay 1998). The financial incentives can be orthodox, like tax relief and attractive credit facilities, or heterodox, such as:

1   compulsory purchases by utilities of the power generated by these plants at avoided costs;
2   purchase, for the public grid, of energy blocks through bidding restricted to some types of these plants; or

3 granting of purchase tariffs above the market rate for the energy generated in these plants (to be paid for by all consumers, or on a voluntary basis, e.g. consumers willingness to pay more for 'green' energy).

During 1996–7, the Brazilian Ministry of Mines and Energy discussed with the interested parties a possible federal government decree that would oblige the utilities to buy surplus power from co-generators, up to a certain share of their market growth, more or less on the same lines as the American PURPA legislation, during its first phase. The project was badly designed and the proposed measures came up against the main directives dealing with the opening up of the Brazilian power sector. As a consequence, many utilities rallied against the project and succeeded in aborting it.

Brazil's most important development bank, Banco Nacional de Desenvolvimento Econômico e Social (BNDES), offers some credit facilities for the building of co-generation plants. Although BNDES conditions are more favourable than the average on the Brazilian credit market, they are much worse than those on the international market. Thus, this credit line has been little used.

So far, the owners of small hydro power stations (up to 30 MW) are the only renewable power producers to enjoy 'heterodox' financial incentives in Brazil. In 1998, Law No. 9,648 granted them access to any consumer with a contracted demand higher than 0.5 MW and relieved them of the payment of half the value of transmission grid use fees. The current limits that define 'free' consumers are 10 MW for existing consumers and 3 MW for new ones.

Bajay (1998) proposes that the issue of financial incentives for co-generation and power generation units employing renewable sources of energy in Brazil be tackled in a comprehensive way. A working group with members from the Ministry of Mines and Energy, ANEEL, utilities, IPPs, self-producers, consumer associations and environmentalist groups should put forward short- and long-term proposals, allowing for feedback according to the results obtained. The proposed measures should not affect competition among the other agents in the new power sector structure and they also should not constitute a heavy burden on consumers.

## 2.6 Biomass and the environment

### 2.6.1 Introduction

Carbon dioxide ($CO_2$), nitrogen oxides ($NO_x$), sulphur oxides ($SO_x$), and metal emissions from the burning of fossil fuels are the more evident sources of air pollution. They affect the quality of urban air, help to form acid rain and are among the main causes of global environmental changes. Thus there is a broad understanding that energy conservation and the transition to potentially renewable and cleaner energy sources (solar, biomass, hydroelectricity and wind power), together with the development of pollution abatement technologies, should direct future energy policies, in order to meet environmental goals. Such needs are even more evident when the short-term view of the problem gives way to longer time horizons, where the needs of future generations are considered.

In the case of Brazil, given the favourable climatic conditions, land availability and the accumulated experience, biomass should have a fundamental role in the search for a sustainable energy mix in the country. The substitution of biomass for fossil fuels, using energy-efficient and environmentally-sound conversion technologies, is an important alternative, that contributes, simultaneously, to the reduction of air pollution and the pressure on the country's non-renewable resources. Naturally, therefore, environmentalist pressures can be seen as an important factor favouring the use of biomass.

Despite its potential environmental benefits, if the production and use of biomass are not sustainable, and without adequate policies and management schemes, numerous and complex adverse effects can result. These may impact economically, socially, environmentally and on the food supply. For instance, the occupation of large, continuous land spaces by monocultures (besides inhibiting other relevant agricultural activities, especially the production of food) can contribute to the suppression of significant native vegetation, affecting wild-life habitats and contributing to the reduction of biodiversity. For this reason, besides providing raw material and energy for industry, agriculture and forestry should also have the function of rehabilitating soils, blocking desertification, providing better drainage and preserving biodiversity. Thus, the sustainability of a biomass-based programme should start from a strategic viewpoint, which incorporates the challenge of making compatible the need to improve the air and to improve the rural spaces in Brazil (Ab'Saber 1990).

### 2.6.2 The environmental effects of extracting energy from biomass in Brazil

#### 2.6.2.1 The sugar and alcohol industry

The use of sugar-cane as an energy source makes Brazil responsible for one of the most important renewable energy programmes in the world. Besides being an important strategic option, the partial substitution of ethanol for gasoline in the country's automobile fleet has a strong environmental appeal: it contributes to improving the air quality in urban areas and to the reduction of emissions of greenhouse-effect gases.

As far as air pollution in large urban centres is concerned, the example of the Metropolitan Region of São Paulo can be mentioned, where the acceptable levels of air quality for particulate matter, CO, troposphere $O_3$ and $NO_x$ are often exceeded. Studies of the São Paulo state company, Compantia de Tecnologia de Saneamento Ambiental (CETESB), charged with monitoring pollution emission standards there, shows that vehicles consuming fossil fuels are the main culprits. Despite an increase in the emission of aldehydes, studies show that an increase in the percentage of vehicles running on neat alcohol in the country's fleet would lead to improvements in air quality, because of reductions on the emissions of CO, hydrocarbons (HC) and $SO_x$ (CETESB 1997). As for the addition of anhydrous ethanol to gasoline, besides the elimination of lead emissions, there is also a reduction in the emissions of other pollutants, as shown in Table 2.9.

In the terms of global air pollution, the use of sugar-cane as an energy source is a mitigating measure, since the $CO_2$ emissions during combustion are compensated by the photosynthetic process, inherent to the growing phase of biomass. The Brazilian sugar-cane agro-industry activities are estimated to be responsible for the annual fixing of 12.74 M tonnes of carbon, as shown in Table 2.10.

*Table 2.9* Relative changes in the emission of pollutants by vehicles, as a function of the percentage of anhydrous ethanol in the mixture with gasoline

| | *Percentage of ethanol in the mixture* | | | |
|---|---|---|---|---|
| *Pollutant* | 22 | 18 | 12 | 0 |
| CO | 100 | 120 | 150 | 200–450 |
| HC | 100 | 105 | 110 | 140 |

*Source:* CETESB (1997)

*Table 2.10* Contribution of the Brazilian sugar-cane agro-industry to reducing $CO_2$ emissions (in $10^6$ t C/year)

| *Activities* | *Balance (Mt C/year)* |
|---|---|
| Use of fossil fuels in the sugar-cane agriculture | +1.28 |
| Methane emissions (sugar-cane burning) | +0.06 |
| $N_2O$ emissions | +0.24 |
| Substitution of ethanol for gasoline | −9.13 |
| Substitution of bagasse for fuel oil (in the chemicals and food industries) | −5.20 |
| Net consumption | −12.74 |

*Source:* Macedo (1998)

In the past few years, the continual fall of oil prices in the international market, together with the lack of clear public policies for PROALCOOL, have provoked a vivid debate about the future of the programme. According to Freitas *et al.* (1996), two positions are evident in this debate. First, is the view, which appears to be strengthened with the end of the oil monopoly of PETROBRAS (the state oil monopoly), which argues simply that ethanol is not competitive with gasoline and depends on subsidies for survival in the market. Secondly, the view, defended mainly by the alcohol producers and reinforced by part of the scientific community and environmentalist groups in the country, that the comparison between the pump price of ethanol and the price paid to the alcohol producers, plus transport and retail costs and taxes is misleading, since it does not take into account the related benefits. These include new jobs in the fields, the reduction of air pollution in large towns, the mitigation of emissions of greenhouse-effect gases, the strategic value of PROALCOOL in the event of a future international oil crisis, the savings in foreign currency with the reduction of oil imports, and the existence of a national technology.

Despite these conflicting positions, recent government statements have had in common the need to place the programme at a new stage. Favourable aspects, like the large-scale use of a renewable source of energy, and the programme's strategic value have been highlighted. Firm actions, however, should be taken to eliminate its social, economic and environmental distortions.

The improved utilisation of the sugar-cane by-products, particularly bagasse, vinasse and barbojo, can contribute substantially to making the sugar-cane agro-industry more competitive. There is a large under-used energy potential that, if it is fully exploited, particularly in co-generation systems, can generate extra income through the sales of surplus power. It can also increase the contribution of the sugar-cane agro-industry to reducing the emissions of greenhouse-effect gases.

### 2.6.2.2 Charcoal-consuming iron and steel plants

In principle, charcoal is more environmentally 'friendly' than coal, which often has a high sulphur content and releases more greenhouse-effect gases. It is estimated, for example, that for each tonne of charcoal consumed, 0.4–1.2 t of $CO_2$ are fixed, compared to 1.86 tonne released by the production of pig-iron from coke (as shown in Table 2.11). For example, in 1992 Brazil produced 6.2 million tonnes of pig-iron using charcoal, 39 per cent of which came from planted forests. This represents a positive balance of at least 1 million tonnes of $CO_2$, compared to the use of coke (Rosillo-Calle et al. 1996).

Despite these potential advantages, the Brazilian charcoal industry is associated with poor working conditions and environmental problems, the most important of which is the destruction of native forests that still supply about 30 per cent of the wood used in charcoal production. In the state of Minas Gerais, with the largest concentration of pig-iron and steel plants in the country, the production of charcoal without any environmental consideration is causing serious deforestation. Native forests in the river valleys of the Doce and Medium São Francisco are affected, as well as part of the prairies in the northeast region of the state. The problem has also reached the state of Goias (Luczynski and Sauer 1996). Deforestation is forcing the local charcoal consumers to invest in reforestation projects with species of rapid growth, to meet their wood demand. The charcoal from such planted forests, however, does not appear so far to be competitive with coke.

The reduction of native forests in Minas Gerais and the availability of wood and iron ore in Carajas, in the Amazon region, has, since the 1980s, attracted the production of pig-iron from charcoal to that region. Although charcoal production there was planned to be environmentally sound, based on sustainable forest management, that is not what is actually happening, except in some pilot experiences, described by Rosillo-Calle et al. (1996). The current production of charcoal in Carajas (nearly 500,000 t/year) is mostly fed by wood residues, obtained free or at low cost from illegal deforestation, carried out mainly by timber dealers and cattle-ranchers. A recent report about forestry policy, from the Secretary of Strategic Affairs (Brazil's Security Agency), concluded that

*Table 2.11* $CO_2$ balance associated with the production of pig-iron from coke and from charcoal produced from planted forests (in kg/t of pig-iron)

| Process | CO₂ balance | |
|---|---|---|
| | *Charcoal* | *Coke* |
| Photosynthesis | −3,814 | – |
| Carbonisation | 1,496 | 160 |
| Sintering | 144 | 114 |
| Reduction | 1,791 | 1,589 |
| Total | −388 | 1,863 |

*Source:* Rosillo-Calle *et al.* (1996)

80 per cent of the wood produced in the Amazon region is extracted illegally, without any forestry sustainable management scheme. In the state of Para alone there are 3,500 timber dealers (Schwartzman 1997).

### 2.6.2.3 The paper and pulp industry

After the second international oil price crisis, the Brazilian paper and pulp industry began a strong substitution process of firewood and black liquor for fuel oil. Although the process was started for strategic reasons, related to the supply of oil, the consequent environmental benefits have been remarkable.

Other changes in this industry that have brought about important environmental gains include:

1 lower specific energy consumption figures, obtained through a fast modernisation process, required by the equally fast-growing exports; and
2 rising share of electricity self-production, mainly in co-generation plants, but also, on a much smaller scale, in small- and medium-sized hydro power plants.

Unlike the situation in the production of charcoal, the quality and the standardisation of the raw material required in the production of pulp and paper restricts the sector's forestry activities to planted forests, usually of eucalyptus or pine. As a consequence, the paper and pulp industry, along with the sugar and alcohol and part of the charcoal industries, are often questioned about the wisdom of planting just one species in large, continuous land areas, inflaming the debate about keeping biodiversity, the dislocation of food cultures, contamination by biopesticides, etc. Nevertheless, it should be mentioned that the practice already exists, at least among the more professional, responsible and far-seeing companies in the country, of using plantations in conjunction with native forests ('mosaics'), together with corridors of native vegetation. These practices ensure good levels of preservation of biodiversity and leave room for the biological control of plagues and diseases.

There is also pressure to minimise the planted areas and to reduce industrial pollution and urban waste generation (in this case, waste paper) that, in the end, forces greater use of recycled paper. This is a world-wide tendency, not just because of environmental requirements, but also to allow a lower demand for pulp and, consequently, less need for investment.

In general terms, it can be observed that the main, long-term strategic goal of the paper and pulp industry is a radical reduction in the emission of pollutants, through the recycling of all the manufacturing process by-products, resulting in a closed system, without effluents, and an environmentally-sound management of planted forests.

### 2.6.3 Limits and potentialities of environmental policy tools to provide incentives for the use of biomass

In the difficult international debate about environmental responsibilities, particularly concerning the problem of global warming, the use of biomass in Brazil is often mentioned both in positive terms (because of the important share of renewable energy sources in the country) and in negative ones (because of deforestation and forest fires, mainly in the Amazon Region). However, despite providing a more environmentally acceptable alternative, the growth and the upkeep of current levels of biomass energy uses, particularly ethanol and charcoal, in Brazil faces a strong barrier of insufficient cost-effectiveness in current market conditions, when compared to fossil fuels. So, the challenge to be faced is how to make feasible an energy supply that is environmentally and socially more acceptable (if sustainably managed), given that the competitors, which cause more local and global pollution, are cheaper, at least in the short term. To meet this challenge, it is necessary to create an adequate regulatory environment, employing both direct government regulation tools and market forces. In the case of market forces, this could take place through financial incentives and/or making use of the opportunities created by environmental certificates of voluntary adoption, which are currently included in several international standards. There are also the enforcement perspectives of the Clean Development Mechanism, provisioned in the United Nations Convention on Global Climatic Change.

### 2.6.3.1 Direct regulation

As far as direct regulation tools (licences, emissions and quality standards, environmental zoning, etc.) are concerned, there is the possibility of implementing in the energy area environmentally-motivated public policies, that could favour the use of biomass. For instance, if the illegal exploitation of native forests could be more controlled and if due taxes were collected, charcoal from reforestation would be more competitive *vis-à-vis* charcoal from native forests. It should also be mentioned that many of the coke-based pig-iron plants would need to make high investments to fully meet the environmental requirements of Brazilian legislation, particularly that concerning emissions of air pollutants. The same reasoning applies to meeting fully the legal air quality standards in large cities, which would tend to increase the market for neat ethanol-fuelled vehicles. Another example, also provisioned in law, is the sustainable management of forests, which is a form of biomass exploitation that can provide economic gains, with manageable environmental impacts.

The problem is that, in practice, public management of the environment in Brazil is inefficient, because the government lacks enforcement powers. Better defined public policies are needed, as are greater manpower and financial resources for the institutions charged with applying the policies. In addition, there is considerable resistance to incorporating the environmental element in other public policies.

In the state of São Paulo, the environmental policies formulated recently by the state government indicate that, gradually, energy policy issues are being put into practice by government actions. The following examples illustrate this:

1  A multi-secretary commission, co-ordinated by the Secretary of the Environment, was created to formulate directives for a sustainable mix of energy sources in the state.
2  Two provisions of the state of São Paulo Law No. 9,509, from 1997, that define the state environmental policy and its enforcement mechanisms, deal with energy issues: 'the public authorities should assure the necessary conditions for sustainable development, among other measures, through guidance in a rational use of natural resources, including incentives to energy conservation; . . . It is CETESB's duty to

set the maximum amount, per unit of time, of each pollutant delivered to the environment'. In the case of air pollution from combustion processes, it is evident that reductions in the allowed emissions of air pollutants will induce the conservation and substitution of fuels.

3   State of São Paulo Law No. 9,472, from 1996, requires that: 'In the evaluation of the manufacturing process of industrial plants, the institution responsible for pollution control should observe, among other things, the management forms and conservation measures taken, concerning the process energy inputs'.

### 2.6.3.2 Environmental certificates

Due to the growing concern of several countries with environmental protection and its impacts on international trade, international environmental certificates (EMAS, ISO 14,000, BS 7,750, FSC, etc.) were created, mainly in the 1990s. This set of environmental standards, of voluntary adoption, seeks to promote the mobilisation of market forces, through conscious consumers and producers, to promote improvements in the environmental quality of products and processes.

As far as the production and use of energy from biomass are concerned, environmental certificates can contribute in the following ways:

1   in the market, through the requirement of energy carriers with reduced environmental impacts;
2   in the manufacturing process, through the use of technologies and procedures that minimise the generation of pollutants and the consumption of materials and energy, per unit of product; and
3   in the supply of raw material, through the requirement of a sustainable management of plantations.

### 2.6.3.3 The Clean Development Mechanism

The consequences of the meetings provisioned in the United Nations Convention on Global Climate Changes are of strategic importance for the future of biomass in Brazil, despite the fact that Brazil's participation in the global emission of greenhouse-effect gases is relatively small and the actions seeking its reduction are not among the government's environmental priorities. Already, within the Convention's scope, the details of the Clean Development Mechanism (CDM) are being finalised. It will allow governments and private companies of the industrialised countries to promote actions in developing countries, aimed at reductions in carbon emissions. These reductions will be accounted for by the industrialised countries as a way to accomplish their reduction targets, since experience has shown that measures to reduce emissions cost less in developing countries.

There are still issues to be settled, concerning the criteria to qualify projects which can collect credits, as well as the reference value above which these credits can be accounted for. Nevertheless, if well conducted, this mechanism can be an excellent opportunity to attract foreign capital to expand the use of energy from biomass in Brazil.

## 2.7 Bibliography

Ab'Saber, A. (1990) 'Um plano diferencial para o Brasil', *Revista Estudos Avançados*, **4** (9): 19–62.

Bajay, S. V. (1998) *Proposta, à ANEEL, de uma Metodologia de Obtenção de Tarifas para o Fornecimento de Energia Elétrica, à Guisa de Reserva, para Autoprodutores*, Relatório Técnico – Versão Preliminar, NIPE/UNICAMP.

Bajay, S. V. and Carvalho, E. B. (1994) 'Biomass energy use in Brazil: trends, problems and solutions', paper presented at BioResources '94, Bongalore, India (October).

Bajay, S. V. and Carvalho, E. B. (1996) 'Reestruturação do setor elétrico: Motivações econômicas, financeiras e políticas', *Anais do 7° Congresso Brasileiro de Energia, v. 2*, Rio de Janeiro: COPPE/UFRJ, 1,188–95.

Bajay, S. V. and Carvalho, E. B. (1998) 'Planejamento indicativo: Pré-requisito para uma boa regulação do setor elétrico', *Anais do 3° Congresso Brasileiro de Planejamento Energético*, São Paulo: UNICAMP/USP/EFEI/SE-SP/SBPE, 324–8.

Bajay, S. V., Walter, A. C. S., Carvalho, E. B. and Guerra, S. M. G. (1993) *Sistemas Energéticos Comparados – Fase I: Informe Preliminar – Brasil*, Relatório para o SGT No. 9 do MERCOSUL – Política Energética, Campinas, SP: NIPE/UNICAMP.

Boas, P. V. (1990) *Energia na Indústria de Papel e Celulose: Outubro – 1990*, São Paulo, SP: Associação Nacional de Fabricantes de Papel e Celulose.

Borges, J. M. M. (1994) 'Energy from BioResources: the case of the Brazilian Alcohol Program (PROALCOOL)', paper presented at BioResources '94, Bangalore, India (October).

BRACELPA (1997a) *Annual Report*, São Paulo, SP: Brazilian Pulp and Paper Association.

BRACELPA (1997b) *Consumo de Energia Elétrica no Setor de Papel e Celulose – 1996*, São Paulo, SP: *Associação Brasileira de Celulose e Papel*.

BRACELPA (1998) 'Resultados de 97 superam expectativas do setor', *Celulose e Papel*, **14** (61): 12–15.

CETESB (1997) *Relatório Anual sobre Qualidade do Ar – 1996*, São Paulo, SP.

Freitas, M. A. V., Cecchi, J. C. and Rosa, L. P. (1996) 'Energia da biomassa no Brasil: A atual situação do setor sucro-alcooleiro', *Anais do 7° Congresso Brasileiro de Energia, v. 2*, Rio de Janeiro: COPEE/UFRJ, 949–62.

Goldemberg, J. (1996) 'The evolution of ethanol costs in Brazil', *Energy Policy*, **24** (12): 1127–31.

Goldemberg, J. and Macedo, I. C. (1994) 'Brazilian alcohol program: an overview', *Energy for Sustainable Development*, 1(1): 17–22.

Luczynski, E. and Sauer, I. L. (1996) 'O impacto ambiental da produção de ferro gusa na Amazônia', *Anais do 7° Congresso Brasileiro de Energia, v. 2*, Rio de Janeiro: COPPE/UFRJ, 812–26.

Macedo, I. (1998) 'Greenhouse gas emissions and energy balances in bio-ethanol production and utilization in Brazil', *Biomass and Bioenergy*, **14** (1): 77–81.

MME (1998) *National Energy Balance – 1998*, Brasilia, DF: Ministry of Mines and Energy.

Rosillo-Calle, F., Rezende, M. A. A., Furtado, P. and Hall, D. O. (1996) *The Charcoal Dilemma*, London: Intermediate Technology Publications.

Schwartzman, S. (1997) 'Queimadas na Região Amazônica aumentaram 28% em 97', in *Revista Parabólicas*, No. 34, year 4, São Paulo: Instituto Socioambiental.

Walter, A. C. S., Bajay, S. V. and Nogueira, L. A. H. (1994) 'Power co-generation from sugarcane by-products: an overview of the Brazilian case', paper presented at BioResources '94, Bangalore, India (October).

# THE POTENTIAL CONTRIBUTION OF TECHNOLOGY ASSESSMENT TO BIOENERGY PROGRAMMES

*Harry Rothman and André Furtado*

## 3.1 Introduction

We are coming to the end of an amazing millennium, during which the human population has increased twenty-fold, from about 300 million to 6 billion. Over the millennium our growth in energy exploitation has been almost beyond measure. For the first 800 years of the millennium we had access to relatively few energy sources and the amount generated was, by current standards, slight but mostly renewable. Since then, the growth of industrial capitalism has called into being, at exponential growth rates, massive energy markets and enormous technical advances in energy sources, conversion and motion; first coal and steam, then oil, electricity, nuclear, and so forth. Such was this growth that in the USA this century the energy required to keep the American economy functioning has grown over a thousand-fold (Lapp 1973). During the twentieth century the energy released by our technology, and used by our productive forces generally, created an unprecedented level of wealth whose benefits, however unevenly, flowed through society and around the world. So dazzling were these benefits that their costs went for a long time unseen, or ignored. Almost against our will these costs forced themselves on our attention. Because most of the growth was due to the use of fossil fuels, localised pollution and extraction damage were first noted. Next, concern was expressed that fossil fuel resources would eventually be depleted (Meadows *et al.* 1972). It seemed, however, that there was always a technological solution – for example, nuclear power; human ingenuity was inexhaustible, and thus growth could continue unabated as it had over the last thousand years. That comforting belief dominated policy thinking (Simon and Kahn 1984) until the extent of climatic anthropogenic change became increasingly apparent over the last decade-and-a-half. Governments began to sit up and take notice and sustainability entered the policy lexicon (World Commission for Environment and Development 1987).

We are entering the new millennium with a big question mark against our current unsustainable technological paradigms. How do we change to sustainable technologies, and to sustainable economies able to serve our current wants whilst ensuring an ecological basis for the future? Fortunately we possess knowledge of energy technologies that are more benign, such as bioenergy, wind and solar power, that might form the basis of an alternative energy paradigm. Such knowledge alone will not, of course, ensure its wide acceptance and dominance, we have to find ways and means of demonstrating their advantages over current practices, and develop appropriate policies. We are not able to avoid the 'technological dilemma', i.e. progress always has a cost – it brings disbenefits as well as benefits. Therefore, we should try to structure our policy thinking so that the technological dilemma is recognised and incorporated into our thinking. That is part of what technology assessment (TA) is about; the other part is a broader democratic concern over social equity with respect to any costs and benefits.

## 3.2 The origins of Technology Assessment

Technology Assessment was defined by Huddle (Medford 1973) as

> the purposeful, timely and iterative search for the unanticipated consequences of an innovation derived from applied science or empirical development, identifying affected parties, evaluating the social, environmental and cultural impacts, considering feasible technological alternatives, and revealing constructive opportunities, with the intent of managing more effectively to achieve societal goals.

Hetman (1973: 57), in his classic study for OECD, also added to the definition by arguing, 'It [TA] is neutral and objective seeking to enrich the information for management decisions. Both 'good' and 'bad' side effects are investigated since missed opportunity for benefit may be detrimental to society just as is an unexpected hazard.'

TA first became institutionalised in the USA on the establishment of the Congressional Office of Technology Assessment (OTA) in 1972. This was the culmination of political struggle and of various trends in technology policy studies. The two decades after the end of World War II saw enormous technological development: the jet engine, computing, nuclear power, synthetic organic agricultural chemicals, to name but a few. Inexorably they changed everyday life. As the Fourth Economic Longwave (Freeman 1984) got underway we saw the consumer-based economy develop first in USA, followed by Western Europe and Japan. On a more global scale, the period saw the end of colonialism and the beginning of attempts to universalise industrialism and modernisation. The Cold War and a permanent technological arms race also dominated it. Consequently, development, modernisation and technology transfer were tainted by the struggle of the super-powers; e.g. the Alliance for Progress Programme in Latin America was a US response to the Cuban Revolution of 1959.

Advances occurred not only in technological hardware, but also in managerial tools. Forecasting and planning techniques developed in World War II began to be applied and expanded on a greater scale, especially within special think-tanks set up to speed up R&D and innovation, particularly for weapons technology and military intelligence. Advancing science and technology were perceived by policy-makers as vital elements of the Cold War, at the same time strengthening the competitive abilities of national economies by accelerating the rate of innovation. Science and technology policy obtained a more important role for government, and this was recognised at an international level by organisations such as OECD.

The advance of technology brought many new problems, which in turn demanded solutions. By the 1960s in the First World there was a perceived deterioration in the 'quality of life' by the general public, despite rising living standards. Many trends lay behind this anomaly. These include, along with the associated social and political reactions:

1  Defects in consumer goods, such as cars, which brought forth the consumerist movement, for example, the role of Ralph Nader in the USA (Nader 1965);
2  Increased awareness of pollution, exemplified by Rachel Carson's book *Silent Spring* (Carson 1962), led to widespread protests and the creation of the modern environmentalist movement;
3  Changes in culture and work changed everyday life in the advanced industrial countries; these included greater sexual freedom, the rise of feminism and civil rights movements (Roszak 1968);
4  Nuclear weapons testing produced widespread radiation fallout and a nuclear arms race. A globalised anti-nuclear campaign movement was created in response to the potential catastrophic outcomes (Katz 1986);
5  Expensive state-funded technology programmes, for example, nuclear power and space programmes, fed political scepticism and calls for greater accountability (Commoner 1966).

The iconic year 1968, and its discontents, can now be seen as a first breath of a new reflexive society; a gradual, if uneven, emergence of a new self-awareness – what Beck (1992) called 'the risk society'.

In the United States such developments helped put the regulation of technology firmly on the political agenda by the mid-1960s. Amongst the outcomes were the National Environmental Policy Act (NEPA) in 1969, and the Office of Technology Assessment (OTA) in 1972. NEPA, amongst other things, included the requirement for an environmental impact statement for any Federal programme. The Act establishing the OTA demanded that 'to the fullest extent possible, the consequences of technological applications be anticipated, understood and considered in determination of public policy on existing and emerging national problems.' The OTA, during its lifetime, carried out over 500 assessments, some of which dealt, as we explain later, with biomass energy and related issues (US Congress OTA 1996). In September 1995 a reactionary Republican Congress killed off the OTA. Ironically, by then the OTA had come to serve as a model for TA agencies around the globe. At least ten European Union members now have some form of TA institution, also many international and national agencies – often dealing with environmental or medical issues – carry out TA without necessarily using that appellation.

## 3.3 Methodological issues

TA has several underlying assumptions (Rothman 1978). The first is that a holistic approach to technological programmes is preferable to a fragmented one. Another is that all technical change will cause second, third and $n^{th}$ order changes in the environmental, economic and cultural spheres, most of which will not have been thought about by the initiators, and may prove to be of greater significance than the original aims. It acknowledges that our application of new knowledge opens up new areas of ignorance. It also assumes that we are able to devise means of forecasting these and their degree of risk. It has a democratic dimension since it implies a right to know and be informed, and finally it has a managerial/policy function.

Having become institutionalised in the OTA, TA during its early formative years was characterised by methodological debates. Many schemas were produced (Porter *et al.* 1980); however, their differences were rarely as significant as their authors proclaimed. Implicit in much of the discussion was the naive view that the 'facts' of any situation could be laid out objectively and speak for themselves. Some early enthusiasts seemed to believe neutral analysts and scientific analysis could overrule vested interests, and that such analysis could provide the basis of social consensus. The naivety of such a position led some critics to dismiss TA as a new 'technocratic rhetoric' (Wynne 1975). Yet despite such inadequacies, which we will discuss later, TA does offer valuable new insights for protecting society and the environment, and furthermore, attempts continue to develop it into a constructive policy approach to the socio-economic role of technology.

The basic thinking behind TA can be seen from an examination of the framework (Figure 3.1) suggested by Coates (1998: 41). The reader should avoid a linear interpretation of the framework, which is meant to represent, however crudely, a dynamic and interconnected process. Looking at the boxes we can see how TA thinking expands the traditional engineering and business modus operandi, which might be confined within boxes 1–3.

Similar schemes are found elsewhere, for example, Braun (1998) merges several of Coates' stages to give a five-stage framework, which he summarises as scope, technology, impacts (beneficial and harmful) and policy (STIP). Basically the aim is to describe: the technology in its broadest context, including alternatives; the society and environment; stakeholders, beneficiaries and losers; the nature and type of impacts; and issues and policy options. There are many potential outcomes of an assessment; some are summarised (Coates 1998) in Table 3.1.

Within an assessment there are many techniques that might be used to generate needed information and analysis, their choice depends upon the nature of the problem and predilections of the analysts. Some of the forecasting techniques were developed in the military think-tanks mentioned earlier, others have emerged from academic disciplines, e.g. cost–benefit analysis from welfare economics, and others developed

*Table 3.1* Potential TA outcomes

| |
|---|
| • Modify project to reduce negative aspects and/or to increase benefits |
| • Identify regulatory or other control needs |
| • Stimulate R&D to clarify uncertainties, to define risks more reliably and forestall anticipated negative effects, and to identify new benefits |
| • Define a surveillance programme for operationalised technology |
| • Identify alternative methods for achieving goals of technology |
| • Identify needed institutional change |
| • Provide sound information to all interested parties |
| • Delay the project |
| • Provide partial or incremental implementation strategies |
| • Prevent the technology from being developed or used |
| |
| *Source:* Coates (1998) |

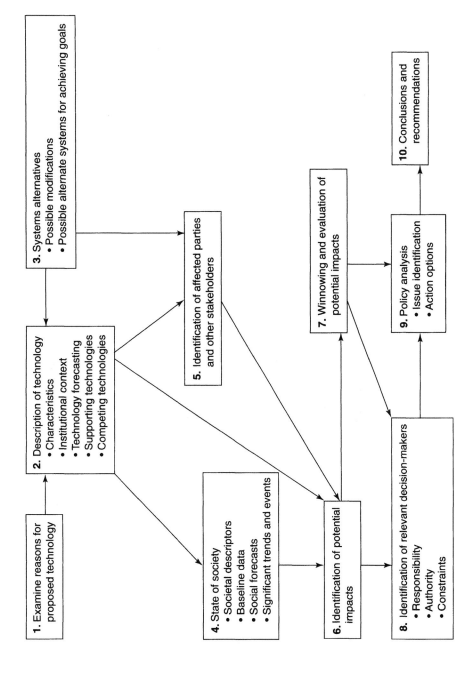

*Figure 3.1* Schema of Technology Assessment
*Source:* Coates (1998)
*Note:* The figure has omitted the feedback loops which connect all boxes

within the context of policy-making, e.g. public enquiries. Porter *et al.* (1980) have provided a comprehensive formal discussion of guidelines for the selection of analytical techniques. Amongst the techniques that they listed are: brainstorming; trend extrapolation; opinion surveys; expert panels and committees; scenarios; checklists; relevance trees; cross-impact matrices; simulation modelling; cost–benefit analyses; and environmental impact statements. Some of the techniques, e.g. sophisticated modelling, described in textbooks and methodological studies, appear too complex, costly and time-consuming for practical purposes relative to any extra value that they might confer upon most studies. Delphi (Office of Science and Technology 1996) and scenario studies (Coates *et al.* 1997), often in combination, seem to be amongst the more popular techniques currently in use. Another methodological trend, which we will deal with in the conclusion to this chapter, is a move to develop ways of encouraging greater public participation.

## 3.4  Case studies of TA exemplars

In this section we will examine some examples of TA studies, which dealt with bioenergy technology issues, to illuminate how they approached key stages in the TA process, such as: the description of technology; identification of impacts, environmental and socio-economic; and policy options. We shall also use material culled from OTA reports, with a few exceptions, to illustrate how contentious issues have been dealt with, and how complex information can be presented in summary form for easier assimilation.

### 3.4.1  Technology description: case study of alternative fuels

The OTA produced a study (US Congress OTA 1990) of the costs and benefits of introducing alternative fuels into the American light duty transport fleet. Three perceived areas of benefit drove the assessment:

1   ozone control in urban air;
2   global warming;
3   national energy security.

We are not concerned in this case study with providing full details of these policy drivers; however, it is well known that fossil fuel combustion and production play an important role in the creation of three of the major greenhouse gases – carbon dioxide, methane and nitrous oxide. This serves to illustrate that it would be advantageous to replace fossil fuels with alternatives that do not produce, or produce reduced amounts of, gases enhancing the greenhouse effect.

The OTA analysts explore which technologies might do this for the light transport fleet. They list three major options: alternative fuels; modifications to existing fossil fuel mixtures; and modifications to the engines. Each of these options also carries with it the need for a series of supporting technologies. The scope of the OTA assessment was limited to examining the potential of alternative fuels only, modification of existing fossils fuels and engines were regarded, in this study, as significant only in so far as they might affect the adoption of alternative fuels.

We shall use the study of alternative fuels to illustrate the role of the 'describe technology' stage of TA. We will also present comparative information of their advantages and disadvantages, and finally present some of the obstacles to their diffusion into the market place.

The examined alternatives were:

1  gasoline;
2  methanol;
3  natural gas;
4  liquid petroleum gas;
5  ethanol;
6  hydrogen;
7  electricity;
8  reformulated gasoline.

The analysts attempt to compare the various advantages and disadvantages of these fuels. We will not present here the detailed analyses; their summarised findings are listed in Table 3.2. Such lists have to be read with care: the findings are not in any way final, since all such comparisons are fraught with uncertainty. Like all good analysts the OTA report's authors stated the major sources of uncertainty that they faced when calculating costs and benefits of introducing any particular fuel alternative. These were:

1  The evolution of current fossil fuel technology, both in the fuel and in vehicles;
2  Few alternative systems had gone beyond the laboratory, or prototypes, their performances in the real world and on a large scale was unknown (the OTA failed to analyse the Brazilian ethanol experience, which might have indicated costings for changing infrastructures);
3  The sensitivity of calculations to numerous future and unpredictable changes in related areas of regulation, manufacturing, marketing, etc.

In the light of these uncertainties the analysts warn 'projections of the costs and benefits of alternative fuels rely on a series of assumptions about technology success, capital charges, feedstock costs, vehicle efficiencies, shipping methods and so forth that are single points in a range of possible values. Changing assumptions to still other plausible values will change the cost–benefit results, sometimes drastically' (US Congress OTA 1990). This warning is one that must be heeded in all such studies.

The report argues that the greatest obstacle to the diffusion of alternative fuels is the entrenchment of gasoline. Enormous sums of capital have been, and are still being, invested in the existing systems by very powerful industrial actors, and by consumers. Furthermore, there is an inextricable tangle of supporting technologies and consumer expectations surrounding the conventional fossil fuel systems that are not easily incorporated into the alternatives. It is easy to make the mistake of viewing a technology, say a specific fuel, in isolation, separate from the myriad interconnections it has developed with other technologies, actors and institutions. Evolutionary economists (Hodgson *et al.* 1994) have attempted to open what they term the black box of technology to reveal and comprehend these by using concepts such as technology systems, national systems of innovation, technological paradigms, etc. The elaboration

*Table 3.2* Pros and cons of alternative fuels

**Advantages**

*Methanol*
- Familiar liquid fuel
- Vehicle development relatively advanced
- Organic emissions (ozone precursors) will have lower reactivity than gasoline emissions
- Lower emissions of toxic pollutants, except formaldehyde
- Engine efficiency should be greater
- Abundant natural gas feedstock
- Less flammable than gasoline
- Can be made from coal or wood (as can gasoline), though at higher cost
- Flexi-fuel 'transition' vehicle available

*Ethanol*
- Familiar liquid fuel
- Organic emissions will have lower reactivity than gasoline emissions (but higher than methanol)
- Lower emissions of toxic pollutants
- Engine efficiency should be greater
- Produced from domestic sources
- Flexifuel 'transition' vehicle available
- Lower CO with gasohol (10 per cent ethanol blend)
- Enzyme-based production from wood being developed

*Natural gas*
- Though imported, probable North American source for moderate supply (1 mmbd or more gasoline *dis*placed)
- Excellent emission characteristics except for potential of somewhat higher $NO_2$ emissions
- Gas is abundant world-wide
- Modest greenhouse advantage
- Can be made from coal

*Electric*
- Fuel is domestically produced and widely available
- Minimal vehicular emissions
- Fuel capacity available (for night-time recharging)
- Big greenhouse advantage if powered by nuclear or solar
- Wide variety of feedstocks in regular commercial use

*Hydrogen*
- Excellent emission characteristics – minimal hydrocarbons
- Would be domestically produced
- Big greenhouse advantage if derived from photovoltaic energy
- Possible fuel cell use

*Reformulated gasoline*
- No infrastructure change except refineries
- Probable small to moderate emission reduction
- Engine modifications not required
- May be available for use by entire fleet, not just new vehicles

*Table 3.2* continued

---

**Disadvantages**

*Methanol*
- Range as much as 1/2 less, or larger fuel tank
- Would probably be imported from overseas
- Formaldehyde emissions a potential problem, especially at higher mileage, requires improved controls
- More toxic than gasoline
- Methanol has non-visible flame, explosive in enclosed tanks. Costs likely to be somewhat higher than gasoline, especially during transition period
- Cold starts, a problem for methanol
- Greenhouse problem if made from coal

*Ethanol*
- Much higher cost than gasoline
- Food/fuel competition at high production levels
- Supply is limited, especially if made from corn
- Range as much as 1/3 less, or larger fuel tanks
- Cold starts, a problem for ethanol

*Natural gas*
- Dedicated vehicles have remaining development needs
- Retail fuel distribution system must be built
- Range quite limited, need large fuel tanks, added costs, reduced space (NPG range not as limited, comparable to methanol)
- Dual fuel 'transition' vehicle has moderate performance, space penalties
- Slower refuelling
- Greenhouse problem if made from coal

*Electric*
- Range, power very limited
- Much battery development required
- Slow refuelling
- Batteries are heavy, bulky, have high replacement costs
- Vehicle space conditioning difficult
- Potential battery disposal problem
- Emissions for power generation can be significant

*Hydrogen*
- Range very limited, need heavy, bulky fuel storage
- Vehicle and total costs high
- Extensive research and development effort required
- Needs new infrastructure

*Reformulated gasoline*
- Emission benefits remain highly uncertain
- Costs uncertain, but will be significant
- No energy security or greenhouse advantage

*Source*: OTA (1990)

of the appropriate technology system could be an extension of the traditional 'technology description stage', which has been generally presented in mechanical fashion. A possible approach has been described by Carlsson (1994), who lists ten dimensions and characteristics by which technological systems can be described:

1  phase of development, in terms of life cycle;
2  future potential;
3  buyer competence;
4  buyer–supplier collaboration;
5  supplier competence;
6  industrial R&D;
7  academic infrastructure;
8  government policy.
9  bridging institutions, able to establish and maintain interactions with actors in the system;
10 gaps/weaknesses.

It is also possible to define a system by national, rather than technical, boundaries, as in the concept of national systems of innovation. Later we describe how Actor Network Theory is being used by some sociologists of science to understand how such systems are created, survive and are destroyed.

### 3.4.2 Environmental impacts: case study of bioenergy crop production

Environmental issues are examined in several chapters of this book, e.g. Chapter 2 looks at environmental policy issues in Brazil, whilst Chapter 4 deals specifically with externalities and sustainability issues. Our aim here is to demonstrate how environmental impact analysis was incorporated into an OTA study prepared as a contribution to the national debate in the USA about the potential benefits and impacts of bioenergy crop programmes (US Congress OTA 1993). Despite its American context, the principles and findings of the study possess a general relevance to illustrating the TA philosophy and approach to environmental impacts and issues.

We shall first look at their analytical framework, which differs in certain respects from the model TA methodology described earlier, as its scope is restricted to environmental issues. The framework is:

1  Nature and kinds of energy crops. (The analysts only dealt with short-rotation woody crops (SRWCs), because they considered herbaceous energy crops (HECs) to be essentially similar to conventional agricultural crops, and therefore offering few novel environmental benefits or disbenefits compared to conventional crops; see Table 3.3 for a summary of environmental impacts of agriculture);
2  Environmental impacts;
3  Rural economic impacts;
4  Federal budget impacts;
5  Trade-balance impacts;
6  Second-order impacts;
7  R&D needs;

8   Demonstration systems;
9   Commercialisation and infrastructures;
10  Institutional issues.

The assessment studied a wide variety of energy crops under development across the US, for example: short-rotation woody crops (hybrid poplars, black locust, silver maple, sweet gum, and eucalyptus), and herbaceous perennials (switchgrass and reed canary grass). Among the advantages such species have over conventional agricultural crops are that they need less intensive farming, are potentially more efficient users of fertilisers and other inputs, and have been selected for high cellulosic biomass production. The analysts recognised that the variety in management regimes required by the different crop species implies differing impact spectra, and that we still do not possess a

*Table 3.3* Environmental impacts of agriculture

---

**Water**
- Water use (irrigated only) that can conflict with other uses or cause ground water mining.
- Leaching of salts and nutrients into surface and ground waters (and runoff into surface waters), which can cause pollution of drinking water supplies for animals and humans, excessive algae growth in streams and ponds, damage to aquatic habitats, and odours.
- Flow of sediments into surface waters, causing increased turbidity, obstruction of streams, filling of reservoirs, destruction of aquatic habitat, increase of flood potential.
- Flow of pesticides into surface and ground waters, potential build-up in food chains, causing both aquatic and terrestrial effects, such as thinning of birds' eggshells.
- Thermal pollution of streams caused by land clearing on stream banks, loss of shade, and thus greater solar heating.

**Air**
- Dust from decreased cover on land, operation of heavy farm machinery.
- Pesticides from aerial spraying or as a component of dust.
- Changed pollen count, human health effects.
- Exhaust emissions from farm machinery.

**Land**
- Erosion and loss of topsoil decreased cover, ploughing, increased water flow because of lower retention; degrading of productivity.
- Displacement of alternative land uses – wilderness, wildlife, aesthetics, etc.
- Change in water retention capabilities of land, increased flooding potential.
- Build-up of pesticide residues in soil, potential damage to soil microbial populations.
- Increase in soil salinity (especially from irrigated agriculture), degrading of soil productivity.
- Depletion of nutrients and organic matter from soil.

**Other**
- Promotion of plant diseases by monocultural cropping practices.
- Occupational health and safety problems associated with operation of heavy machinery, close contact with pesticide residues and involvement in spraying operations.

*Source:* OTA (1990)

---

sufficiently deep scientific and technical understanding of their agro-ecosystems. Nevertheless, in what is essentially a literature desk study, they felt able to present the broad environmental issues posed by biomass crops across the range of environmental sub-systems that included:

1   soil quality: physical, chemical, biological, nutrient cycling, and soil erosion;
2   water quality;
3   air quality;
4   habitats: agriculture, forestry, riparian zones and wetlands, energy crops and habitat;
5   greenhouse gases;
6   biomass as a carbon sink;
7   cropping practices.

Since many of the issues concerning impacts on these environmental systems are discussed elsewhere we shall confine ourselves to summary illustrations. The report's discussion of impacts on water quality deals with both negative and positive potential impacts of high-productivity energy crops, where water demand can be up to 300–1,000 tonnes per tonne of biomass grown. Clearly, in some circumstances there may be damaging effects on local ground water supplies, an issue which they considered 'may pose substantial challenges' (US Congress OTA 1993: 137); on the other hand, in poorly drained or flood-prone areas, bioenergy crops may help with water management problems. There is, says the report, a need for managers of energy crop projects to pay special attention to specific local circumstances, and to recognise the limitations of our current knowledge base. For example, there are enormous uncertainties still about the fate of agricultural chemicals entering the ground water or surface waters. Some of the concerns and issues surrounding water quality impacts of chemical and other pollutant runoff from land are summarised in Table 3.4.

One lesson that emerges clearly from this OTA assessment is that TA involves identifying and resolving trade-offs in benefits and disbenefits. Thus, by using biomass energy we hope to alleviate global warming, which would be of great benefit, but the resultant monoculture may lead to loss of biodiversity, which is now recognised as a serious problem. This is not unique to bioenergy crops, of course, since the development of agriculture generally has had profound and disturbing effects on many species, causing population declines and extinction, as well as population explosions of pests. Changes within established systems of agriculture impacting on wildlife include accelerating mechanisation, greater use of agro-chemicals, and removal of hedgerows to increase field size. Sometimes it has proved possible to alleviate these damaging effects by improvements in agricultural management schemes, such the 'mosaics' and natural vegetation 'corridors' described in Chapter 2. Such schemes are, unfortunately, still far from widely understood and practised.

Part of the solution to resolving the issues posed by such trade-offs lies in a deeper and wider public understanding of the concept of biodiversity, which is more complex than is popularly imagined (see Table 3.5). The commonplace understanding of biodiversity is that it is the number and variety of life forms. Technically, however, diversity is measured at three fundamental hierarchically organised levels of biological organisation: genetic, species, and ecosystems. Until recently public concerns tended to

*Table 3.4*   Impacts of non-point water pollution

---

**Non-point water pollution**, whether from agriculture or other activities, has a variety of impacts on US water resources and fish and other wildlife.

**Increased sedimentation** of streams and other bodies of water, primarily from erosion, may destroy fish feeding and breeding areas. Streams may become broader and shallower so that water temperatures rise, affecting the composition of species the stream will support. Riparian wildlife habitats change, generally reducing species diversity.

**Pollutants and nutrients** associated with eroded sediments can have adverse impacts on aquatic environments. Concentrations of toxic substances may kill aquatic life, whereas nutrients in the runoff can accelerate growth of aquatic flora. This can aggravate the sedimentation problem and lead to accelerated eutrophication of water bodies.

**Eutrophication** is a process that usually begins with the increased production of algae and plants. As they die and settle to the bottom, the micro-organisms that degrade them use up the dissolved oxygen. Sedimentation also contributes to exhausting the oxygen supply, especially in streams and rivers by reducing water turbulence. Thus, the aquatic ecosystem changes dramatically.

**Phosphorus and nitrogen** are the major nutrients that regulate plant growth. Soil nitrogen is frequently leached or runs off into water supplies. Phosphorus, on the other hand, is leached in the soil, so runoff typically contains relatively small amounts. Under normal conditions, therefore, phosphorus is more likely to be the limiting factor in aquatic plant growth. Since phosphorus (along with potassium, calcium, magnesium, sulphur, and the trace elements) is held by colloid material, however, it is abundant in waters receiving large amounts of eroded soil. This can lead to eutrophication.

**Natural eutrophication** is generally a slow process, but man-induced eutrophication can be extremely rapid and can produce nuisance blooms of algae, kill aquatic life by depleting dissolved oxygen, and render water unfit for recreation. Replenishing the oxygen supply is a costly remedy because of the energy and equipment investment on the scale required.

*Source:* OTA (1993)

---

focus on the species level. However, as the recent debate over saving traditional agricultural crop varieties demonstrates, there is now some public perception of the genetic level, whilst public concerns at the ecosystem level have emerged in campaigns, such as those over saving tropical rainforests.

Despite past and current losses in biodiversity, a degree of optimism is possible in the light of some countervailing contemporary social trends. These include:

1   the growth in membership of wildlife and environmental organisations (in Britain the membership of the Royal Society for the Protection of Birds is far larger than that of any political party) often with a global reach, for example, via Northern customer boycotts of 'non-green' products from the South;
2   a new awareness among some scientists and planners of the value of traditional knowledge of plants and animals possessed by native peoples and farmers;
3   an increase in the range and number of national conservation policies and legislation, and global treaties.

*Table 3.5*  What is biological diversity?

---

**Biological diversity**: refers to the variety and variability among living organisms and the ecological complexes in which they occur. Diversity can be defined as the number of different items and their relative frequency. For biological diversity, these items are organised at many levels, ranging from complete ecosystems to the chemical structures that are the molecular basis of heredity. Thus, the term encompasses different ecosystems, species, genes, and their relative abundance; it also encompasses behaviour patterns and interactions.

Diversity varies within ecosystems, species, and genetic levels. For example:

- **Ecosystem diversity**: A landscape interspersed with croplands, grasslands, and woodlands has more diversity than a landscape with most of the woodlands converted to grasslands and croplands.
- **Species diversity**: A rangeland with 100 species of annual and perennial grasses and shrubs has more diversity than the same rangeland after heavy grazing has eliminated or greatly reduced the frequency of the perennial grass species.
- **Genetic diversity**: Economically useful crops are developed from wild plants by selecting valuable inheritable characteristics. Thus, many wild ancestor plants contain genes not found in today's crop plants. An environment that includes both the domestic varieties of a crop (such as corn) and the crop's wild ancestors has more diversity than an environment with wild ancestors eliminated to make way for domestic crops.

Concerns over the loss of biological diversity to date have been defined almost exclusively in terms of species extinction. Although extinction is perhaps the most dramatic aspect of the problem, it is by no means the whole problem. Other aspects include consideration of: species having large habitat requirements of relatively pristine ecological condition; species whose movement is easily prevented with the slightest anthropogenic changes in the landscape; unique communities of species; and many others. These are just a few of the aspects of biological diversity that should be considered. Means of coping with these many aspects of biological diversity, in the context of our lack of knowledge of biological diversity, are being developed. 'Fine filter' approaches deal with the potential loss of individual species; 'coarse filters' focus on maintaining the integrity of entire ecosystems. Energy crops may offer an additional tool at the regional landscape level to assist such strategies.

*Source:* OTA (1993)

---

We would expect in the light of such changes in public awareness, and consequent stricter legislation, developers of future biomass energy projects will have to consider more deeply conservation and local habitat issues.

The biodiversity issue is not a simple one, as we have observed already. Species do not live in isolation, they form complex sets of relationships of living organisms and the physical environment, which are not generally recognised by the layperson. These are studied within the ecosystem concept, a central tenet of modern ecology (Dickinson and Murphy 1998). To illustrate this concept the authors of the OTA assessment present a novel short study, 'The value of a dead tree'. For most people a dead tree is something not deserving of attention, other than for its utility or nuisance value. Far from being something to be necessarily eliminated and removed, they show that a dead tree can, on the contrary, play a most valuable function for an ecosystem and should, therefore, sometimes be left in place. They demonstrate how modern research shows that dead trees play a key part in the ecology and health of forests by providing a wide range of ecological niches for many species of animal, plant, fungi and bacteria, and, after their decomposition, organic inputs to the forest soil.

Finally, the OTA report stresses many of the benefits that may accrue from energy crops, some of which have been listed above, or discussed in other chapters. Nevertheless, its analysis recognises that, as far as wildlife is concerned, there is no substitute for natural habitat and that, as we mention above, a trade-off is necessary between the concerns for conservation and for greenhouse gases. We need, the OTA analysts argued, to utilise ecological principles in the design of energy-cropping systems. To that end they drew up a set of ecology-driven guidelines, which are summarised in Table 3.6. On studying these we cannot but be impressed by our general ignorance and the need for far more research, especially field trials involving the interaction of natural and energy crops systems, a time-consuming and costly activity. At the same time our technology moves onward, creating new problems and issues not covered in the guidelines, for example, the potential negative impacts of genetically-modified energy plant species (Robinson 1999).

### 3.4.3 Social and economic impacts: case study of short-rotation eucalyptus plantations in Brazil

In this section we shall examine the social and economic impacts associated with the development of vast areas of eucalyptus plantations created in Brazil over the past thirty-five years as a resource for biomass energy and paper pulp. Brazil had a long tradition of eucalyptus plantations. From 1909 to 1965 some 470,000 hectares of eucalyptus had been planted in Brazil. However, after the accession of the military in 1964, new policies were implemented, as part of a broader modernisation strategy, which encouraged enormous growth in the distribution and area devoted to plantations. As a result today Brazil possesses the second largest area of eucalytus plantations in the world, about 6m hectares supplying about half the world's eucalyptus pulp. However, this progress has not been an unmitigated success; initially many schemes failed economically and technically and, furthermore, little or no attention was paid to the negative impacts that these plantations might have on the environment or social conditions of the local population.

The scientific debates and political struggles over the impacts, both positive and negative, of eucalyptus plantation growing illustrate the difficulties in coming to any final assessment about the value of technological programmes. This is particularly so where scientific knowledge is insufficiently developed, and political and economic power is massively unequally distributed, an all too common situation everywhere – but especially in the economies of the South. This view is reinforced by an examination of two reports seeking to review the field under consideration, 'Short-rotation Eucalyptus Plantations in Brazil: Social and Environmental Issues' (Couto and Betters 1996) and 'Pulping the South: Industrial Tree Plantations and the World Paper Economy' (Carrere and Lohmann 1996). The former argues that on balance science shows the plantations have not had too negative an effect, whilst the latter claims that research shows that plantation monoculture is environmentally and socially damaging.

In all technology assessments we need to examine the reasons for existence of the technology. In this case, why monoculture plantations, and why eucalyptus plantations?

*Table 3.6*  Prototype ecology-driven guidelines for structuring energy crops

Plant species under consideration for use as bioenergy crops are primarily native species that evolved in the regions where they may be used. These crops can provide greater structural diversity on a landscape level than typical agricultural crops, and thus can enhance wildlife habitat. The extent to which such habitat benefits are realised, however, depends on the careful application of ecological principles, as outlined in prototype guidelines below. These guidelines, however, should be considered only a starting point, requiring much further research. Furthermore, these guidelines are based on principles drawn from studies of natural ecosystems and of highly simplified agricultural systems; there are few or no empirical data for energy crops themselves. Conducting dedicated field-trial research on the ecological interactions of natural systems with energy crops would be useful in order to guide the development of large-scale energy cropping. Ecology-driven guidelines for structuring energy crops might include the following:

- **Site:** Energy crops should be concentrated on current, idled, or former agricultural, pasture, or other 'simplified' or 'marginal' lands. Energy crops should not be grown on naturally structured primary growth forest land, wetlands, prairie, or other natural lands.
- **Species:** Energy crops should combine two or more species in various ways in order to improve species diversity. This would preferably include the use of leguminous species or others with nitrogen-fixing capabilities to reduce the need for artificial fertilisers, and other combinations to reduce potential losses from disease or insects and thus reduce pesticide use. Non-invasive species, which will not escape from cultivated plots, are also preferred.
- **Structure:** Energy crops should combine multiple vegetative structures to enhance landscape diversity as needed by particular species. This could include various combinations of short-rotation woody crops, perennial grasses, and other dedicated energy crops, leaving small to large woody debris and other ground cover, as well as inclusions of natural habitat, as needed. These energy crops could also be used to provide structure to conventional agricultural monoculture through the addition of shelter-belts and hedgerow plantings. Similarly, monoculture of energy crops should have shelter-belts or hedgerows of other types of vegetation.
- **Lifetime:** Landscape structure can also be made more diverse by harvesting adjacent stands on different rotation cycles, including leaving some stands for much longer periods, if possible.
- **Non-indigenous species:** Energy crops should use locally native species to the extent possible. Native species or close relatives will harbour richer insect life and other faunas.
- **Chemicals:** Crops should be chosen to minimise application of agricultural chemicals such as herbicides, insecticides, fungicides, and fertilisers, as discussed above.
- **Unique features:** Unique habitats and features such as small natural wetlands, riparian or other corridors, 'old-growth' incisions, and shelter belts should be preserved and enhanced by the energy crop.
- **Habitat assistance:** Artificial nesting structures and other additions to, or supplements of, habitat features should be provided where appropriate.
- **Research:** Energy crops should be studied carefully at all appropriate scales and on a long-term basis to better understand the best means of improving appropriate habitats for desired species, for the energy crop itself as well as for related agricultural, managed forest, and natural lands. This should also be done on a regional basis, as appropriate.

*Source:* OTA (1993)

The linear model of innovation, whether in its science/technology push or market pull versions, is insufficient to explain the existence of specific technologies. Nowadays theorists often prefer to speak of the 'social shaping of technology' (MacKenzie and Wajcman 1985), a view which stresses the ways in which a society may shape a technology. This is a complex task since any technology owes much to existing scientific and technological knowledge, political, economic and cultural factors, as well as to Nature. How these are brought together in any specific case will be unique, at least in the details; this has been studied recently by sociologists of the actor network theory school (Callon 1995). We need, according to this approach, to visualise a situation in which various actors are struggling and competing for resources to create and control networks – of actors and resources – with the goal of creating technological projects, which themselves may be part of a programme for gaining profit or political power. Actor network theorists argue that the only rationality in this struggle is to be found by following the actors in the network, and by treating it as a study of social power (Latour 1988). These ideas are only just beginning to enter the world of TA, which has historically been somewhat technocratic in its belief that there is an optimal solution in all assessments. It is only possible here to indicate how we might apply these ideas to our case study.

In the following section we will examine the environmental, economic and social impacts associated with the development of eucalyptus plantations. Ideally all TA exercises should endeavour to identify all affected parties and involve them in the assessment exercise. In the development of the Brazilian eucalyptus programmes no such exercise occurred, changes and improvements in plantation policy were made only after political struggle and economic losses.

By definition plantations do not plant themselves! Their creation requires trees, capital, and land. How these are obtained is a complex social process involving the cooperation and the collaboration of many actors, and the overcoming of actors in opposition to the plantation project. Actor networks are created to obtain and concentrate necessary resources, which also need to be cut off from potential or actual opponents. One step in an analysis is to identify the actors and their potential roles. Table 3.7 classifies actors associated with pulp and paper plantations. Many of these potentially wield enormous economic and political power, since they represent large industrial and commercial organisations; other actors are relatively weak, and must rely on collective strength and mobilisation of public opinion. Note that our list includes Nature – who shall speak on behalf of Nature? (One of the characteristics of actor network theory (ANT) is the inclusion of non-human actors in the network.) Nature, we suggest, may speak in the form of environmental changes that force themselves on our attention (e.g. global warming), or via proxy spokespersons, who establish themselves, perhaps, in the form of environmental NGOs. The extent of ANT's value for TA remains to be demonstrated; we argue that it has at least served to demonstrate the futility of looking at technologies in isolation from the social forces which create them.

Direct and indirect social benefits that have been claimed for short-rotation plantations include: job creation; wealth creation; flood control; erosion control; soil improvements; watershed protection; improved water quality; creation of leisure and recreational facilities; and improving the $CO_2$ balance. This has led to claims that in certain circumstances low social costs make plantations a better investment than

*Table 3.7*   Plantation development actors

- Pulp and paper firms, able to buy the fibre from the plantations, e.g. Champion International Corporation, a US company with annual sales in excess of US$5 billion, owning 390,000 ha of timberlands in Brazil
- Consulting companies which provide strategic planning, project development, and other services for plantation owners, e.g. the Finnish company Jaakko Poyry, the world leader, with offices in Brazil
- Technology suppliers
- Industrial associations and alliances, e.g. Associacao Brazileira de Exportadores de Cellulose
- Bilateral aid agencies, e.g. Sweden's SIDA, which has funded plantation planning in the South
- State investment and export credit agencies, e.g. the UK Commonwealth Development Corporation
- Multilateral agencies providing direct and indirect subsidies, e.g. The World Bank funded a project in Minas Gerais encouraging local farmers to grow eucalyptus
- National government provides subsidies and other forms of support, e.g. Brazil subsidised plantations in the period 1966–87
- Research institutes and NGOs that provide scientific and technological support
- Trade unions seeking to protect employment and improve working conditions
- People's landlessness movements, e.g. in Brazil the Movimento dos Trabalhadures Rurais Sem Terra (MST)
- Small farmers
- Environmentalists

*Source:* Carrere and Lohmann (1996)

farming. However, the literature also contains many reports of environmental and social costs, amongst which are environmental-related impacts, high employee accident rates from logging, damage to road systems, and indigenous population displacement. Table 3.8 lists potential environmental effects of logging and forestry.

Economic benefits are generally more easily accounted for than the costs of negative impacts, which may go unobserved for a time, or are even covered up. Couto and Betters' literature review of economic benefits of the forestry sector in Brazil provides a positive overview. The sector generated 3.9 per cent of national GDP, US$18.8 billion by the early 1990s, and this was predicted to increase. They note that a single company, Aracruz Celulose SA, paid in 1992 US$18 million in taxes to the state of Espirito Santo (where its mill is located), spent US$15 million in social investments, allocated US$14.7 million to its forest farmer programme, US$12.5 million to forest extension, US$600,000 to R&D, and paid US$30 million in wages, and bought US$50 million in goods from local suppliers. The generation of economic benefits for Espirito Santo for 1992 was calculated to be US$120.8 million. Couto and Betters do not provide monetary data for any environmental and social costs that might have been incurred; it appears nobody has calculated them. However, the large-scale presence of such negative costs can be attested by the scale of protests by environmentalist, trade union and native groups over the last twenty years in Brazil, which have forced major changes in technical practices.

The nature and extent of social impacts seem equally contentious. Sargent and Bass

*Table 3.8*  Potential environmental effects of logging and forestry

**Water**
- Increased flow of nutrients into surface waters from logging erosion (especially from roads and skid trails)
- Clogging of streams from logging residue
- Leaching of nutrients into surface and ground waters
- Potential improvement of water quality, and more even flow, from forestation of depleted or mined lands
- Herbicide/pesticide pollution from runoff and aerial application (from a small percentage of forested acreage)
- Warming of streams from loss of shading when vegetation adjacent to streams is removed

**Air**
- Fugitive dust, primarily from roads and skid trails
- Emissions from harvesting and transport equipment
- Effects on atmospheric $CO_2$ concentration, especially if forested land is permanently converted to cropland or other (lower biomass) use or vice versa
- Air pollution from prescribed burning

**Land**
- Compaction of soils from roads and heavy equipment leading to following two impacts: surface erosion of forest soils from roads, skid trails, other disturbances;
- Loss of some long-term water storage capacity of forest, increased flooding potential (or increased water availability downstream) until re-vegetation occurs
- Changes in fire hazard, especially from debris
- Possible loss of forest to alternative use or to regenerative failure
- Possible reduction in soil quality/nutrient and organic level from short rotations and/or residue removal (inadequately understood)
- Positive effects of reforestation – reduced erosion increase in water retention, rehabilitation of strip-mined land, drastically improved aesthetic quality, etc.
- Slumps and landslides from loss of root support or improper road design.
- Temporary degrading of aesthetic quality

**Ecological**
- Changes in wildlife from transient effect of cutting and changes in forest type
- Temporary degradation of aquatic ecosystems
- Change in forest type or improved forest from stand conversion

*Source:* OTA (1980)

(1992) argue that plantations have been beneficial to local people. For example, Aracruz Celulose SA, which owns more than 200,000 ha in Espirito Santo and Bahia, they say has improved local social and environmental conditions. This sanguine view is, however, contradicted by other sources. 'Aracruz's tenure has had deleterious effects both on local people and their livelihood and on the soil, water and forests of the region' claim Carrere and Lohmann (1996: 151). They also make a series of serious claims regarding malpractices and social damage resulting from plantation programmes in Brazil that we have summarised in Table 3.9.

We have drawn information from two contrasting reports, Couto and Betters is a consultant's assessment, whereas Carrere and Lohmann are providing an investigative

71

*Table 3.9*   Social impacts of plantations in Brazil

- Expulsion of indigenous populations and small farmers
- Use of violence, actual and symbolic
- Corruption of local and state officials
- Undermining of subsistence farming
- Replacement of agricultural food and fruit growing
- Erosion of agricultural land
- Drying up of wells and streams
- Increasing urban migration
- Weakening of local autonomy and social ties
- Reduction of plantation employment by mechanisation
- Concentration of land and power at the expense of the majority

*Source:* Carrere and Lohmann (1996)

social critique. Each report has sought to marshal the scientific literature to support its case and yet the flavour of their conclusions differs markedly. The former sees the plantations as beneficial on balance, whereas Carrere and Lohmann present a far more negative view. Carrere and Lohmann (1991: 61) are reflective about the seeming contradictions of the scientific literature and speak of tree plantation research as falling into 'two libraries', asking 'why this divergence of views?' They provide three main reasons in answer to their question.

1   Science, they argue, is a social construct and one would therefore expect any scientific work to be 'coloured by the experience of the author and linked to his or her scale of values and vision of the world . . . [and] often directly affected by scientists' material interests' (p. 62). This view is supported by research in the sociology of science (Cozzens and Woodhouse 1995).

2   Silviculture, like the majority of disciplines, has a reductionist tendency dividing reality into parts whilst maintaining it still speaks to the whole. Silviculture cannot answer, on its own, the questions posed by the impacts resulting from plantation monoculture.

3   Research findings regarding environmental and social impacts are not always generalisable, since plantations are found in diverse environmental and socio-economic circumstances.

These differing interpretations serve to illustrate a fundamental problem of technology assessment, that we have to live with and accept such seeming contradictions. The experts do not always agree. The school of constructive technology assessment (Rip *et al.* 1995), therefore, argues that part of the function of the assessors should be to bring such expert conflict to the attention of interested parties rather than brush it aside. Analysts should also note the existence of another, and opposite, problem recorded by sociologists of science. That is a tendency of some scientific experts when faced with a lay audience to down-play uncertainties, and emphasise what is known rather than what is unknown about potential risks and uncertainties of new technology (Jasanoff 1995).

### 3.4.4 Policy analysis: case study of energy technologies for developing countries

The final stage of an assessment involves identifying the requisite decision-makers, their institutional location, responsibility, authority and constraints. Porter *et al.* (1980: 60) state that the 'Assumptions and values held by the assessors should be set forth as explicitly as possible . . . procedures should be clear to the user of the assessment to allow for acceptance or rejection of conclusions drawn.' The assessors at this point need to go beyond their evaluations to identify the key issues and suggest the various options for policy and actions.

'Fuelling Development: Energy Technologies for Developing Countries' (US Congress OTA 1992) states that the main barriers to biomass energy include:

1 inadequate R&D and demonstration;
2 direct or indirect subsidies to other energy supplies that may discourage investment;
3 high land and infrastructure (notably roads) costs;
4 lack of credit, which may be less readily available for biomass than for more conventional supplies.

Their policy options can be divided into several categories: regulatory; planning and evaluation; financial and market; R&D, demonstrations and training. A preferred recommendation is to remove market distortions in energy markets produced by a complex pattern of regulations and subsidies. Another favoured policy is to allow market factors to determine prices more readily by privatising state-owned energy utilities, this market approach can act against biomass energy if no means are found to internalise the externalities of conventional energy sources. In the following section we explore some Brazilian biomass policy experiences.

We must remember that biomass technology is a set of technologies, rather than a single one, consequently there are at least three kinds of technology that policy-makers need to consider when seeking to promote biomass as an energy source or as a raw material. These are:

1 efficient biomass crops that are able to compete with fossil fuels in their energy and raw material purpose (wood, sugar-cane, sorghum, etc.);
2 efficient converters of biomass into secondary energy (alcohol, charcoal, and residues)
3 efficient end-use converters of biomass primary or secondary energy into industrial or domestic services (efficient stoves, alcohol motors, etc.).

Generally we find that these technologies form a technological system, described earlier, in which they are linked together as interrelated minor and major innovations. The success of any one technology thus relies on another's success, so that if one of the technologies is discarded, the same could happen to a whole chain of others independently from their efficiency and fitness (Stewart 1987). Policy-makers, therefore, need to consider this systemic complexity of biomass in their policies.

Technological systems, when they are well established, also rely on a complex network of social and economic interests. Every technology has a set of characteristics at productive or consumption level that favours, more or less, some specific social group (Stewart 1987). Thus, most of the time the success of a new technology depends on the strength of the social groups that lie behind it. Governments have to take into account social power relations in policy decisions if they want to be successful; in that sense we can, as mentioned above, speak of national systems of innovation.

Public policies oriented to foster biomass-efficient technologies that are able to compete with fossil fuel and/or centralised sources will have to overcome serious obstacles. Biomass is an old source of energy, which is being discarded 'naturally' by industrialisation because of its lower productivity, and simultaneously an emergent technology. The obstacles that biomass faces are of several kinds: technological, social and economic.

The Brazilian experience in promoting biomass as a source of energy is very illuminating about the opportunities and the limits of public policies in developing countries. A biomass technological system is situated in, and dependent upon, a specific social context and complex network of actors. However, most of the time, policies are not carried out consistently within the technological system, by fostering at the right time complementary technologies and knowledge, by giving the right signals to political actors, and by introducing the necessary institutional changes.

### 3.4.4.1 The FINEP experience

We shall take as our first example FINEP (the Federal Research Project Funding Agency), that promoted alternative sources of energy during the 1970s and 1980s. FINEP created an Energy Department and launched five energy programmes during the second half of the 1970s (solar, hydrogen, coal, biomass and energy efficiency). The purpose of these programmes was to finance, through non-reimbursable funds, the establishment of research groups that would be able to intervene in new renewable technologies, alongside better-informed energy planning in Brazil. This early move into new energy technologies was a correct choice on FINEP's part, since after the second oil shock energy became the first national security priority. Important academic groups in energy were created and nurtured by FINEP support. In fact, these programmes financed most of the research on alternative sources of energy during the 1970s, creating a technological basis for future development in this field.

After the second oil shock, FINEP funds directed towards alternative energy increased significantly. Without abandoning its funding position to universities and research groups, the federal agency broadened its policy scope. The new emphasis was to finance industrial projects with commercial returns by low interest rate loans. This change proved to be much less successful.

Biomass represented 38 per cent of projects and 28 per cent of the funds designated to energy in 1982–93, about one-third of which was directed towards universities and research laboratories, and two-thirds to firms. The loans were scattered in 176 projects with an average value of US$190,000 (La Rovère 1994). Public funding suffered a great setback during the second half of the 1980s, when government reduced their prioritisation of energy as a national goal, and the economic crisis became more intense.

The local impact of some biomass projects was significant, as in the case of semi-continuous fermentation for alcohol, bagasse dryers, vinasse biodigestors, methane recycling of urban residues, wood gasifiers, and charcoal production. However, none of them had a really significant impact for a specific biomass source. One of the limitations of the strategy was its weak articulation among technologies, and between technologies and market. For example, the FINEP contribution had a limited impact on technological advance for alcohol production. Much of the technological efforts that underlay alcohol productivity advances during the 1980s and the 1990s were financed and executed by the industry itself. In other fields, such as vegetable oils or integrated biomass and energy systems, FINEP had only an exploratory function.

A linear approach to technology policy and the distribution of a limited volume of resources in a heterogeneous group of technologies, without real demand from industry and market, compounded by the 'counter-oil shock' and the low prices of oil, along with the success of well-established industries like oil and hydroelectricity, frustrated the attempts of FINEP to promote alternative energies.

### 3.4.4.2 The Alcohol Program (PROALCOOL)

PROALCOOL, created in 1975, was a much more successful case of government policy to promote biomass (Rothman *et al.* 1983). Alcohol needed to cope with the three levels of technology, that we described before. The scale of the programme was great enough to intervene at all levels:

1    by financing, at negative interest rates, productive activities (plantations and distilleries);
2    by creating a captive market for alcohol as fuel for cars (first obliging the adoption of gasohol, and then of pure alcohol) and a storage capacity (assumed by PETROBRAS);
3    by guaranteeing advantageous and competitive prices for the industry.

PROALCOOL could start with such a well-articulated array of actions because the state apparatus had strong interests links with the sugar-cane industry. Since the 1930s government had created a variety of mechanisms to promote and protect this industry in the country, preserving regional interests (Szmrcsanyi 1979). This institutional learning was used to build the alcohol program. Furthermore, the sugar industry found in alcohol production a new dynamic market when it was suffering an over-capacity problem. Thus, from the start, there were converging institutional mechanisms and social interests underlying the programme implementation.

However, this well-established interest network had a very negative effect at the social level (land ownership concentration, exclusion of subsistence crops) and a very low rate of technological dynamism (CNPq 1979; Furtado 1983). The incorporated technical equipment had a low degree of efficiency, and sugar-cane plantations produced very low yields.

When, at the end of the 1970s and the beginning of the 1980s, alcohol production had to reach a larger market (pure alcohol-fuelled cars), the programme faced a serious challenge because the automobile industry demonstrated little interest to develop and produce efficient alcohol motors. The existing institutional arrangement proved inadequate in encouraging changes in the automobile industry. It was only after very hard pressure from government, associated with briskly falling internal sales, that the automobile industry was induced to develop and put on the market reliable pure alcohol-fuelled cars. The technological success of these cars guaranteed almost 90 per cent of total market share at the end of the 1980s.

The 1986 oil counter-shock generated a new and harsher environment for the sugar-cane industry. The drop in crude oil international prices produced a drop in fuel prices at national level and, of course, of alcohol prices. This new context induced rapid technological changes. There was a sharp increase in sugar-cane crop yields, allied with substantive incremental improvements in the industrial process. The sugar-cane crushing process could significantly improve its productivity through the introduction of minor technological innovations. This technological dynamism allowed the healthiest part of the sugar-cane industry to survive the harsh turn in its environment, while weaker distilleries were closed.

In technology assessment, it is necessary to consider the social and economic context of technological generation and diffusion. The success or failure of biomass as an alternative to fossil fuels is connected with the strength and the weakness of technological opportunities and to social support for each one of these sources of energy. PROALCOOL was a successful energy programme in Brazil because it was supported by the powerful São Paulo agro-industry.

In the beginning federal government involvement by financing investment and guaranteeing a price and market for alcohol was decisive for the programme's take-off. However, since the mid-1980s government had an increasing propensity to play a hands-off role in the programme, reducing subsides and letting alcohol prices fall. Alcohol production was able to survive in this adverse environment because the sugar-cane innovation network became stronger as a result of the learning process. A well-articulated innovation network at scientific, technological and productive level proved important in facing this challenge, and surviving.

### 3.4.4.3 Innovation barriers and policy

The success of a technology is related to its own dynamics and to the evolution of the competition with others. Fossil fuels and centralised sources of energy are well-established technologies that benefit from large-scale economies and increasing returns due to learning. Biomass as a re-emerging technology system has to compete on a whole set of levels with the dominant technological system. There are a variety of barriers that need to be overcome. These barriers are technical, financial/economic, social and institutional. As we mentioned before, biomass requires a parallel and continuous development of a whole set of complementary technologies to be successful. All these technologies, when they differ from the dominant system, have to face these barriers.

Biomass technologies will only succeed if they are able, during the development phase, to acquire enough social support, to be technically feasible, economically viable and have the adapted institutions. This success relies on the effective contribution of a public policy to encourage the constitution of a convergent network of actors.

Policy options depend on how accurate the government is in identifying new or alternative technologies, and how well it is able to promote, implement, and co-ordinate a complex set of incentives and signals that induce and orient actors towards innovation. These options need also to take into account the insertion of the new or alternative technology in the framework of the existing technologies. The greater the changes required, the greater are the needed levels of intervention of government action. Most of the time, new energy technologies can be successful only when they are able to blend with the existing ones (Bhalla 1994).

One of the paths to encourage biomass technology is to take advantage of its present use in some sectors of Third World economies. Rural areas and a lot of small towns in the so-called traditional sector employ biomass frequently for domestic uses. However, in countries like Brazil, with large stocks of natural resources and land, biomass is very frequently used in industry. There the modern sector, i.e. industry and transport, is responsible for 81.8 per cent of biomass energy consumption – mostly used as an alternative to fossil fuels. Thus, biomass can occupy some space in the energy supply of the modern sector without having to remove the infrastructure of the existing technological system. Consequently, we believe there already exist market niche opportunities for biomass which need political engagement to be explored. However, excluding alcohol, at the moment most of these biomass sources of energy do not have any articulated policy. They are also exhibiting strong social problems, and are not environmentally sustainable, for example, in the case of natural wood extraction and wood production.

Government should encourage a favourable environment for investment in each one of these biomass niches. These technologies need to be locally reliable and economically attractive to a network of actors with convergent interests. These actors are research centres and laboratories, government banks, engineering expertise bureaux, equipment suppliers, equipment users, biomass producers, transportation companies, workers, consumers, etc. Government, through the executive branch, must try to co-ordinate all these actors towards innovation.

## 3.5 Conclusions

Most of the examples we have examined issued from the work of the OTA and so reflected American Congressional concerns, values and political commitments. That is not to argue that they have no value out of that context; we think we have shown that those assessments contained information and issues of general concern relevant to problems in Brazil. This raises the questions of where the locus should be for any specifically Brazilian TA study, and to whom it should be directed. What impact, if any would TA have in Brazil? It is not easy to answer these questions because there are problematic issues within TA itself which we need to explore, as well as with the political system of Brazil. First let us explore, briefly, some of the currently perceived problems with TA.

Industrialists are often opposed to TA, which they might rename 'Technology Arrestment', seeing it as a critical threat to their business interests. Certainly this is a regrettable situation, but understandable, especially if business works in an environment that favours short- over long-term interests. Political will is required to change this, and it needs to be stressed that good TA can open up as well as close down business options. Proponents of TA, and sustainable technologies, will have to overcome scepticism and show that it is possible to use TA constructively to improve the level of technical decision-making, and that it is in the national interest. The enormity of this task cannot be overestimated; as the sad spectacle of the closing down of the American OTA shows, it is an ongoing one and there is no final victory.

### 3.5.1  Constructive TA

Over the last decade an interesting development has occurred in the European Union, primarily in the Netherlands and Denmark, namely the attempt to establish what has been termed 'constructive technology assessment' (CTA). Just how influential CTA will become, or even whether it represents a new TA paradigm (Rip *et al.* 1995), is too soon to judge. CTA reflects certain European political and intellectual trends: SETI (Science, Engineering, Technology and Innovation) policy at the national level and EU level; the experience, often bitter and divisive, of technological controversies; and theoretical trends in the sociology of science and technology – some which we have already touched upon above.

Within the EU, a long-running DG XII programme, Forecasting and Assessment in Science and Technology (FAST), was influential in spreading the TA concept (de Hoo *et al.* 1987). There were various national TA offices established as well as supranational organisations and networks, for example, in the European Parliament Scientific and Technological Assessment (STOA), and the European Parliamentary Technology Network. Much of this, and a growing interest by politicians in SETI and Foresight programmes, can be related to widely believed concerns that Europe was falling behind the US and Japan in emerging technological fields, such as IT, communications and biotechnology. Such developments expanded the space across Europe for academic and consultancy researchers in the area, and new ideas, previously confined to academic ghettos, entered the SETI and TA policy arena (Rip 1994). Some of these we have mentioned already: a deeper understanding of the nature of technological innovation, controversy studies, and the application of constructivist ideas to technology (SCOT – Social Construction of Technology); drawing on these CTA emerged as a revision of the original TA concepts.

The basic idea of CTA, says Arie Rip (1994), one of its leading spokespersons, is

> to shift the focus of TA from assessing fully articulated technologies, and introduce the anticipation of technology impacts at an early stage in the development. Actors within the world of technology become an important target group then . . . [and they] are not the only ones to be involved . . . the preferred strategy for CTA is to broaden the aspects and the actors that are taken into account.

The CTA claims to be placing more emphasis on the initial design stages of technological innovation so as to increase the opportunities of identifying impacts prior to a technology becoming entrenched. Is this significantly different from what the TA pioneers were seeking? Is it, with reference to Wynne's initial concern's (1975), an even more sophisticated technocratic tool for enforcing consent from populations that are both bemused and scared of new technologies? There may also be an unwarranted assumption that the professional consultancy and critical scholarly roles of the CTA proponents are compatible in the world of power politics and business.

On the positive side, CTA may offer the opportunity to open up some possibilities for greater democracy, and what Rip (1994) terms 'societal learning', in dealing with technology. This emerges from CTA's constructivist perspective that rejects the traditional image of science and technology as value free, objective, and embodied in consensual experts. As far as technology is concerned, autonomous deterministic images are rejected in favour of images of social shaping. This of course opens up the question of what social factors are involved in the shaping process. This is by no means obvious, and the CTA analyst needs to cast the net widely and creatively to draw in all the likely actors. The original TA steps of finding affected parties and stakeholders might appear to cover this, and early commentators dealt especially with the need to actively seek such actors (Gibbons and Voyer 1974). In practice, however, this has not been very successful if we are to judge by the large number of anti-technology struggles and protests occurring *after* the technology has emerged. If we take, for example, the case of genetically-engineered food, we see that it first reached the shops, and then, afterwards, the protest campaigns began. A CTA approach would have called in the first place, as the technology was being developed, for a concurrent study of potential societal responses.

Sometimes members of the public protest but still feel impotent in the face of technology and powerful corporate owners (Goldsmith 1999), perceiving the technology as something obdurate and entrenched, beyond their control. Constructivists might argue that such a view is unduly pessimistic and that their perspective offers a way to a more democratic politics of technology, which CTA may embody, by showing that social analysis of technology makes it possible to demonstrate both the sources of technological inflexibility and potential for changing it. Such recognition of the possibility of alternatives is a necessary step in political action. Bijker (1996) has sought to order various models of public involvement according to whether they build on the traditional image of science and technology or on a more constructivist image. Travelling that spectrum from traditional to constructivist his list includes: expert advisory committees; hearings; advisory commissions; public inquiry; referendums; consensus conferences; CTA; and citizen's juries. Their relative role and importance within a democracy is not obvious and it is clear that we need more experience (Bijker 1996; Coote and Lenaghan 1997; Dunkerley and Glasner 1998).

CTA, in principle, appears more open than traditional TA to accepting erosion of the traditional boundary between experts and lay-public. From a broader perspective the advisability of such a development has been convincingly argued by Alan Irwin (1995: 7)

> any kind of citizenship which neglects the knowledges held by citizen groups will be restricted in its practical possibilities. Such a limited approach to citizenship will restrict the 'social learning' between science, technology and public groups which . . . is essential to the process of sustainable development.

### 3.5.2 Final words

There is a need to ensure that the question of vested interests is faced. A TA of the PROALCOOL's future, the Third Stage, carried out by the sugar industry or the petroleum industry might be seen as somewhat partisan. Of course, that is not to say an industry or a business should eschew TA; on the contrary, it should grasp the opportunity of discovering novel strategic insights that might be forthcoming from such assessments carried out on its behalf (Braun 1998). It is necessary to find an institutional locus able to bring together interested parties, identify and prioritise issues of conflict, and overcome public mistrust of experts – which may well be founded on bitter past experience. Perhaps there is a space to experiment with some of the citizen-based approaches mentioned above, currently being tried in Europe.

The case for TA at a national level is easier where there a science, engineering, technology and innovation policy clearly linked to political goals. Where these are confused and unclear nobody will wish to take responsibility for prioritising specific technologies. If, for example, Brazil, at national and state levels, is honestly committed to goals of sustainable development, full employment, social equity and education, it is possible to see TA playing a key role in improving policy decisions and democratic participation. That is, of course, a utopian vision, given current asymmetries of economic and political power; in practice TA could be quite partisan and un-representative. Nevertheless, contemporary CTA thinking does contain elements of a critical emancipatory approach to the problems of identifying solutions in society's management of technical change. These are: the multi-disciplinary holistic approach through concurrent studies, the search for affected parties, alternative technologies, and environmental and social impacts; the willingness to deal with difficult questions of technology and equity, technology and gender, and technology and values; and its rejection of any traditional scientistic and technocratic elements within the original TA conceptions.

## 3.6 Bibliography

Beck, U. (1992) *The Risk Society*, London: Sage Publications.

Bhalla, A. (1994) 'Technology choice and development' in Salomon, J. J., Sagaski, F. and Sachs-Jeantet, C. (eds) *The Uncertain Quest: Science, Technology and Development*, Tokyo, United Nations University Press, pp. 412–45.

Bijker, W. (1996) 'Democratization of technology: who are the experts?', *The World Series on Culture and Technology, http://www.desk.nl/~acsi/WS/speakers/bijker.htm*

Braun, E. (1998) *Technology in Context: Technology Assessment for Managers*, London and New York: Routledge.

Callon, M. (1995) 'Technological conception and adoption network: lessons for the CTA prac-titioner', in Rip, A., Misa, T. J. and Johan Schot, J. (eds) *Managing Technology in Society: The Approach of Constructive Technology Assessment*, London: Pinter.

Carlsson, B. (1994) 'Technological Systems and economic performance' in Dodgson, M. and Rothwell, R. (eds) *The Handbook of Industrial Innovation*, Aldershot, Edward Elgar.

Carrere, R. and Lohmann, L. (1996) *Pulping the South: Industrial Tree Plantations and the World Paper Economy*, London and New Jersey: Zed Books.

Carson, R. (1962) *Silent Spring*, New York: Fawcett Crest.

CNPq (1979) *Avaliação Tecnológica do Álcool Etílico*, Brasília: CNPq.

Coates, J. F. (1998) 'Technology assessment as guidance to government management of new technologies in developing countries', *Technology Forecasting and Social Change* 58, 35–46.

Coates, J. F., Machaffie, J. B. and Hines, A. (1997) *2025: Scenarios of US and Global Society Reshaped by Science and Technology*, Greenboro, NC: Oakhill Press.

Commoner, B. (1966) *Science and Survival*, London: Gollancz.

Coote, A. and Lenaghan, J. (1997) *Citizens' Juries: Theory into Practice*, London: Institute for Policy Research, p. 3.

Couto, L. and Betters, D. R. (1996) *Short-rotation Eucalyptus Plantations in Brazil: Social and Environmental Issues*, ORNL/TM-12846, Oak Ridge, TN: Oak Ridge National Laboratory, US Department of Energy.

Cozzens, S. E. and Woodhouse, E. J. (1995) 'Science, government and the politics of knowledge' in Jasanoff, S., Markle, G. E., Peterson, J. C. and Pinch, T. (eds) *Handbook of Science and Technology Studies*, Thousand Oaks: Sage.

de Hoo, S. C., Smits, R. E. H. M. and Petrella, R. (eds) (1987) *Technology Assessment: An Opportunity for Europe*, the Hague: Netherlands Organisation for Technology Assessment (NOTA).

Dickinson, G. and Murphy, K. (1998) *Ecosystems*, London: Routledge.

Dunkerley, D. and Glasner, P. (1998) 'Empowering the public? Citizens' juries and the new genetic technologies', *Critical Public Health* 8 (3), 181–92.

Freeman, C. (ed.) (1984) *Long Waves in the World Economy*, London: Pinter.

Furtado A. (1983) *Biomasse et Style de Développement – les leçons du Programme PROÁLCOOL au Brésil*, Thèse pour le Doctorat de Troisième Cycle, Université de Paris I.

Gibbons, M. and Voyer, R. (1974) *A Technology Assessment System: A Case Study of East Coast Offshore Petroleum Exploration*, Ottawa: The Science Council of Canada.

Goldsmith, Z. (1999) 'The Monsanto test', *The Ecologist* 29 (1), 5–8.

Hetman, F. (1973) *Society and the Assessment of Technology*, Paris: OECD.

Hodgson, G. M., Samuels, W. J. and Tool, M. R. (eds) (1994) *The Elgar Companion to Institutional and Evolutionary Economics*, 2 vols, Aldershot: Edward Elgar.

Irwin, A. (1995) *Citizen Science: A Study of People, Expertise and Sustainable Development*, London: Routledge.

Jasanoff, S. (1995) 'Product, process, or programme: three cultures and the regulation of biotechnology' in Bauer, M. (ed.) *Resistance to New Technologies: Nuclear Power, Information Technology, and Biotechnology*, Cambridge: Cambridge University Press.

Katz, M. S. (1986) *Ban the Bomb: A History of SANE, the Committee for a Sane Nuclear Policy, 1957–85*, Westport, CT: Greenwood.

La Rovère, E. M. (1994) *Energia: atuação e tendências*, Série Especial FINEP, Departamento de Transporte e Energia, Rio de Janeiro.

Lapp, R. E. (1973) *The Logarithmic Century*, Engelwood Cliffs: Prentice Hall.

Latour, B. (1988) *Science in Action: How to Follow Scientists and Engineers Through Society*, Milton Keynes: Open University Press.

MacKenzie, D. and Wajcman, J. (eds) (1985) *The Social Shaping of Technology*, Milton Keynes: Open University Press.

Meadows, D. H., Meadows, D. L., Randers, J. and Behrens, W. W. (1972) *The Limits to Growth*, London: Earth Island.

Medford, D. (1973) *Environmental Harassment or Technology Assessment?* London: Elsevier.

Nader, R. (1965) *Unsafe at Any Speed*, New York: Grossman.

Office of Science and Technology (1996) *Winning Through Foresight – A Strategy Taking the Foresight Programme to the Millennium*, London: OST/Department of Trade and Industry.

Porter, A., Rossini, F. A., Carpenter, S. R. and Roper, A. T. (1980) *A Guidebook for Technology Assessment and Impact Analysis*, New York and Oxford: North Holland.

Rip, A. (1994) 'Science & technology studies and constructive technology assessment', *EASST Review* 13 (3).

Rip, A., Misa, T. J. and Johan Schot, J. (eds) (1995) *Managing Technology in Society: The Approach of Constructive Technology Assessment*, London: Pinter.

Robinson, C. (1999) 'Making forest biotechnology a commercial reality', *Nature/Biotechnology* 17 (1) 27–30.

Roszak, T. (1968) *The Making of a Counter-culture*, New York: Doubleday.

Rothman, H. (1978) 'Technology assessment and the unintended consequences of technology', in Sharma, K. D. and Qureshi, M. A. (eds) *Science, Technology and Development*, New Delhi: Sterling Publishers.

Rothman, H., Greenshields, R. and Rosillo-Calle, F. (1983) *The Alcohol Economy: Fuel Ethanol and the Brazilian Experience*, London: Pinter.

Sargent, C. and Bass, S. (eds) (1992) *Plantation Politics: Forest Plantations in Development*, London: Earthscan.

Simon, J. L. and Kahn, H. (1984) *The Resourceful Earth*, Oxford: Blackwell.

Stewart, F. (1978) *Technology and Underdevelopment*, London: Macmillan Press.

Stewart, F. (1987) 'Macro-policies for appropriate technology: an introductory classification', in Stewart, F. (ed.) *Macro-Policies for Appropriate Technology in Developing Countries*, Westview Special Studies, London: Westview Press.

Szmrcsanyi, T. (1979) *O planejamento da agroindustria canavieira do Brasil: (1930–1975)*, São Paulo: Hucitec and Universidade Estadual de Campinas.

US Congress, Office of Technology Assessment (1980) *Energy from Biological Processes: Volume II – Technical and Environmental Analyses*, Washington, DC: US Government Printing Office.

US Congress, Office of Technology Assessment (1990) *Replacing Gasoline: Alternative Fuels for Light-Duty Vehicles*, OTA-E-364, Washington, DC: US Government Printing Office.

US Congress, Office of Technology Assessment (1992) *Fueling Development: Energy Technologies for Developing Countries*, OTA-E-516, Washington, DC: US Government Printing Office.

US Congress, Office of Technology Assessment (1993) *Potential Environmental Impacts of Bioenergy Crop Production*, OTA-BP-E-118, Washington, DC: US Government Printing Office.

US Congress, Office of Technology Assessment (1996) *The OTA Legacy 1972–1995*, Washington, DC: US Government Printing Office.

World Commission for Environment and Development (1987) *Our Common Future (The Bruntland Report)*, Oxford and New York: Oxford University Press.

Wynne, B. (1975) 'The rhetoric of consensus politics: a critical review of Technology Assessment' *Research Policy* 4, 1–53.

# 4

# ASSESSMENT OF EXTERNALITIES
# AND SUSTAINABILITY ISSUES

*Ausilio Bauen*
*(with contributions from Arnaldo Walter and André Faaij)*

## 4.1 Introduction

Brazil, a country rich in natural and human resources, is struggling to develop economically. While strong economic growth is a clear necessity for Brazil, its implications for the nation's well-being need careful consideration. More than mere growth expressed in terms of Gross Domestic Product (GDP), Brazil is in need of economic development going beyond traditional economic growth, and which should include all factors leading to advances in social welfare, such as environmental quality and enhancement of the individual. Economic development should be part of a sustainable development strategy which 'meets the needs of the present without compromising the ability of future generations to meet their own needs' (WCED 1987).

The full costs and benefits of economic activities are frequently not accounted for. Decision-making processes will consider private costs and benefits, whilst often ignoring a series of additional costs and benefits, known as externalities, which are then borne by society as a whole. A blatant example of externality is the damage caused by pollution. Accounting for the full costs of economic activities is one possible step towards strategies aimed at sustainable development.

Energy is a cornerstone of economic growth and of social development. However, rapid economic growth can be achieved with great detriment to environmental quality and energy may be one of its greatest causes. Energy generation and its use in the residential, commercial, industrial and transport sectors have severe consequences on the local, regional and global environment. Smog, acidification and climate change are some of the environmental concerns associated with energy-related activities, and to these we can add concerns regarding resource depletion, energy security and the drain on the economic resources of countries poor in primary energy sources. All these and more may translate into external costs which are generally not accounted for when choosing amongst different energy supply and end-use options.

Externalities valuation has received little attention to date in Brazil, in particular in relation to energy. This chapter aims at providing an understanding of externalities and their valuation, a discussion on the current state of energy externalities and their role in sustainable development strategies, and a discussion of biomass fuel cycle externalities, with emphasis on industrial uses of biomass energy in Brazil.

## 4.2 Externalities and their internalisation

### 4.2.1 Neoclassical foundations of externalities

The concept of externality is part of the neo-classical theory of welfare economics and was first established by Arthur Pigou (Pigou 1920). Externalities are the result of market failures which lead to certain effects of economic activities not being accounted for in economic transactions. The consideration of these effects is important as it accounts for the effects of economic activities on society as a whole, as opposed to only the parties involved in the economic transaction, and therefore reflects social preferences as opposed to individual preferences. The social costs of an economic activity consist of the sum of the private (internal) and external costs, and the aim of social costs is for products and services to reflect their true costs to society. Externality adders can be added to the private costs of goods and services to reflect their true costs. Social costs, as opposed to private costs, should allow for an economically efficient allocation of resources such that the economic welfare of society and individuals is maximised simultaneously. This relies on the assumption of perfect information, perfect markets and rational (i.e. aiming at maximising individual utility or profit) behaviour on the part of the players. In fact, there are shortcomings to all three of the previous assumptions, making optimal allocation only possible in theory.

According to the neoclassical theory of welfare economics, the optimal level of pollution abatement (Q) is such that the marginal cost of emissions abatement is equal to the marginal cost of pollution damage. For all values of emissions different than Q, the theory implies that there is a welfare loss. For values of abatement below Q this welfare loss consists of an excess of damage costs, and for values of abatement above Q it consists of an excess of abatement costs. In both cases the allocation of resources is not considered optimal, with resources being used which could, in theory, find better use elsewhere in the economy.

Distinction is often made between environmental and non-environmental externalities. Environmental externalities are meant to consider all effects of human economic activity on the natural and man-made environment. However, the concept of environmental externality is strongly anthropocentric, focusing on direct and indirect effects on human health and amenity. Non-environmental externalities are broader in scope and may include issues such as value added to the economy, employment, resource use, and security of supply. Careful consideration is required when assessing if an effect of an economic activity is an externality, as this may not be evident, especially in the case of non-environmental externalities.

The main reasons for a monetary valuation of impacts are:

1   The internalisation of externalities to eliminate market distortions and achieve a more efficient allocation of resources, improving thus economic efficiency.
2   The introduction of market-based mechanisms (e.g. environmental taxes) as a more economically efficient way of achieving environmental objectives than command-and-control measures (Pearce *et al.* 1989).
3   The achievement of weak sustainability through the consideration of social costs and benefits. (Weak sustainability assumes that man-made and natural capital are perfect substitutes and strong sustainability assumes that the ecosystem possesses

fundamental life-sustaining functions which cannot be substituted by man-made capital.)

4   The use of money as a common measuring rod, with which different players are familiar, for the quantification of impacts.

### 4.2.2 Steps towards the internalisation of externalities

To date, the internalisation of environmental externalities has been accomplished mainly through command-and-control measures, which remain the most common regulatory tool adopted for environmental protection. They mainly consist in imposing emission limits on specific activities and represent a reactive approach, favouring end-of-pipe solutions and offering little flexibility. They are generally a costly solution both for the regulated and the regulator. Furthermore, emission limits being fixed, command-and-control measures offer no incentives for improvements beyond those set by the limits. Nevertheless, they may be desirable or even necessary in some cases, as for example in limiting point sources of toxic pollutants. Issues of economic efficiency and of flexibility in achieving environmental protection and the economic burden on industry and government resulting from an increasing number of ever-stricter command-and-control measures have driven the increased interest in market-based mechanisms, in particular directed at certain pollutants (e.g. sulphur emissions). The trend has also been assisted by an increasing liberalisation of western economies, increasing environmental awareness on the part of the public, improved understanding of the environmental damages of pollutants, the refining of methods to quantify the damages of pollution, and the greater availability of technological options to abate emissions.

A variety of economic instruments, such as taxes, subsidies, tradable permits, and also tax credits and deposit refund schemes, can be used as an alternative to command-and-control measures to meet environmental targets. Taxes and subsidies act in a similar way, the first will act by penalising a given polluting activity and the second will act by inciting cleaner activities. For example, $CO_2$ abatement could be achieved by introducing a $CO_2$ tax or by subsidising technologies which reduce emissions. In the case of taxes and subsidies, the level of the economic incentive is fixed and the market will determine the level of emissions. However, the effectiveness of the measures may vary. Tradable permits act differently. The maximum level of emissions is imposed and permits are issued accordingly. Then, following an initial distribution, the permits are traded in the market place and it is the market which will decide on their value. Economic instruments are more flexible than command-and-control measures, reduce the level of government intervention and incite enterprises to make improvements as long as there is an economic benefit. In the case of taxation based on damage cost estimates, there is concern that economic benefit could cease at a level of environmental impact higher than what would be the critical level for the environment. For this reason tradable permits are often seen as an option more in tune with sustainability considerations.

For market-based instruments to be most effective, they must be applied to a level playing field and, before thinking of implementing them, it may be beneficial to eliminate taxes and subsidies in place which bear a negative effect on the environment (i.e. correct government intervention failure) (Roodman 1998). Also, fiscal systems are being reviewed in many countries to shift the burden of taxation away from drivers of economic activity, such as labour, towards unsustainable practices, such as waste and pollution.

## 4.3 The externalities of energy

The energy sector is a major source of environmental and non-environmental externalities (Table 4.1). Consideration of the externalities in decision- and policy-making with regard to energy is fundamental to reduce its negative impacts and move towards a more sustainable energy supply and use.

*Table 4.1* Examples of impact categories leading to potential externalities

| Environmental | Non-environmental |
| --- | --- |
| • Human health<br>• Ecotoxicity (impacts of noxious substances on flora and fauna)<br>• Acidification<br>• Eutrophication<br>• Soil quality<br>• Climate change<br>• Amenity (e.g. noise, odours and visual impacts)<br>• Biodiversity | • Resource use<br>• Employment<br>• Security and reliability of supply<br>• Effects on Gross Domestic Product<br>• Rural development |

Externalities occur at all stages of a fuel cycle. (A fuel cycle is defined as consisting of all activities involved in the supply of thermal or electrical energy to an end-user and consists of the following principal activity groups: primary fuel production and transport, conversion to heat and electricity, and electricity and heat distribution. In the case of energy supply to the transport sector, the fuel cycle consists of primary fuel production, transport and refining, fuel distribution, and conversion to mechanical power. Renewable fuel cycles, such as wind, solar and hydro, do not possess upstream fuel production and transport stages and the fuel cycle consists uniquely of conversion and transport and distribution stages. Energy-saving measures could be considered as a fictitious fuel cycle in which energy, instead of being generated, is saved (negawatts).) The externalities of energy can be reduced by improving fuel cycles, switching between fuel cycles, a more efficient end-use of energy, and reductions in energy consumption. The ultimate goal of externalities valuation is to achieve an economically efficient allocation of resources through the integration of externalities in energy prices. Given the current situation relating to externalities valuation, we are still far from being able to use externalities in search of Pareto optimal solutions (assuming that markets operate perfectly!). However, the valuation of externalities (and the process of assessing externalities generally) is useful for providing an indication of damages/benefits associated with different energy options, for assessing trade-offs between different energy options, and for ranking energy options, and it can serve as a basis for the introduction of economic instruments to reflect the social costs of energy.

The 'monetary valuation of environmental externalities now seems to be the dominant paradigm in the comparative environmental appraisal of energy options' (Stirling 1997). However, the path to assessing externalities is mined with difficulties and uncertainties.

### 4.3.1 Approaches to externalities assessment

The determination of the external costs and benefits of fuel cycles is characterised by three main stages: identification, quantification and monetisation of the impacts.

Two methodologies are commonly used to determine the externalities associated with fuel cycles and are based on top–down or bottom–up approaches. Most of the earlier externalities studies employ a top–down approach where generic damage costs are estimated at a national level for different impact categories (e.g. damage to forests) and are then attributed to various emissions (e.g. $SO_2$) to determine, based on an emissions inventory, an average external cost per unit of emission. The external cost per unit of energy is finally obtained on the basis of generic emissions from different fuel cycles (Hohmeyer 1988; Friedrich and Voss 1993; Pearce 1995; Ott 1997). The top–down approach is generally based on highly aggregated data for damages and emissions. It may be suitable to provide a first indication of the environmental externalities of energy, where sufficient data are available on the state of the environment to estimate specific impacts resulting from emissions of pollutants to the environment. It does not, however, allow for the assessment of the marginal effects of additional energy supply, which are usually of interest for decision-making and planning purposes.

The bottom–up approach is also known as *impact-pathway approach* or *damage-function approach* (DFA) and it allows for the calculation of marginal external costs. The approach can be generally applied to all sorts of impacts for which an impact-pathway can be defined. In the case of pollutants the approach begins with determining the quantity of emissions from a defined source, then makes use of dispersion models and exposure–response functions to determine the marginal damages resulting from the emissions. The final step consists of multiplying the marginal damages by their estimated monetary value. DFA studies are site specific and the marginal external costs obtained are in principle not transferable. The application of this methodology requires large quantities of data and is time-consuming. The results of past studies have shown that externalities calculated using a bottom–up approach tend to be lower that those calculated using top–down approaches. In part this difference appears to be due to the limited consideration of synergistic effects between pollutants and the adoption of linear exposure–response functions in bottom–up studies. The more recent studies use this approach (RCG/Tellus 1995; ORNL/RFF 1994; CEC 1995 and 1998a).

A series of valuation techniques are used to assign monetary values to environmental impacts. *Market prices* can be used for the direct valuation of damages or benefits to commodities which are traded (e.g. damages to forests lead to the loss of timber which can be quantified based on the price at which it is traded on the market). For environmental goods and services for which no direct market exists, economists have had to devise other valuation tools. A direct method consists of the *contingent valuation method* (CVM), in which individuals are asked about their *willingness to pay* (WTP) for improved environmental quality or the *willingness to accept compensation* (WTA) for environmental damage, creating thus a fictitious market for the goods and services considered. Non-market items can also be valued *indirectly*, by examining changes in prices of traded commodities which are linked to them. *Hedonic valuation* looks at differences in prices of market-based goods (e.g. housing prices) to determine individuals' willingness to pay to avoid certain impacts. The *revealed preference method* infers what value individuals place on goods and services by observing their behaviour. For example, *travel-cost valuation* looks at individuals' expenditure to travel to places where a desirable environment may be experienced.

Where damage costs are difficult to determine using the above valuation techniques, or if the uncertainty of the values is judged to be too large, control costs have been proposed in some cases as a proxy for damage costs. Control costs can be determined by assessing the costs of achieving emissions reductions to specific levels (or also costs incurred for mitigating the damages). They do not give an indication of the externality but of what society would have to pay to avoid it. This may be useful in relation to impacts which are characterised by a high degree of uncertainty, as is the case with climate change.

### 4.3.2 The ExternE methodology

The most exhaustive study to date on the external costs of energy is the ExternE project which began as a collaborative effort between the EC and the US in 1991, and of which the European side has completed a third phase in 1998 (CEC 1995 and 1998a). The ExternE methodology (CEC 1995) uses a bottom–up approach to determine the environmental external costs of fuel cycles (Figure 4.1). The project has been principally concerned with the determination of impacts and externalities of air emissions from conventional thermal power plants, as these are likely to cause the most significant (i.e. priority) impacts in the case of conventional fossil fuel cycles. The first step in the methodology is to provide a fuel cycle inventory and impact matrix, based on which a set of priority impacts is identified by expert judgement for further study. To determine the damages of atmospheric pollution, the dispersion and transformation of pollutants is modelled, based on a short-range and long-range atmospheric dispersion model. The local atmospheric dispersion model calculates the pollution increments for one hundred $10 \times 10$ km grid cells around the emission source. The regional atmospheric dispersion model calculates the pollution increments for $100 \times 100$ km grid cells across Europe. The pollution increments can be translated into impacts via exposure–response functions (ERFs). ExternE has selected a large number of ERFs relating impacts to the polluting species considered (e.g. effect of exposure to particulate concentration on acute mortality). The ERFs are the result of an extensive literature survey and are mostly based on recent epidemiological studies carried out across Europe (ExternE Phase III, CEC 1998a). It is important to note that the ERFs used are linear. The economic valuation of the physical impacts is carried out, based on a database of monetary values associated with the different impacts. The monetary values are based on different valuation techniques and have been obtained through a literature survey.

The ECOSENSE software developed within the framework of the ExternE project performs the external costs calculations for short-range and long-range atmospheric pollution from point sources. ECOSENSE requires input in the form of plant characteristics and location, emissions per unit flue gas volume, and meteorological data for the short-range dispersion model. The results are provided as a range of low- and high-cost estimates for damages to human health, damages to forestry and crops and damages to building material.

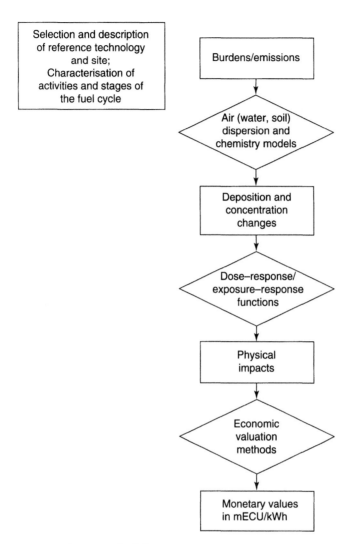

*Figure 4.1* Impact-pathway methodology (CEC 1998b)

The ExternE methodology stresses three principles which are important in externality valuation. They are:

1   transparency (i.e. clear description of method, assumptions and data used);
2   comprehensiveness (i.e. consideration of all significant impacts and full account of their spatial and temporal effects); and
3   consistency (i.e. allow for comparisons between different fuel cycles and sites).

### 4.3.3 *Review of externalities of energy*

The previous sections have made reference to a number of studies which have addressed the externalities of energy, and Table 4.2 provides the ranges for the externalities obtained by some prominent studies since the late 1980s for different fuel cycles (the value of the externality associated with climate change impacts is given in parentheses). The review which follows provides an indication of the impacts on which valuation has focused to date, the magnitude of energy externalities, the differences in values between studies and the relative importance of different impacts.

Studies have mainly focused on the impacts of fuel cycles on human health, and these generally represent the most significant contribution to the value of the externality. In cases where estimates of damage from climate change caused by the emission of greenhouse gases are considered, they often overwhelm other externality values. The range of externality values are also wider where damages from climate change are considered because of even larger uncertainties over the impacts. Apart from the uncertainties surrounding the physical impacts of climate change, further uncertainty is added by different economic assumptions made in valuing potential damages. Typically, the level of discounting is the cause of considerable controversy and an important ethical issue. Small changes in the discount rate cause large variations in damage estimates because of the long-term effects of climate change.

Hohmeyer (1988) uses a top–down approach to value damages associated with environmental impacts of fossil fuel generation on flora, fauna, humans and materials, and considers climate change impacts. A single externality has been attributed to electricity from fossil fuels in general, expressed per unit of electricity generated. However, this cost is likely for the most part to be attributable to coal, in particular old coal plants, which should account for most of the damage. The external costs of nuclear energy are found to be large and are attributed to impacts on human health from normal operation and accidents and to resource depletion. Climate change accounts for just a small part of the externalities valued in this study, representing less than 1 per cent of the estimate.

The study also considers a number of non-environmental externalities such as the depletion of non-renewable resources and government subsidies, with the first contributing between a quarter and half of the externality estimate for fossil fuels and between one-third and two-thirds of the value for nuclear. The externalities estimated are believed to represent only the tip of the iceberg and consideration of further externalities would further strengthen the stance of renewables. The net benefits of wind and solar energy result from economic effects such as gross value added, savings and employment.

Hohmeyer's study, being one of the first to attempt the quantification of the externalities of energy, attracted great attention, particularly because it indicated that the externalities of conventional generation are significant and similar in magnitude to the price of electricity. As a response, Friedrich and Voss (1993) carried out a similar study, which resulted in much lower externalities for fossil and nuclear fuel cycles. The study rejects most of the non-environmental externalities claimed by Hohmeyer (1988) (e.g. employment), estimates the cost of utilisation of non-renewable resources as being small and possibly internalised, and estimates R&D expenditure and public subsidies as being significant externalities. Expenditure on R&D accounts for most of the externality estimate in the case of wind and solar.

The PACE study (Ottinger *et al.* 1990) is based on a literature review of environmental impacts based on bottom–up studies. The externalities valued refer to damages of air pollution. The damage cost of climate change impacts accounts for a large portion of the externality associated with the fossil fuel cycles. The bulk (80 per cent) of the externality of the nuclear cycle is associated with the risk of accidental emissions. The externalities associated with renewable energy are mainly a result of toxic emissions

Table 4.2 Review of externalities of energy[a, b] (UScents/kWh)

| | Hohmeyer (1988) | Friedrich and Voss (1993) | Ottinger et al. (1990) | RCG/Tellus (1995) | Masuhr and Ott (1994) | ORNL/RFF (1994)[c] | Pearce (1995) | CEC (1998a) |
|---|---|---|---|---|---|---|---|---|
| Coal | 4.1–9.3 | 0.4–2.0 | 2.9–6.7 (1.7) | 0.3 | – | 0.06–0.13 | 1.5–8.1 (0.7–0.8) | 0.8–31.4 (0.5–18.0) |
| Oil | – | – | 2.9–6.8 (1.2) | 0.2 | 6.0–88.0 (3.1–85.4) | 0.018–0.024 | 5.7–9.1 (0.7) | 2.0–24.8 (0.4–15.8) |
| Gas | – | – | 0.8–1.2 (0.79) | 0.02 | 3.3–61.0 (2.2–59.8) | 0.0013–0.024 | 0.6–0.7 (0.3) | 0.3–10.5 (0.2–9.8) |
| Nuclear | 10.2–21.9 | 0.03–0.6 | 3.4 | 0.01 | 0.3–3.0 (0.1–2.8) | 0.022–0.034 | 0.07–0.5 (0.02) | 0.3–1.0 (0.01–0.04) |
| Biomass | – | – | 0–0.8 | 0.3 | – | 0.19 | 0.4 (0.04) | 0.1–4.2 (0.08–0.3) |
| Hydro | – | – | – | – | 0.2–1.2 | 0–0.017 | 0.06 (0.008) | 1.0–0.9 |
| Solar | 7.2–18.0 | 0.05–1.2 | 0–0.5 | – | – | – | 0.1 (0.005) | 0.08–1.1 (0.03–1.0) |
| Wind | 5.9–13.0 | 0.02–0.4 | 0–0.1 | 0.001 | – | – | 0.02–0.07 (0.005) | 0.05–0.5 |

Notes
[a] Values in parentheses indicate contribution of climate change externality to the externality value provided
[b] Italicized values denote an environmental benefit
[c] Data summarised in Lee (1996)

from the manufacturing process in the case of photovoltaics, of noise in the case of wind, and of atmospheric emissions in the case of biomass.

Masuhr and Ott (1994) and Ott (1997) discuss a top–down approach applied to Switzerland. The externalities of the fossil fuel cycles account for the damages of air pollution to human health, buildings, agriculture and forestry. The nuclear energy externality accounts only for estimated deaths caused by normal plant operation. The principal externalities associated with hydropower are a result of the impairment of natural landscapes and the impacts on water systems. The externality values are based on willingness to pay surveys on conservation and biodiversity and on the valuation of the recreational function of natural landscapes. The costs of climate change are valued in terms of damage cost estimate ranges and average avoidance costs for Switzerland (damage cost estimates are shown in Table 4.2).

Pearce (1995) estimates externality adders for UK power generation based on a literature review of externalities associated with different pollutants and on a range of emissions for different generating technologies. The estimates account for air pollution and climate change impacts. The climate change damage cost is based on an estimate by Fankhauser (1995). The externality adder for nuclear energy is largely a result of damage estimates for accidental emissions. The externality estimates for hydro and wind account only for damages from emissions of pollutants from equipment production and from the construction stage, and do not include – although they are mentioned – more site specific effects such as noise, landscape changes and effects on fauna, which may be dominant for such generating systems.

The RCG/Tellus (1995) study, also known as the New York State Externalities Study, is based on a bottom–up approach. The study considers the impacts of air, water and soil pollution. For fossil fuel cycles, air pollution impacts are the only ones of significance. The study does not account for climate change impacts. Impacts of water pollution appear to be significant in the particular biomass case considered. The nuclear energy externalities are dominated by radiation exposure impacts from normal operation. The wind energy externalities are a result of impacts on the landscape. The externality adders calculated are lower than those obtained by the previous studies. However, the results for the Sterling, NY, site are particularly low, and within the same study, the siting of fossil facilities (natural gas and oil) at other sites has resulted in increases in externality adders of up to a factor of eight. Ottinger (1996) has criticised the study as suffering from serious omissions and undervaluations and his criticism extends to other bottom–up studies. Also, the low health damages calculated may be explained by the low population densities exposed to the pollutants and by a lack of adequate air dispersion modelling (CEC 1998b).

The Oak Ridge National Laboratory (ORNL) and Resources for the Future (RFF) study (ORNL/RFF 1994) has been carried out as part of the EC/US External Costs of Fuel Cycles study, the first phase of the ExternE project (CEC 1995). The study, based on the bottom–up approach, focuses on the impacts on human health of atmospheric emissions from power-generating facilities. Like the RCG/Tellus study, the externalities calculated are significantly lower than those of other studies.

The most recent and extensive effort to value the externalities of energy is provided by the ExternE project. The third phase of the project (CEC 1998a) has assessed the externalities of fossil, nuclear and renewable fuel cycles across the European Union member states. For the fossil fuel cycles the range of externalities is strongly influenced by the technology chosen for the case studies and by their location. For example, a similar facility sited in Sweden and in Germany is likely to present lower externality values for Sweden because of the likely lower population which may be exposed to pollution. Such site-specific effects may lead to different priorities with regard to the impacts of fuel cycles at different sites. In the case of the nuclear fuel cycle, the external costs associated with the risk of accidental emissions are very small. However, the study admits that much controversy exists on how public perception of risk should be included in the analysis. Most of the damages are attributed to radioactive emissions of abandoned mill tailings and to climate change impacts of the emissions from reprocessing stages. The externalities of the biomass fuel cycles are generally lower than those of the best fossil fuel cycle considered. The external benefit obtained for hydropower reflects the Austrian case study where only benefits of protection from flooding and effects on navigation have been considered. The site dependency of externality estimates is also likely to be great for hydropower because of the strong influence of local amenity and ecological issues. The externalities of both nuclear and renewables are small, but the uncertainties over the risks associated with nuclear are much larger.

The recent BioCosts study (CEC 1998b) has focused on the externalities of biomass fuel cycles and their comparison to reference fuel cycles at different sites within the European Union. The study is based on the ExternE methodology and the ranges of externalities obtained are provided in Table 4.3 and are discussed in more detail in Section 4.4.1.

Few studies on the externalities of energy have been carried out outside Europe and the US. A study by Carnevali and Suarez (1993) assessed the effects of Argentinean energy policies of the 1970s and 1980s on air pollution emissions and emissions control costs. It is estimated that fuel switches avoided a capital expenditure on emissions control of over US$1.5 billion. Van Horen (1996) carried out an assessment of the externalities of coal and nuclear energy for South Africa. The externalities of coal consider mining injuries and deaths, health impacts from air pollution and climate change impacts, and they range between 0.6 and 3.4 UScents/kWh (0.16 and 0.24 UScents/kWh excluding climate change impacts). The externalities for nuclear consider exclusively fiscal subsidies and range between 0.9 and 3.1 UScents/kWh. Furtado (1996) carried out a contingent valuation study to assess the WTP to avoid environmental impacts from hydro, coal and nuclear power in Brazil (Table 4.4). The study which related to three specific facilities showed public preferences to favour hydropower, followed by coal and finally nuclear. After comparison with externalities determined in other European and US studies, Furtado found the values to be sufficiently reliable for use in cost–benefit analysis of energy generation options. However, for the facilities considered, the inclusion of the external costs considered would not have influenced their ranking based on private costs. Furtado's study is a pioneer in the valuation of energy externalities in Brazil. The study, though, relies on contingent valuation alone, with all impacts aggregated in a unique value, and lacks specificity with regard to technology and knowledge of actual impacts.

The review of the externalities of energy illustrates the wide range of values found in the literature. The assumptions and methods vary greatly for the different studies, and many results are strongly site-dependent. However, it can generally be concluded that the externalities of energy are most likely to be significant in relation to the current price

Table 4.3 BioCosts study results (UScents/kWh) (CEC 1998b)

| | Biomass | | | | Fossil | | | |
| | Original site | | Lauffen DE | | Original site | | Lauffen DE | |
| | Conversion | All stages | Conversion | All stages | Conversion | All stages | Conversion | All stages |
|---|---|---|---|---|---|---|---|---|
| Nassjo SE[a] | 0.11 | 0.14 | 0.51 | 0.67 | 0.46 | 0.60 | 1.96 | 2.62 |
| Mangualde PT[b] | 1.14 | 1.16 | 1.96 | 2.09 | 14.39 | 14.39 | 23.54 | 23.54 |
| Hashoj DK[c] | 0.90 | 1.06 | 2.49 | 3.01 | 0.82 | 0.82 | 2.35 | 2.35 |
| Varnamo SE[d] | 0.22 | 0.27 | 1.05 | 1.29 | 0.48 | 0.63 | 2.09 | 2.62 |
| Eggborough UK[e] | 0.29 | 0.61 | 0.69 | 1.44 | 3.27 | 3.27 | 7.85 | 7.98 |
| Weissenburg DE[f] | 18.31 | 18.31 | 19.62 | 20.93 | 19.62 | 20.93 | 23.54 | 23.54 |

Notes
[a] Utilisation of forestry residues in the Nässjö circulating fluidised bed combustion plant, Sweden, versus the use of Polish coal in the same plant.
[b] Utilisation of woody biomass for industrial combined heat and power production in Mangualde, Portugal, versus the use of fuel oil in an engine generating heat and power.
[c] Production of biogas from slurry for municipal combined heat and power generation at Hashöj, Denmark, versus the use of Danish natural gas in the same engine.
[d] High-pressure gasification of forestry residues for combined heat and power generation at Värnamo, Sweden, versus the use of coal in the Nässjö plant.
[e] Atmospheric pressure gasification of short-rotation coppice for electricity generation at Eggborough, UK, versus the use of coal in a power plant.
[f] Production of cold-pressed rape-seed oil and its use in a co-generation plant at Weissenburg, Germany, versus the use of diesel fuel in a similar engine.

*Table 4.4* CVM estimates of externalities of energy in Brazil (Furtado 1996)

| Conversion facility | Externality (UScents/kWh) |
| --- | --- |
| Hydro | 0.38–0.81 |
| Coal | 1.34–2.81 |
| Nuclear | 2.97–5.95 |

of energy. The difference in externality between fossil and renewable sources is also likely to be significant, in particular when considering $CO_2$ emissions. Greatest benefits of renewables appear when comparing old coal technology to wind, while the benefits are reduced when comparing natural gas to biomass, where the benefit may largely be attributed to reduced $CO_2$ emissions. The case of natural gas and biomass fuel cycles will be discussed in greater detail later. The range of externalities of nuclear energy is large, mainly due to difficulties in assessing the risk of nuclear accidents. Nuclear energy also presents difficulties (e.g. disposal of radioactive waste material) which raise questions about the sustainability of the fuel cycle.

### 4.3.4 The limitations of externalities

Most of the externalities studies carried out to date acknowledge fundamental problems due to lack of scientific knowledge, uncertainties at various stages of the valuation process, biases in valuation, differences in economic assumptions and ethical issues. However, there is also general agreement that because of omissions in the quantification of impacts, the externalities presented are in most cases believed to underestimate the actual level of externalities.

There are considerable differences between the values obtained by the studies reviewed here. They are mainly due to the variety of methodological approaches used, to differences in the impacts considered, in the emissions estimates for the fuel cycles, in the specific damages attributed to emissions, and in the assumptions underlying the risk of nuclear energy, and, where climate change is considered, to the wide range of damage estimates calculated (Lee 1996).

Most of the earlier studies, based on top–down approaches, obtain higher externality values compared to more recent studies based on impact-pathway approaches. The extent of the effects considered is also very important. For example, the inclusion of climate change damage estimates generally leads to much higher externalities being attributed to fossil fuel cycles and has a significant influence on the nuclear cycle. The consideration of non-environmental externalities may also affect significantly the externality estimates. The technologies considered in the fuel cycle are also a cause of differences in estimates. For example, noxious emissions from a coal-based fuel cycle using integrated gasification combined cycle technology (IGCC) are much lower than those from a coal-based fuel cycle using old coal-fired boilers with no emissions control. Also, differences in generic damage estimates or in exposure–response functions used lead to significant differences in externality estimates, and so do assumptions made with regard to the risks associated with nuclear energy, for example with regard to the probability of severe accident, releases and exposure in the case of severe accident, and risk perception.

Many of the problems affecting the reliability of externality studies can be mitigated or solved through methodological refinements and improvements in scientific knowledge. Addressing the variability of results in externality studies, the US Office of Technology Assessment (OTA) stated that 'many differences can be addressed through further research and analysis. Some critical agreements over methodology, however, mask deeper disputes over values, basic policy goals, and the intended role of environmental cost studies. It is unlikely that these disputes can be resolved by technical analysis or scientific research.' (OTA 1994).

There remain a number of limitations associated with externalities values which raise questions about their usefulness in decision-making processes. Stirling (1997) asserts that externality valuation suffers the same drawback as other aggregated quantitative techniques, that is the 'failure to address the multidimensional nature of environmental appraisal'.

Some important issues concern the distribution of environmental effects. They are: the predominantly local effects of certain fuel cycles as opposed to the predominantly regional and global effects of others; the question of how to deal with intragenerational and intergenerational equity (e.g. how impacts are distributed among the population and how to address long-term impacts such as climate change); and the anthropocentrism which characterises environmental valuation and which may not attribute the necessary relevance to the diversity of ecological systems.

Questions can also be raised as to the way monetary valuation addresses environmental effects in terms of severity (e.g. deaths as opposed to serious injuries), immediacy (e.g. injury as opposed to disease), gravity (e.g. the high probability of small impacts of fossil generation as opposed to the low probability of large impacts of nuclear generation), and reversibility (e.g. the irreversibility of climate change and radiation impacts as opposed to the reversibility of changes in landscape of certain renewables such as wind).

Monetary values may also give a false sense of objectivity in aggregating impacts over which those affected have different degrees of voluntariness and control (e.g. the health impacts of air pollution as opposed to the right to a pristine landscape). Monetary valuation is also undermined by issues of comprehensiveness, emphasis being mainly on more readily monetisable impacts, and by issues of reliability in the techniques used in estimating impacts and monetary values, which affect the uncertainty of externalities. These are principally due to lack of sufficient knowledge, data quality, complexity of some of the effects, and diversity of empirical and theoretical models used. The variety of influences affecting the uncertainty of externalities renders their treatment by orthodox probabilistic approaches a difficult task. The best way to deal with uncertainty appears to be to make use of ranges of values and sensitivity analysis. It is fundamental, given the current state of the valuation of external effects, to specify, at different stages of the process leading to the monetisation of the impacts, the degree of confidence in the data and models used.

### 4.3.5 The internalisation of externalities and sustainability

The difficulties experienced with externalities valuation have considerably hindered their application in decision- and policy-making. Although they are critical with respect to the correction of pricing mechanisms, these difficulties should be relativised when externalities are to be used as an indication of the potential costs or benefits of fuel cycles, as a measuring rod for comparing fuel cycles, and as a tool supporting rational market-based instruments aiming at the correction of market distortions. Under present circumstances externalities appear best suited as inputs for policy formulation, rather than as corrections to market prices.

In essence, the current scope of externalities of energy renders them insufficient as unique criteria, in association with private costs, on which to base decision- and policy-making for energy options. Other considerations, not satisfactorily addressed in terms of externalities, need to be taken into account. The inability to express a variety of impacts in terms of externalities, uncertainties governing the values of those impacts which can be monetised, and the risk that externalities alone will not ensure that life-supporting functions are conserved over time cause concern with regard to the achievement of sustainability. It is then fundamental to address the sustainability of fuel cycles and to consider the role of externalities in achieving sustainable energy systems.

## 4.4 The externalities of biomass

Biomass is a renewable energy source with a large potential for exploitation, in particular in countries like Brazil. The social cost of biomass energy as opposed to energy from other sources needs to be considered for future energy sector decision- and policy-making if a move towards more sustainable energy systems is desired. Hall and Scrase (1998) discuss a series of issues associated with biomass energy, ranging from misconceptions to real problems, which need to be addressed if biomass is to become 'the environmentally-friendly fuel of the future'.

A biomass fuel cycle for power generation, like a fossil fuel cycle, consists of a fuel production stage, a fuel transportation stage, a fuel conversion stage and a final stage consisting of waste disposal and recycling. There are, though, some fundamental differences in terms of impacts and their distribution. Impacts of fuel production and fuel conversion may arise at distant locations in the case of fossil fuels, but are generally close for biomass plants. In the case of fossil fuel cycles, the impacts from the fuel production stage are considered as having little significance compared to the impacts from the generation stage, and externalities estimates have focused on the damages of emissions to air from the conversion stage. It is most likely, though, that the focus on the conversion stage has tended to underestimate the externalities of upstream activities. In the case of biomass, fuel production activities generally play a more significant role. In particular, when biomass fuel cycles make use of purpose-grown energy crops, the effects of their production may be diverse and significant. Also, impacts from the transport of biomass fuels, solid fuels in particular, may be significant because of the relatively low density of the fuel and the fact that transport is often carried out by road. The type of conversion technology, as in the case of fossil fuels, is a key factor influencing the impacts of the fuel cycle and the relative importance of the different stages. Impacts from waste disposal and recycling will generally be low. Possible burdens, receptors and impact categories associated with the different stages of biomass fuel cycles for power generation are shown in Table 4.5.

The list of impacts may not be exhaustive, and other specific impacts may occur depending on the particular biomass fuel cycle considered. The diversity of biomass fuel cycles available for power generation makes it difficult to provide a general list of

Table 4.5 Biomass fuel cycle impact pathways

| Fuel cycle stage | Burden | Receptors | Impact categories |
|---|---|---|---|
| *Biomass production* | Air emissions from machinery use | Public | Human health, acidification, climate change |
| | Noise from machinery use | Workers, public | Human health, rural amenity |
| | Occupational hazards from machinery use | Workers | Human health |
| | Fuel consumption from machinery use | | Resource use |
| | Labour requirement of biomass production | | Employment |
| | Emission of toxics to air, water and soil from herbicide and pesticide application | Workers, public, flora and fauna | Human health, ecotoxicity |
| | Nutrient leaching (N, P, K) from fertiliser application | Public, fauna and flora | Human health, eutrophication |
| | Emission to air of nitrogen compounds | | Climate change |
| | Heavy metals in soil, pathogens in air, water, soil from sewage sludge application | Workers, public, fauna and flora | Human health, ecotoxicity |
| | Odours from sewage sludge application | Public | Rural amenity |
| | Nutrient leaching (P, K) from ash application | Public, fauna and flora | Human health, eutrophication |
| | Heavy metals in soil from ash application | Public, fauna and flora | Human health, ecotoxicity |
| | Particulates in air from ash application | Workers, public | Human health, rural amenity |

| | | | |
|---|---|---|---|
| | Soil erosion as result of exploitation of biomass energy | | Soil quality, eutrophication |
| | Landscape change as result of exploitation of biomass energy | Public | Rural amenity |
| | Water consumption as result of exploitation of biomass energy | | Resource use |
| *Biomass transport* | Air emissions | Public | Human health, acidification, climate change |
| | Road accidents from vehicle use | Workers, public | Human health |
| | Fuel use | | Resource use |
| | Labour requirement | | Employment |
| | Road use | | Rural amenity (e.g. road congestion and damage, noise) |
| *Biomass conversion* | | | |
| Biomass storage | Fire hazard | Workers | Human health |
| | Biological degradation | Workers | Human health |
| Biomass drying and processing | Emission of organic compounds | Workers, public, fauna and flora | Human health, eutrophication, rural amenity (odours) |
| | Dust emissions | Workers, public | Human health, rural amenity |
| Power generation | Emissions in flue gas | Public, fauna and flora | Human health, acidification, climate change |
| | Waste water emissions | Public, fauna and flora | Human health, ecotoxicity |
| | Labour requirement for plant operation and maintenance | | Employment |
| | Use of renewable and endogenous energy resource | | Resource use, national balance of payments, security of supply |

Table 4.5 continued

| Fuel cycle stage | Burden | Receptors | Impact categories |
|---|---|---|---|
| Plant construction and decommissioning | Emissions to air from construction equipment | Public | Human health, acidification, climate change |
| | Fossil fuel consumption | | Resource use, national balance of payments, security of supply |
| | Occupational hazard | Workers | Human health |
| Waste disposal and recycling | Air emissions from transport of ashes | Public | Human health, acidification, climate change |
| | Road accidents from vehicle use | Workers, public | Human health |
| | Labour requirement for vehicle operation | | Employment |
| | Road use | | Rural amenity (e.g. road congestion and damage, noise) |
| | Landfill of ashes | Public | Human health (leaching of noxious substances) |
| | Recycling of ashes | | Soil quality Human health (leaching of noxious substances) |

*Note*
The biomass production stage impacts consider also situations in which sewage slude and ash may be used as fertilisers.

activities and impacts, and the biomass production impacts listed in Table 4.5 apply rather to a biomass fuel cycle based on energy crops.

The choice of the system boundaries is critical in the assessment of a fuel cycle. It will determine which activities are believed to have impacts with potentially significant externalities. System boundaries may account for direct effects (i.e. those derived from activities which are part of the fuel cycle) and indirect effects (i.e. those which derive from activities which act as inputs to fuel cycle activities). An example of indirect effects which may be significant in the assessment of biomass fuel cycles are those resulting from the production of fertilisers used in the fuel production stage. System boundaries must comprise all activities leading to potentially significant impacts.

An exhaustive list of potential impacts which could result from the activities considered within the system boundaries needs then to be provided. An attempt at a detailed assessment of all impacts would prove a strenuous, perhaps impossible and certainly not necessary task, which is why the next step consists of the identification of priority impacts, those impacts which are believed to be significant. The choice of priority impacts is based on expert judgement and on information available in the literature. It is likely that there may be a tendency to focus on those impacts which are more readily monetisable. Other impacts which are judged to have significant impacts but which are more difficult to translate into monetary terms need equal consideration. While many priority impact categories may be common to different biomass fuel cycles, the variety and strong site specificity of biomass fuel cycles lead to priority impacts specific to the system and location considered.

### 4.4.1 Recent studies on the externalities of biomass energy

There has been a steady evolution in the attention dedicated to the externalities of biomass fuel cycles since the late 1980s. Past studies provide an insight into externalities which may prove useful when addressing biomass fuel cycles in the Brazilian context.

A summary of biomass energy externality estimates is provided in Table 4.2. Ottinger et al. (1990) provide an estimate for the externalities of biomass energy based on the results of a previous study (ECO Northwest 1986). The estimate is based on specific US biomass conventional co-generation facilities fuelled with pulp and paper mill waste, waste liquor and forest residues. The effects considered in this early study were health damages from particulates and CO emissions, and visibility improvement as a result of avoided open burning of slash.

The Oak Ridge National Laboratory (ORNL) and Resources for the Future (RFF) study (ORNL/RFF 1994) focused on the use of wood residues for electricity generation using combustion and gasification technology at a specific site in the US (the externality in Table 4.2 relates to combustion technology). Another study, using a methodology similar to that of the ExternE study and focusing on human health impacts of atmospheric emissions, has been carried out as part of the New York State Externalities Study (RCG/Tellus 1995).

Two studies have been carried out in Europe as part of the second phase of the ExternE project (CEC 1995) by the Centro de Estudos em Economia da Energia, dos Transportes e do Ambiente (CEEETA 1993 and Fernandes 1995) and by the National Technical University of Athens (NTUA 1995 and Diamantidis et al. 1996). These studies considered the use of forestry plantations, wood residues and energy crops at specific sites in Portugal and Greece. The ORNL/RFF, CEEETA and NTUA studies focused on the following priority impacts: public health, ground water contamination from nitrogen leaching, soil erosion, occupational health, and road damage.

Another European effort to study the externalities of biomass energy has been carried out as part of the EU-APAS programme (Faaij *et al.* 1998 and Saez *et al.* 1998), and includes certain macroeconomic effects as externalities. More recently, the third phase of the ExternE project has included numerous biomass fuel cycle studies as part of the National Implementation studies (CEC 1998a) and another EU project, known as the BioCosts project (CEC 1998b), has focused on the economic and environmental performance of different biomass fuel cycles for power and automotive applications across the European Union (see Tables 4.2 and 4.3).

The BioCosts study identified four categories of priority impact: impacts of atmospheric emissions on human health; impacts of greenhouse gases on climate change; impacts of the fuel production stage on the local biosphere; and impacts on rural amenity.

As in the case of fossil fuel cycles, the human health effects of atmospheric emissions appear to dominate the externalities of biomass fuel cycles. This is in part due to the fact that human health impacts from atmospheric emissions are those on which most monetisation efforts have concentrated to date. In this respect, the externalities of the biomass production cycle associated with the local biosphere may in many cases be underestimated. These, however, need not necessarily be negative, for example biomass plantations for energy may provide an increase in biodiversity compared to alternative land uses.

Effects other than health impacts, where they are considered, are generally found to be small (CEC 1998b). Such is the case of impacts of air emissions on agriculture, forests and materials (less than 0.013 UScents/kWh in the BioCosts study and less than 0.13 UScents/kWh in other studies cited in it), ground water contamination from nitrogen leaching, soil erosion (only the ORNL/RFF study finds an effect estimated at about 0.13 UScents/kWh of the same order of magnitude as health damages from air pollution), occupational health (except for the CEEETA study which arrives at damages from road accidents, estimated at 0.18–0.31 UScents/kWh, of the same order of magnitude as health damages from atmospheric emissions) and road damages (although it is debatable whether this is actually an externality).

The range of external costs of nitrogen leaching from short-rotation coppice appears significant (0.1–3.9 UScents/kWh) in a study analysing Dutch conditions (Faaij *et al.* 1998). However, the net effect of a biomass for energy plantation will depend on the alternative land use, and it may be beneficial if the alternative land use is for food crops or detrimental if the alternative land use is fallow land. The same study estimates the external costs of herbicide and pesticide use at 0.1–1.3 UScents/kWh. The net effect of biomass for energy plantations will again depend on the alternative land use. These values do not represent actual damage costs of fertiliser and agrochemical use, but are based on the WTP to reduce levels of nitrates in water and on a shadow price for herbicide and pesticide use based on estimates of reductions in yields associated with lower use. Other studies show much lower values for externalities associated to agrochemicals use (e.g. Saez *et al.* (1998) estimate them at 0.008–0.05 UScents/kWh). Impacts of soil erosion can be significant (Saez *et al.* 1998), but the net effects of biomass plantations for energy with respect to alternative land use may be positive.

The variety of possible biomass sources, variations in practices used for the procurement of biomass and the influence of site-specific considerations impose a careful examination of potential effects on soil, water, biodiversity and rural amenity. In the case of energy crops, potentially negative effects on soil and water can in many cases be avoided if attention is paid to site-specific considerations (e.g. nitrogen-sensitive areas) and adequate agricultural practices are followed. The use of biomass for energy purposes is also not likely to negatively affect biodiversity. When considering energy crops such as short-rotation coppice, it is believed that biodiversity is likely to benefit compared to alternative land uses. Rural amenity, principally in terms of visual amenity, can be a major issue. Where biomass energy schemes imply landscape changes, they may receive strong opposition from the public, which translates to high individual monetary preferences against biomass energy. However, opposition to the schemes can be dealt with through information and involvement of the local population. While the impacts of small-scale exploitation of biomass are likely to be of little significance, large-scale exploitation of biomass for energy would require more careful consideration.

Environmental externalities have dominated the scene, but some studies have also addressed non-environmental issues. Faaij *et al.* (1998) consider the externalities associated with effects of fuel cycles on GDP and employment. The consideration of the effects on GDP of biomass (based on energy crops) and coal for the Netherlands leads to a significant net benefit for biomass (biomass GDP increment: 0.8–2.0 UScents/kWh; coal GDP increment: (–0.9)–(–1.0) UScents/kWh). Also, with regard to employment, biomass presents benefits over coal in the Dutch situation (biomass employment benefit: 0.11–0.53 UScents/kWh; coal employment benefits: 0.04–0.2 UScents/kWh). Employment generation is often hailed as a potential benefit of renewable energy, biomass in particular. Overall, there is reason to believe that there is a net positive effect on employment from the use of biomass energy (CEC 1998b), in particular for countries which rely heavily on energy imports and which possess a high rate of long-term unemployment.

The BioCosts study shows that the environmental externalities considered, excluding the impacts of greenhouse gas emissions, are of the same order of magnitude as the internal costs, but vary considerably according to location. For the cleaner technologies such as circulating fluidised bed combustion and gasification, they are usually lower than the internal costs, but can range between a few percent and values close to those of the internal costs. They are generally significantly lower than the externalities of the reference fossil fuel cycles. However, only in a few cases would the internalisation of the external costs calculated make the biomass fuel cycles the economically preferred option, because of the generally higher internal costs of the biomass systems. It must be noted, though, that some biomass technologies are in the early stages of commercialisation and can achieve important reductions in costs. The study shows that the externalities estimates are strongly site-specific, with large increases in the values calculated for all case studies between the original sites and a reference site in Germany. The difference is mainly due to the exposure of a larger population to emissions for plants at the German site. While the ratio between the externalities of the biomass and fossil fuel cycles is constant for different sites, the difference between the biomass and fossil fuel cycles externalities increases with the externality. The net benefit of biomass use appears then more important for sites which result in greater environmental impacts.

The externalities of biomass energy vary greatly, depending on the biomass fuel cycle, the conversion technology and the site considered. They appear to be generally low, about an order of magnitude lower than typical energy costs, and significantly lower than the externalities of most fossil fuel cycles. However, some studies present ranges with external costs similar in magnitude to the cost of energy on to the externality associated with the better-performing fossil fuel cycles (e.g. natural gas). As for fossil fuel cycles, the dominant externality is attributed to health impacts of atmospheric emissions from the fuel cycle, in particular the conversion stage. Some studies have pointed out the potential significance of nitrogen leaching from fertiliser application, herbicide and pesticide use, soil erosion and road accidents. Potentially significant macroeconomic benefits have been attributed to biomass energy in some cases.

### 4.4.2 Dealing with greenhouse gas emissions

Greenhouse gas emissions can be reduced typically by a factor of 10 or more by using biomass instead of fossil fuels. However, the external costs of these emissions and the benefits of their avoidance, compared to the reference fossil fuel cycles, are difficult to estimate and are uncertain. Estimates of damage costs for greenhouse gas emissions vary by orders of magnitude in the literature. Variations in the estimates are a result of uncertainty over the type and magnitude of impacts, the impact categories considered by different valuation studies, the valuation method employed and the underlying economic assumptions. Damage cost estimates of the studies reviewed by the IPCC (1996) range between US$6 and US$221/tC for the period 1991–2030. Hohmeyer (1996) has shown that differences in the countries covered, in the value of a statistical life and in the discount rate, cause monetary estimates of potential damage to vary by several orders of magnitude. The valuation of non-market damage and discounting remain the great unresolved issues. Categories and estimates of non-market damages vary greatly in the literature. Discounting of damages which may occur years from now and affect future generations remains an unresolved economic issue and a fundamental ethical question (e.g. problems of intra- and inter-generational equity).

The uncertainties and value judgements surrounding the valuation of climate change damages indicate that a standard economic approach, based on the comparison of the marginal damage costs of climate change with the marginal benefits of actions taken to prevent them in the search for an economically optimal solution, is most likely not to be applicable nor appropriate. In the words of the IPCC Working Group III, 'both the costs and the benefits may be hard, sometimes impossible to assess. This may be due to large uncertainties, possible catastrophes with very small probabilities, or simply lack of consistent methodology for monetising the effects.' (IPCC 1996).

The approach to dealing with climate change may be better sought by applying principles of ecological economics (Daly 1996), advocating strong sustainability, rather than those of neoclassical economics. As a result, issues like climate change, for which there is evidence that impacts on future generations could be considerable, need to be treated in a cautionary fashion, by application of the precautionary principle. This implies that actions should be taken, although not justifiable in terms of traditional cost–benefit analysis, where there is sufficient reason to believe that the consequences

104

of not taking action could be severe. With regard to climate change, the idea of a 'safe minimum standard' for greenhouse gas emissions could be applied (Hohmeyer 1996), and emissions could then be equitably distributed with market-based instruments used to efficiently allocate resources under the constraints.

The Kyoto Protocol represents a move in this direction, although much work is yet to be done to achieve consensus over the risks associated with climate change and 'safe' levels of greenhouse gas emissions. The reductions in emissions envisaged by the Kyoto Protocol may not be sufficient to eliminate the risk of significant damage from climate change; in fact it has been estimated that to mitigate but not completely avoid climate change, current emissions should be reduced by 50–60 per cent globally (IPCC 1996). The question then is to what extent reductions in $CO_2$ emissions can be achieved efficiently, that is at a minimum cost. Some reductions in $CO_2$ emissions may even be achieved at a negative cost, as is the case of some energy-efficiency measures. Also, in many cases, reductions in $CO_2$ emissions are likely to be accompanied by additional benefits, such as reductions in other air pollutants. Reductions in greenhouse gas emissions are technically feasible and their costs likely to be affordable, in particular, if proper policies are implemented aiming at the elimination of market distortions (e.g. subsidising polluting activities) and at the implementation of a fiscal policy aimed at shifting the burden of taxation away from drivers of economic activity, such as labour, towards unsustainable practices, such as pollution.

The BioCosts study has assessed the avoidance costs for biomass fuel cycles by comparing their internal costs and $CO_2$-equivalent emissions to those of reference fossil fuel cycles. Figure 4.2 shows the avoidance costs based uniquely on the internal costs of the fuel cycles and based on the internal costs net of the external costs quantified for the original site and the reference site in Germany. It can be noted that abatement costs range from negative to high positive values. Some of the biomass fuel cycles present high internal costs because of the early commercial stages of the technologies involved and important cost reductions are expected as deployment of the technology continues. In particular, technologies like gasification (Bauen *et al.* 1998a), which present abatement costs in the middle range and which are believed to have considerable potential for costs reductions, appear as an interesting candidate for R&D and market penetration support. Figure 4.2 also shows estimates of the damage costs of climate change to provide an indication of how they compare with avoidance costs and to possibly provide supporting arguments for action.

## 4.5 The Brazilian energy sector

Renewable energy plays an important role in Brazil. It provides about 70 per cent of the country's primary energy needs, mainly in the form of hydropower, the remaining 30 per cent being satisfied by fossil fuels. Wood and sugar-cane products account for about 12 per cent and 14 per cent of the primary energy, respectively (BEN 1998). The pulp and paper, agro-industry (including sugar and alcohol industry) and the steel industry are the major consumers of biomass energy. In 1997, alcohol production in Brazil provided 14.4 per cent of the primary energy used in the transport sector, corresponding to about 6.7 Mtoe and representing about 40 per cent of the fuel energy used in light vehicles (BEN 1998). Although the absolute contribution of alcohol has been gradually increasing, its relative contribution has been decreasing.

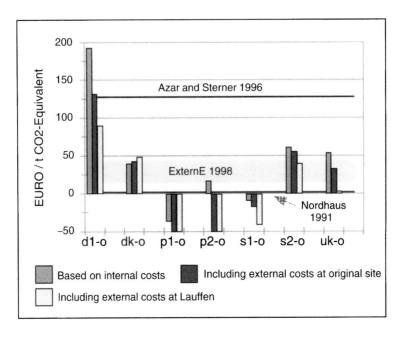

Legend:

d1-o: Production of cold-pressed rape-seed oil and its use in a co-generation plant at Weissenburg, Germany, versus the use of diesel fuel in a similar engine.

dk-o: Production of biogas from slurry for municipal combined heat and power generation at Hashöj, Denmark, versus the use of Danish natural gas in the same engine.

p1-o: Utilisation of forestry residues for industrial combined heat and power production in Mangualde, Portugal, versus the use of fuel oil in an engine generating heat and power.

p2-o: Utilisation of short-rotation coppice for industrial combined heat and power production in Mangualde, Portugal, versus the use of fuel oil in an engine generating heat and power.

s1-o: Utilisation of forestry residues in the Nässjö circulating fluidised bed combustion plant, Sweden, versus the use of Polish coal in the same plant.

s2-o: High-pressure gasification of forestry residues for combined heat and power generation at Värnamo, Sweden, versus the use of coal in the Nässjö plant.

uk-o: Atmospheric pressure gasification of short-rotation coppice for electricity generation at Eggborough, UK, versus the use of coal in a power plant.

Note: 1 EURO = US$1.3 (1995)

*Figure 4.2* $CO_2$ abatement costs of BioCosts case studies with and without considering damages to human health by conventional air pollutants; including damage costs given by Nordhaus (1991), Azar and Sterner (1996) and ExternE (CEC 1998a; Eyre *et al.* 1997). (Source: CEC 1998b)

Energy demand is rising in Brazil, requiring increased inputs of primary energy and the installation of additional generating capacity. Deregulation, privatisation and market forces may in the long term favour the penetration of renewable energy other than large hydro, which has been so far the main renewable energy source. However, the short-term tendency is towards an increased use of fossil fuels (e.g. natural gas) because of the lower cost associated with power generation from fossil plants compared to alternatives (e.g. hydro, nuclear, renewables). In particular, plans for several large hydroelectric schemes have been delayed. Hydropower potential is large in Brazil, with only about 20 per cent of it being exploited. About two-thirds of the remaining potential is estimated to be situated in the Amazon region and far away from the main energy consumption centres.

The expansion of the power-generating capacity in Brazil using fossil fuels (natural gas and coal) would result in significant externalities, in particular related to impacts on human health deriving from emissions of pollutants at the conversion stage and to the impacts of greenhouse gas emissions. The magnitude of the externalities would depend on the fuel and technology used and on the siting of the facilities.

Additional large hydroelectric schemes are also likely to result in significant externalities. In the Amazon region, 'hydroelectric flooding' is considered a potentially large source of greenhouse gas emissions, resulting from biomass decay. However, Rosa and Schaeffer (1995) arrive at the conclusion that although hydropower plants may be important sources of greenhouse gas emissions, they remain in most cases a better option than fossil-based thermoelectric power generation with respect to climate change. Fearnside (1996) believes that the picture, in particular in the Amazon region, is likely to be worse than that presented by Rosa and Schaeffer, and criticises their study because of its neglect of $CO_2$ emissions (only methane is considered), lack of discounting and supposedly optimistic power outputs per unit of flooded area. Large hydropower schemes may also have significant environmental impacts on terrestrial and aquatic ecosystems, as well as social impacts (e.g. displacement of indigenous populations) (Moreira and Poole 1993; Fearnside and Barbosa 1996a and 1996b; Rosa *et al.* 1988). Therefore, there is reason to believe that, besides the high investment costs, large hydroelectric schemes, particularly in regions such as the Amazon, may also have significant external costs. Brazil also possesses a large potential for small hydropower schemes which may present fewer economic and environmental problems compared to large schemes.

There is a large untapped biomass potential in Brazil. The use of biomass for power generation, in particular through the use of modern conversion technologies, such as circulating fluidised bed combustion and steam turbine (CFB/ST) systems and biomass integrated gasification and gas turbine combined cycle (BIG-CC) systems, could contribute significantly to satisfying future power needs (Larson *et al.* 1989; Walter 1994; Bauen *et al.* 1998b). Modern biomass conversion technologies emit low levels of pollutants and biomass could provide an indigenous renewable fuel with little or no $CO_2$ emissions. The use of biomass in the pulp and paper, agro-industry (including sugar and alcohol industry) and steel industry already contributes significantly to reduced $CO_2$ emissions. A more efficient use of the biomass fuel, in particular in association with co-generation aiming at the sale of surplus electricity to the transmission grid, could result in additional benefits associated with displaced generation. The use of alcohol in transport, as an unblended fuel, or blended with petrol and eventually with diesel, serves to mitigate air pollution and greenhouse gas emissions from transport.

Brazil is a large contributor of greenhouse gas emissions, as a result mainly of deforestation. Deforestation-related $CO_2$ emissions in 1991 were estimated at 150–220 Mt of carbon per year, while 1990 energy consumption-related $CO_2$ emissions were estimated at 73 Mt of carbon per year (La Rovère *et al.* 1994). However, $CO_2$ emissions from the energy and transport sector on a per capita basis are low compared to other countries because of Brazil's level of economic development and because of the important renewable energy contribution to primary energy use. However, energy demand in Brazil, in particular in the form of electricity and transport fuels, is likely to grow considerably, and so will $CO_2$ emissions, given the current trend to install fossil fuel-based generating capacity and the stagnation of alcohol use in vehicles. The renewable resources are nevertheless there, which could avoid a considerable portion of the emissions.

Energy from biomass in the pulp and paper, sugar and alcohol, and steel industries results in impacts leading to externalities. However, the impacts of the biomass fuel cycles need to be discussed in comparison to those of competing fuel cycles. Electricity generation in the pulp and paper industry, based on waste generated at the industrial site (see Chapters 2 and 7), only needs to consider the impacts of emissions from the conversion stage and possibly emissions associated with any disposal of waste from the conversion stage. In the case where additional biomass is supplied to the site for energy purposes, the potential impacts of the biomass production and transport activities need to be considered.

The use of charcoal in the steel industry needs to consider impacts from the production of biomass for use in charcoal production, the impacts from the conversion of biomass to charcoal, and the impacts of charcoal conversion to useful energy for steel production. Medeiros (1995) has attempted a preliminary evaluation of the environmental costs of charcoal production and use for steel production in Brazil. The environmental cost estimates for charcoal are high and range between US$57/t of charcoal from native forests to US$66/t of charcoal from planted forests. Most of the externality in the case of charcoal from native forests is associated with $CO_2$ releases and in the case of planted forests with soil erosion. The high value obtained for charcoal from planted forests would deserve more careful consideration, in particular, in view of potential benefits which could be associated with the use of charcoal in steel production compared to coke (see Chapters 2 and 8; Rosillo-Calle *et al.* 1996).

The following section discusses the externalities of alcohol and electricity from sugar-cane.

### 4.5.1 *Externalities of alcohol and electricity from sugar-cane*

Sugar-cane is a major crop in Brazil and a source of both food and energy products. A large part of the 5 Mha of sugar-cane is grown as an energy crop to produce alcohol (ethanol) mainly for use in vehicles. Production of alcohol could still be increased in Brazil by extending sugar-cane cultures (in particular in regions such as the Northeast), by increasing yield, and by using more efficient technologies for the conversion of sugar and molasses to alcohol. Also, it may be possible to use bagasse and sugar-cane residues from the fields to produce additional alcohol by hydrolysis processes.

At the beginning, the main drive behind the alcohol programme originated from issues such as energy security, employment and reduced expenditure on oil imports. While these concerns remain today, the environment has, over the years, become one of the main drivers. Also, the lobbying strength of the powerful sugar-cane producers and industry is certainly not to be underestimated in the past, present and future of PROALCOOL. At present, proponents of the PROALCOOL programme are pushing for its revival, claiming important environmental, economic and social benefits;

however, questions can be raised about the programme on the same grounds (e.g. ecological impacts of sugar-cane cultures and use, economic inefficiency of alcohol production and high subsidisation of the industry, land distribution and working conditions) (see Chapters 2 and 3; Paixao 1996).

Bagasse, resulting from the processing of sugar-cane, is widely exploited in the Brazilian sugar and alcohol industry to satisfy its mechanical and electrical energy, as well as process steam requirements. However, in Brazil, as in the majority of sugar-cane producing countries, bagasse is generally converted to mechanical and electrical power with low efficiency, due to the large quantities of bagasse available and to the need for its disposal. Much scope exists for an enhanced valorisation of this residue.

The harvesting of unburned sugar-cane results in additional residues, consisting of dry and green leaves and tops, which we will refer to as trash. Their quantity per tonne of cane is roughly equal to that of bagasse. Currently, most of these residues go up in smoke or are deposited in the fields as ashes due to the widespread burning of the plantations to ease manual harvesting. Recent legislation in the Brazilian state of São Paulo has decreed the suspension of pre-harvest burning by 2005 in areas judged suitable for mechanical harvesting (estimated to be about 50 per cent of the planted area) and by 2012 in the remaining areas. Legislation and an inexorable move towards the mechanisation of the harvesting process will lead to large quantities of trash, which can be potentially exploited for energy use (Bauen *et al.* 1998b; Braunbeck *et al.* 1999). Sugar-cane trash could be used to complement bagasse as an economically viable fuel (Bauen *et al.* 1998b). Electricity sales could provide an additional source of income for the industry, which could contribute to reducing the private costs of alcohol. Williams and Larson (1993) discuss the potential economic benefits of generating surplus electricity in alcohol plants.

### 4.5.1.1 *Potential impacts of alcohol and electricity from sugar-cane*

When considering the externalities of alcohol and electricity from sugar-cane, different system boundaries may apply. In the case of sugar-cane, which is purpose-grown for the production of alcohol, the possible impacts of sugar-cane cultivation need to be considered. This stage may be neglected when considering the generation of electricity from sugar-cane waste products such as bagasse and trash. Similar reasoning would apply if alcohol were produced from sugar-cane waste products.

The urban and global environments have benefited from the use of alcohol because of reduced greenhouse gas and noxious pollutant emissions from alcohol-fuelled vehicles, compared to conventional petrol- and diesel-fuelled vehicles (Table 4.6). However, the environmental effects of the alcohol fuel cycle are more difficult to assess when the sugar-cane and alcohol production stages are considered. Since 1975 the sugar-cane planted area has increased significantly, in particular in the state of São Paulo. The monoculture may have impacts on soil quality, on biodiversity and on the landscape, apart from other environmental impacts which depend on agricultural practice. Sparovek and Lepsch (1993), in a study carried out in the Piracicaba region in the state of São Paulo, indicate that land dedicated to sugar-cane culture has increased by about 15 per cent since the early 1960s and that it represents about 50 per cent of land use in the region. They express concern over the fact that about 48,000 ha (27 per cent of the total area comprised in the region) are exposed to a high risk of degradation because of over-exploitation, mainly from sugar-cane plantations. Herbicides and pesticides may have impacts on the environment but these are likely not to be significant if good agricultural practice is followed. Also, harvesting has resulted in adverse environmental effects in the form of deteriorated air quality due to pre-harvest burning of cane.

The ethanol production process itself results in emissions from the burning of bagasse to supply energy to the plant, and produces liquid effluents from cane washing and ethanol distillation. Bagasse boiler efficiency is generally low, but there is no evident economic interest to improve it, and emissions are poorly controlled. If co-generation were to be more widely exploited to increase income from electricity sales, then efficiency would become a more important issue. Surplus electricity could be produced, using sugar-cane bagasse and trash, and exported to the electricity grid to displace electricity generation from sources with possibly greater impacts on the environment. More modern installations would also result in lower emissions per unit of energy generated.

A significant amount of water is used at the factory. Water is used for washing out soil carried over from harvesting, for juice extraction in the milling process and for BRIX (soluble solids degree) adjustment in the dilution process preceding fermentation. Water abstraction for use in the mills and liquid effluents from the mills contribute to the depletion and pollution of water courses in the dry season when cane is processed (Cortez *et al.* 1998).

The important water quantities used in the process and the water content of sugar-cane lead to a large volume of liquid effluent from the ethanol distillation process, known as vinasse. Vinasse, which is produced at a volume rate 10 times larger than ethanol itself, is disposed of in cultivated areas to both irrigate and fertilise the cane crop. Much of this residue is transported to the fields by truck or by pipeline. This has been the only economically sound and successful large-scale method for its disposal, although significant research has been conducted to investigate alternatives (Cortez *et al.* 1998). The runoff and percolation of vinasse may affect water bodies through excess concentrations of nutrients such as N, P and K, and lead to impacts on human health and to the eutrophication of water bodies. The chemical stability of the soil and the life of organisms, from micro-organisms to insects and other small animals, are also affected. The effects on water bodies may be more evident than the long-term effects on soil quality and biodiversity, however all impacts are difficult to assess and are very much dependent on site-specific characteristics and on agricultural and industrial practices.

Technological improvements within the industry have been slow and important changes are expected in the coming years, driven by industrial competitiveness and regulation. Also, greater public awareness has been influencing government policies towards changes in practices, in particular concerning cane harvesting and washing.

With regard to its end-use, ethanol is believed to have helped abate pollution in Brazil's urban areas (Szwarc 1995). Table 4.6 provides estimates of emissions from light vehicles.

### 4.5.1.2 Priority impacts of alcohol from sugar-cane

The priority impacts of the sugar-cane production stage are likely to be those resulting from the pre-harvest burning of the fields and the application of vinasse to the fields. Where the cultivation of sugar-cane is extensive and intensive, impacts on soil quality,

*Table 4.6* Comparison of emissions from light vehicles (CETESB 1997)

| Fuel type | Emissions (g/km) | | | | |
|---|---|---|---|---|---|
| | CO | HC | $NO_x$ | $SO_x$ | PM |
| Gasohol (with 22% ethanol) | 19.9 | 1.7 | 1.1 | 0.16 | 0.08 |
| Alcohol | 16.3 | 1.9 | 1.2 | – | – |
| Diesel | 17.8 | 2.9 | 13.0 | 1.13 | 0.81 |

biodiversity and landscape may become priorities. Other impacts from the sugar-cane production stage should not present significant impacts, and good agricultural practices should reduce most potential impacts below levels of concern. Impacts from the road transport of sugar-cane may require greater attention.

The impact of the pre-harvest burning of the fields is most likely to lead to significant externalities related to human health and amenity (e.g. visibility and damage to buildings), in particular in the more densely populated areas of the country, such as the state of São Paulo. Recent legislation aimed at phasing out the burning of the fields should proceed to avoid the externality imposed by the current harvesting practice. In this case, the avoidance of the externality may not imply a rise in the internal costs of sugar-cane production, since the mechanical harvesting of unburned fields may be more cost-effective than manual harvesting (Braunbeck *et al.* 1999), and could lead to a win–win situation, where environmental and economic gains occur simultaneously. The social impacts of mechanisation need careful consideration, as many seasonal jobs will be lost (some 70,000 jobs in the state of São Paulo alone, assuming mechanical harvesting of half of the sugar-cane fields (Cortez *et al.* 1998)). The occupational conditions of manual cane cutters are generally poor, and the preservation of such jobs is not necessarily socially beneficial. What is needed is a gradual transition to mechanical harvesting, accompanied by investments in other economic sectors which can absorb the workforce and offer better occupational conditions.

The potential impacts of vinasse application are complex to assess. It is applied to the fields to provide irrigation and nutrients, but also to dispose of an undesired waste product from the alcohol production process. Restrictions in its application to the fields would most probably result in additional costs of alternative disposal means, in particular as regulation on water quality becomes stricter. The question of vinasse is one that needs to be examined more carefully. Given the site- and practice-specific nature of the issue and the current lack of information, it is difficult to produce any valuation of the impact of vinasse produced in alcohol production.

The Brazilian sugar and alcohol industry is on average self-sufficient with regard to energy through the use of bagasse. The use of bagasse as a fuel has considerable benefits as it is a renewable fuel and it can be considered neutral in terms of greenhouse gas emissions. Also, bagasse has a negligible sulphur content. However, because bagasse is used in relatively old and low-efficiency conversion systems, emissions of other conventional pollutants such as $NO_x$, particulates and CO are likely to be significant and possibly higher than modern conversion systems fuelled with fossil fuels. The externalities per unit of energy generated may therefore be significant for the current use of bagasse.

Ethanol-fuelled vehicles emit pollutants which are likely to impact on human health; however, the use of ethanol, pure and blended with gasoline, is believed to be beneficial compared to the use of unblended gasoline and diesel (Szwarc 1995). More research on the actual benefits is, however, desirable. The benefits of using ethanol with regard to reductions in non-renewable resource use and greenhouse gas emissions are more evident.

### 4.5.1.3 Priority impacts of heat and electricity from sugar-cane

There is considerable scope to increase power generation within the industry and reduce the externalities per unit of energy generated through the use of more efficient systems. In the case of bagasse, it appears suitable to define the system boundaries so that no impacts from the sugar-cane production cycle are considered, all impacts being attributed to the production of sugar and alcohol. In the case of sugar-cane trash, the system boundaries will consider all activities and impacts related to its collection and transport. It is still a matter of discussion what portion of sugar-cane trash should be left in the field for agronomic reasons. Both an excessive and insufficient removal of trash may have negative impacts on soil and water and on sugar-cane growth. Good practice in the use of the machinery for the collection of trash is likely to avoid any significant impacts (e.g. soil compacting). The impacts of the road transport of trash need to be considered carefully because of the important quantities of sugar-cane already transported, and impacts could be mitigated by transporting trash outside the harvesting season. Emissions from the collection and transport of trash may be significant. In particular, collection and transport will contribute some net $CO_2$ emissions to the fuel cycle.

The viability of power exports would be an incentive for the industry or third parties to invest in modern, cleaner and more efficient generating equipment. The viability of using trash as a fuel may also accelerate the phase out of the pre-harvest burning of sugar-cane fields. Multiple benefits could result from such schemes aimed at generating power for export outside the industry. The externalities of alcohol production would be reduced, and the electricity exported is most likely to present lower externalities compared to electricity from other generating sources.

### 4.5.1.4 Potential external benefits of electricity from sugar-cane

The priority impacts of greatest significance in the production of electricity from sugar-cane are likely to be those associated with emissions to air from the collection, transport and conversion activities.

Given that no pollutant-specific externality values are available for Brazil, the present analysis wishes to provide an indication of the magnitude of the externalities of electricity from sugar-cane based on externality values specific to European Union countries. Table 4.7 shows the ranges of values for pollutant-specific externalities derived from site-specific externality valuations in five European Union countries (Sweden, UK, Portugal, Denmark and Germany) (CEC 1998b). The externalities are

*Table 4.7* Ranges of pollutant-specific externalities for Europe (CEC 1998b)

| Pollutant | Externality (US$/kg) |
|-----------|----------------------|
| $NO_x$ | 3.90–23.2 |
| $SO_2$ | 8.03–16.5 |
| PM | 5.06–30.6 |

mainly a result of impacts on human health of the long-range dispersion of pollutants. Damage costs for $CO_2$ vary widely and to provide an indication of the potential benefits of using biomass we have chosen the damage cost range 19.6–54.9 US$/t of $CO_2$ calculated in the ExternE project (CEC 1998a).

Table 4.8 compares emissions from a sugar-cane residues fuel cycle for electricity generation using BIG-CC and from a natural gas fuel cycle for electricity generation using a CCGT. It is then possible to provide an indication of the external costs and benefits of using biomass compared to natural gas, based on differences in fuel cycle $NO_x$, $SO_2$, PM and $CO_2$ emissions (Table 4.9).

Electricity from bagasse and trash could be generated at an average-sized mill in Brazil (350 tc/h) in the early commercialisation stages of the BIG-CC technology at a cost of 5.0–5.6 UScents/kWh, assuming a 10 per cent discount rate. As a comparison, the

*Table 4.8* Comparison of sugar-cane residues fuel cycle and natural gas fuel cycle emissions

|  | Production (mg/kWh) | | Conversion (mg/kWh) | | Total (mg/kWh) | |
|--|---------|-------------|---------|-------------|---------|-------------|
|  | Biomass | Natural gas | Biomass | Natural gas | Biomass | Natural gas |
| $NO_x$ | 21.1 | 28.8 | 213 | 787 | 234 | 816 |
| $SO_2$ | 0.442 | 16.9 | 5.56 | 0 | 6.01 | 16.9 |
| PM | 3.09 | 0 | 11.8 | 0 | 14.9 | 0 |
| CO | 8.36 | 9.72 | 556 | 393 | 563 | 403 |
| NMHC | 3.68 | 11.9 | 0 | 11.0 | 3.68 | 22.9 |
| $CO_2$ | 1355 | 2700 | 0 | 392870 | 1355 | 395570 |

*Table 4.9* Indicative external costs and benefits of using sugar-cane residues for electricity compared to natural gas

|  | External cost difference (UScents/kWh) |
|--|----------------------------------------|
| $NO_x$ | 0.2–1.4 |
| $SO_2$ | 0.009–0.02 |
| PM | (−0.05)–(−0.008) |
| $CO_2$ | 0.8–2.2 |
| Net cost/benefit (excluding GHGs) | 0.2–1.3 |
| Net cost/benefit (including GHGs) | 1.0–3.5 |

*Note*
Positive values denote a net benefit and negative values denote a net cost associated with the biomass fuel cycle compared to the natural gas fuel cycle.

South/Southeast Brazil electricity generation marginal expansion cost is estimated by ELECTROBRAS at 3.8–4.5 UScents/kWh (likely to cover the costs of electricity from natural gas) (ELECTROBRAS 1996; Coelho and Zylbersztajn 1996).

It appears then, from the illustrative external costs/benefits, that externalities may significantly reduce any private costs-based competitive advantage of power generation from more conventional technologies (e.g. natural gas). The above calculations are meant only as an example to show how externalities may influence the total costs of energy from different sources and thereby question the ranking of energy options based on current private costs considerations.

The actual determination of the externalities of the sugar-cane residues and natural gas fuel cycles would require a more careful examination of all the stages of the fuel cycles and the determination of externality values specific to the Brazilian context. A series of non-environmental externalities would also require consideration in both the production of alcohol and electricity from sugar-cane. For example, alcohol could contribute positively to the national balance of payments compared to oil imports, and the same would apply to the use of indigenous sugar-cane residues for electricity generation, as opposed to natural gas imports. Also, apart from potential benefits which may result from direct employment, positive indirect economic and employment effects may result from the establishment of an indigenous industry, as opposed to the reliance on imported goods.

## 4.6 Conclusion

This chapter has provided a discussion on the current state of the externalities of energy and extended the discussion to externalities associated with industrial uses of biomass energy in Brazil.

The international literature reviewed demonstrates how externalities may be significant and influential in the ranking of energy options based on social costs. Biomass energy appears generally to possess lower external costs compared to other conventional energy sources; in particular, if the externalities associated with climate change are considered. The externalities valued and the valuation techniques used differ across the literature, providing an indication of the significance of various impacts and examples of valuation techniques which could be useful in the Brazilian context. However, the review also emphasises the limitations and uncertainties underlying externalities studies, which need to be considered in future externalities studies applied to Brazil.

There is a lack of, and a need for, research on the externalities of energy and on sustainable energy options in Brazil. Efforts are, then, required in the collection and analysis of data related to fuel cycles and to the state of the environment. Work is also required in the quantification and monetisation of impacts in the Brazilian context.

A general discussion has been provided on the impacts of biomass fuel cycles, which can serve as a starting point for the analysis of specific fuel cycles. A more detailed qualitative analysis of the alcohol and electricity from sugar-cane fuel cycles points to their likely priority impacts and to various aspects of the fuel cycles which would require more careful consideration. An illustrative calculation of the external costs of electricity from sugar-cane and their comparison with those of a natural gas fuel cycle reveals the likely significance of the externality values and provides an indication of the possible benefits of the biomass fuel cycle.

More work on the impacts and externalities of fuel cycles is desirable in Brazil to limit negative impacts, to compare energy options, and to lay the foundations for discussing sustainable energy options for the future.

## 4.7 Bibliography

Azar, C. and Sterner, T. (1996) 'Discounting and distributional considerations in the context of global warming', *Ecological Economics*, 19, 169–84.

Bauen, A., Hall, D. O., Rosillo-Calle, F. and Scrase, J. I. (1998a) 'Biomass gasification for heat and electricity generation', Internal BioCosts case study report, King's College London.

Bauen, A., Cortez, L., Rosillo-Calle, F. and Bajay, S. (1998b) 'Electricity from sugar-cane in Brazil', in Kopetz, H. *et al.* (eds), *Biomass for Energy and Industry – Proceedings of the 10th European Biomass Conference*, June 8–11, Würzburg, Germany, CARMEN, 341–4.

BEN (1998) 'Balanço Energético Nacional – Ano base 1997', Ministerio de Minas e Energia, Brasilia.

Braunbeck, O., Bauen, A., Rosillo-Calle, F. and Cortez, L. (1999) 'Prospects for green cane harvesting and cane residue use in Brazil', *Biomass and Bioenergy*, (in press).

Carnevali, D. and Suarez, C. (1993) 'Electricity and the environment: Air pollutant emissions in Argentina', *Energy Policy*, 21 (1).

CEC (Commission of the European Communities), DG XII (ed.) (1995) *ExternE: Externalities of Energy*, vols 1–6, Office for Official Publications of the European Communities, Luxembourg.

CEC (Commission of the European Communities), DG XII (ed.) (1998a) *Final Reports of ExternE Phase III*, forthcoming (preliminary versions available at http://ExternE.jrc.es).

CEC (Commission of the European Communities), DG XII (ed.) (1998b) *Total Costs and Benefits of Biomass in Selected Regions of the European Union. Final Report of the BioCosts Project*, Contract JOR3-CT95-0006, Brussels.

CEEETA (Centro de Estudos em Economia da Energia, dos Transportes e do Ambiente) (1993) *External Costs of Biomass Fuel Cycle. Final Report of the EU-JOULE Project*, JOU2-CT92-0106, Brussels.

CETESB (Companhia de Tecnologia de Saneamento Ambiental) (1997) *Relatorio Anual de Qualidade do Ar – 1996*, São Paulo: CETESB.

Coelho, S. T. and Zylbersztajn, D. (1996) 'A preliminary analysis of mechanisms to improve biomass-origin co-generation in Brazil', in Chartier, P. *et al.* (eds) *Biomass for Energy and the Environment*, vol. 1, Oxford: Elsevier Science, 446–58.

Cortez, L. A. B., Braunbeck, O., Rosillo-Calle, F. and Bauen, A. (1998) 'Environmental aspects of the alcohol program in Brazil', *Proceedings of the International Conference on Agricultural Engineering*, Oslo, August 24–7.

Daly, H. (1996) *Beyond Growth – The Economics of Sustainable Development*, Boston: Beacon Press.

Diamantidis, N. D., Arvelakis, S. and Koukios, E. G. (1996) 'External costs of the biomass fuel cycle under Greek conditions', in Chartier, P. *et al.* (eds) *Biomass for Energy and the Environment*, vol. 3, Oxford: Elsevier Science, 1683–94.

ECO Northwest Ltd (1986) *Estimating Environmental Costs and Benefits for Five Generating Resources*, report prepared for Bonneville Power Administration (BPA), Portland, OR.

ELECTROBRAS (1996) *Plano decenal de expansao – 1996–2005*, Brasilia.

Eyre, N., Downing, T., Hoekstra, R., Rennings, K. and Tol, R. S. J. (1997) *ExternE – Global warming damages. Final Report of the Global Warming Sub-Task of the ExternE Project*, JOS3-CT95-0002, Brussels.

Faaij, A., Meuleman, B., Turkenburg, W., Wijk, A., Bauen, A., Rosillo-Calle, R. and Hall, D. O.

(1998) 'Externalities of biomass based electricity production compared to power generation from coal in the Netherlands', *Biomass and Bioenergy*, **14** (2), 125–47.

Fankhauser, S. (1995) *Valuing Climate Change – The Economics of the Greenhouse*, London: Earthscan Publications.

Fearnside, P. M. (1996) 'Hydroelectric dams in Brazilian Amazonia: response to Rosa Schaeffer and dos Santos', *Environmental Conservation*, **23** (2), 105–8.

Fearnside, P. M. and Barbosa, R. I. (1996a) 'Political benefits as barriers to assessment of environmental costs in Brazil's Amazonian development planning: the example of the Jatapu dam in Roraima', *Environmental Management*, **20** (5), 615–30.

Fearnside, P. M. and Barbosa, R. I. (1996b) 'The Cotingo dam as a test of Brazil's system for evaluating proposed developments in Amazonia', *Environmental Management*, **20** (5), 631–48.

Fernandes, M. (1995) 'External costs and benefits of the biomass fuel cycle', *European Network for Energy Economic Research (ENER) Bulletin* 17.95, 77–92.

Friedrich, R. and Voss, A. (1993) 'External costs of electricity generation', *Energy Policy*, February, 114–22.

Furtado, R. C. (1996) 'The incorporation of environmental costs into power system planning in Brazil', PhD thesis, Imperial College, London.

Hall, D. O. and Scrase J. I. (1998) 'Will biomass be the environmentally friendly fuel of the future?', *Biomass and Bioenergy*, **15** (4/5), 357–67.

Hohmeyer, O. (1988) *Social Costs of Energy Consumption – External Effects of Electricity Generation in the Federal Republic of Germany*, Berlin: Springer-Verlag.

Hohmeyer, O. (1996) 'Social costs of climate change – strong sustainability and social costs', in Hohmeyer, O., Ottinger, R. L. and Rennings, K. (eds) *Social Costs and Sustainability – Valuation and Implementation in the Energy and Transport Sector*, Berlin: Springer-Verlag, 61–83.

IPCC (Intergovernmental Panel on Climate Change) (1996) *Climate Change 1995. Economic and Social Dimensions of Climate Change – Contribution of Working Group III to the Second Assessment Report of the IPCC*, Cambridge: Cambridge University Press.

La Rovère, E. L., Legey, L. F. L. and Miguez J. D. G. (1994) 'Alternative energy strategies for abatement of carbon in Brazil – a cost–benefit analysis', *Energy Policy*, **22** (11), 914–24.

Larson, E. D., Svenningsson, P. and Bjerle, I. (1989) 'Biomass gasification for gas turbine power generation', in Johansson, T. B., Bodlund, B. and Williams, R. H. (eds) *Electricity: Efficient End Use And New Generation Technologies, and their Planning Implications*, Lund: Lund University Press.

Lee, R. (1996) 'Externalities studies: why are the numbers different?', in Hohmeyer, O., Ottinger, R. L. and Rennings, K. (eds) *Social Costs and Sustainability – Valuation and Implementation in the Energy and Transport Sector*, Berlin: Springer-Verlag, 13–28.

Masuhr, K. P. and Ott, W. (1994) *Couts externes et surcouts inventories du prix de l'energie dans le domaine de l'electricite et de la chaleur*, Berne: Office federal des questions conjoncturelles.

Medeiros, J. X. (1995) 'Aspectos Economico-Ecologicos da Produçâo e Utilizaçâo do Carvâo Vegetal na Siderurgia Brasileira', in May, P. H. (ed.) *Economia Ecologica – Aplicaçôes no Brasil*, Rio de Janeiro: Editora Campus Ltda, 83–114.

Moreira, J. R. and Poole, A. D. (1993) 'Hydropower and its constraints', in Johansson, T. B., Kelly, H., Reddy, A. K. N. and Williams, R. H. (eds) *Renewable Energy – Sources for Fuels and Electricity*, Washington DC: Island Press, 73–120.

Nordhaus, W. D. (1991) 'To slow or not to slow. The economics of the greenhouse effect', *Economic Journal*, **101**, 920–37.

NTUA (National Technical University of Athens) (1995) 'ExternE. External costs of fuel cycles – the case of biomass in Greece', final report, Athens.

ORNL/RFF (Oak Ridge National Laboratory and Resources for the Future Inc.) (1994)

*External Costs and Benefits of Fuel Cycles. A Study for the US Department of Energy and the Commission of the European Communities*, Report No. 1–8, Washington DC: McGraw-Hill/Utility Data Institute.

OTA (Office of Technology Assessment) (1994) *Studies of the Environmental Costs of Electricity – Background Paper*, OTA-ETI-134, Washington: US Government Printing Office.

Ott, W. (1997) 'External costs and external price addings in the Swiss energy sector', in Hohmeyer, O., Ottinger, R. L. and Rennings, K. (eds) *Social Costs and Sustainability – Valuation and Implementation in the Energy and Transport Sector*, Berlin: Springer-Verlag, 176–83.

Ottinger, R. L. (1996) 'Have recent studies rendered environmental externality valuation irrelevant?', in Hohmeyer, O., Ottinger, R. L. and Rennings, K. (eds) *Social Costs and Sustainability – Valuation and Implementation in the Energy and Transport Sector*, Berlin: Springer-Verlag, 29–43.

Ottinger, R. L., Wooley, D. R., Robinson, N. A., Hodas, D. R. and Babb, S. E. (1990) *Environmental Costs of Electricity*, Pace University Center for Environmental Legal Studies, New York: Oceana Publications.

Paixao, M. (1996) *Os vinte anos do Proalcool: As controversias de um programa energetico de biomassa*, Serie ' Brasil: Sustentabilidade e Democracia', FASE – Federaçâo de Orgâos para Assistencia Social e Educacional, Brasil.

Pearce, D. (1995) 'The development of externality adders in the United Kingdom', prepared for the EC/IEA/OECD workshop on 'The External Costs of Energy', CSERGE, University College London.

Pearce, D., Markandya, A. and Barbier, A. (1989) *Blueprint for a Green Economy*, London: Earthscan.

Pearce, D., Bann, C. and Georgiou, S. (1992) 'The Social Costs of Energy', Centre for Social and Economic Research on the Global Environment, University College London and University of East Anglia.

Pigou A (1920) *The Economics of Welfare*, London: Macmillan.

RCG/Tellus (RCG, Hagler, Bailly, Inc. and Tellus Institute) (1995) 'New York State Environmental Externalities Cost Study', Final Reports EP 91–50, Reports 1–4, prepared for the Empire State Electric Energy Research Corporation, Albany, New York.

Roodman, D. (1998) *The Natural Wealth of Nations*, New York: W.W. Norton & Company.

Rosa, L. P. and Schaeffer, R. (1995), 'Global warming potentials – the case of emissions from dams', *Energy Policy*, **23** (2), 149–58.

Rosa, L. P., Sigaud, L. and Mielnik, O. (eds) (1988) *Impactos de grandes projetos hidreletricos e nucleares*, Rio de Janeiro: AIE/COPPE/Marco Zero/CNPq.

Rosillo-Calle, F., de Rezende, M. A. A., Furtado, P. and Hall, D. O. (1996) *The Charcoal Dilemma – Finding a Sustainable Solution for Brazilian Industry*, London: Intermediate Technology Publications

Saez, R. M., Linares, P. and Leal, J. (1998) 'Assessment of the externalities of biomass energy, and a comparison of its full costs with coal', *Biomass and Bioenergy*, **14** (5/6), 469–78.

Sparovek, G. and Lepsch, I. F. (1993) 'Diagnostico de uso e aptidao das terras agricoals de Piracicaba', in Tauk-Tornisielo, N. G. *et al.* (eds) *Analise Ambiental: Estrategias e Açôes*, São Paulo: Queiroz.

Stirling, A. (1997) 'Limits to the value of external costs', *Energy Policy*, **25**, 517–40.

Szwarc, A. (1995) 'Emissôes de Poluentes em Veiculos Automotivos', in Fernandes, E. S. L. and Coelho, S. T. (eds) *Perspectivas do Alcool Combustivel no Brasil*, São Paulo: IEE/USP, 101–4.

van Horen, C. (1996) 'Counting the social costs – electricity and externalities in South Africa', University of Cape Town: Elan Press/UCT Press.

Walter, A. C. (1994) 'Viabilidade e perspectivas da cogeraçâo termoéletrica junto ao setor sucroalcoleiro', PhD thesis, University of Campinas, Brazil.

Williams, R. H. and Larson, E. D. (1993) 'Advanced gasification-based biomass power generation', in Johansson, T. B., Kelly, H., Reddy, A. K. N. and Williams, R. H. (eds) *Renewable Energy – Sources for Fuels and Electricity*, Washington DC: Island Press, 729–86.

World Commission on Environment and Development (WCED) (1987) *Our Common Future*, Oxford: Oxford University Press.

# 5

# SUGAR-CANE CULTURE AND USE OF RESIDUES

*Oscar A. Braunbeck*
*and Luís A. B. Cortez*

## 5.1 Introduction

### 5.1.1 The history of sugar-cane production in Brazil

Sugar-cane (*Saccharum officinarum*) was introduced by the Portuguese in Brazil in the early sixteenth century in two different regions: the northeastern part of the state of Pernambuco and the southeastern part of the state of São Paulo. The Pernambuco region presented better responses to sugar-cane than São Paulo and the crop was responsible for the second economic cycle in the colony, after wood had declined.

Brazilian sugar-cane experienced its first shock after the crop was introduced in the Caribbean region by the Dutch, French, Spanish and also the British. Better results experienced in the Caribbean region, but especially in Cuba, favoured the collapse in Brazil. However, this turmoil was not sufficient to eliminate the crop. The basis for this culture and the related industry was well rooted and it became an important activity, particularly in the region named *zona da mata*, a 150 km land slab along the coast in the Northeast region.

Another setback was felt when the French implemented in Europe the technology to obtain sugar from beet. The new crop offered a significant independence from imported sugar; several decades were necessary before a new equilibrium was reached between sugar-cane and beet sugar production.

The Brazilian economy learned, somehow, to live with the constant fluctuations of the world sugar market and throughout the last centuries other economic commodity cycles occurred, chiefly gold in the seventeenth and eighteenth centuries and coffee in the late nineteenth and early twentieth centuries.

At the beginning of the twentieth century Brazil experienced a fast growth in the production of natural rubber, but coffee was largely the most important cash-crop, leading the country's economy. Meanwhile sugar-cane was becoming more important in Southern Brazil, particularly in the state of São Paulo, where recent European immigration had started to play a significant role in developing an industrial basis. This fact gave an important boost for the metal industry, which was very important for building modern mills.

In 1930, at the beginning of the Vargas regime, the Sugar and Alcohol Institute (IAA) was created, which was to play a very important part in the Brazilian sugar business.

The first Vargas Government also started experimenting with the use of cane ethanol in official vehicles.

### 5.1.2 Sugar-cane importance in the current Brazilian economy

This situation did not change radically until the early 1970s, when the Geisel administration, a military government, created the Brazilian Alcohol Program (PROALCOOL), which significantly increased the sugar-cane area in Brazil, which today covers more than 4 million ha.

Although the current figures demonstrate that Brazil has achieved an important world-wide position in the sugar-cane business, the activity is very much concentrated in the São Paulo region, where most of the sugar and ethanol is produced. This predominance of the state of São Paulo and parts of the state of Alagoas over the other producing areas is due mainly to the following factors: better land, providing higher agricultural yields; more organised entrepreneurs; and more intensive use of available technology, in both agriculture and industry.

Today, the sugar-cane industry is directly responsible for employing nearly 600,000 people in the several activities, from agriculture to industry. Brazil is now the world's largest sugar-cane producer and alone this sector generates nearly US$5 billion of income, helping the country to export and to decrease its petroleum dependence. It is a sector almost entirely owned by local entrepreneurs, with important potential to increase its participation in the country's economy through more intensive use of its by-products.

### 5.1.3 Development of technology in sugar-cane agriculture

Sugar-cane is a crop planted in Brazil for two main reasons: the production of sugar and of ethanol. Cane is cultivated utilising the so-called 'ratoon system', in which the first cut is made 18 months after planting and then annually for a four- or five-year period, with decreasing yields. The warm Brazilian climate, with rainy summers and clear skies in the winter, helps the cane to build a strong fibre structure during its growing phase and then to increase the sugar content in the winter.

The sugar-cane crops have benefited from results obtained at experimental stations in the state of São Paulo, particularly in the Agronomic Institute of Campinas (IAC). There, agronomists have conducted long-term experiments developing and adapting new varieties, even though the most important contribution came from a foreign variety, NA-5679, developed in Argentina. Other varieties, developed by the Co-operative of Sugar and Ethanol Producers of the state of São Paulo (COPERSUCAR), are the SP 70-1143, SP 71-1406, SP 71-6163 and SP 79-1011. The 'National Plan for Sugar-cane' (PLANALSUCAR), managed by IAA, developed varieties such as the RB 72454, RB 765418 and RB 785148, which are also well known for their contribution to the sector. However, in the last thirty years the entire agricultural research apparatus in São Paulo, including research stations and extension activities, has been submitted to constant dismantling, which slowed down sugar-cane productivity at a time when oil prices were falling and making ethanol less competitive. At the same time, the federal government put into practice the policy of eliminating state intervention in the sugar business, by closing down the IAA and ceasing the activities of PLANALSUCAR, among other things. This particularly affected the development of new varieties and what was previously done by the state has now become private responsibility.

COPERSUCAR had to concentrate its efforts on its programme of new varieties. The cooperative set up its Centre of Technology (CTC), in Piracicaba, SP, in 1979. It also has an experimental station in the state of Bahia, for conducting tests for more resistant and productive varieties. Another constant concern in COPERSUCAR's research in this area is to extend the crushing season by developing early ripening, increased yield and disease-resistant varieties. The sugar-cane breeding programme has also incorporated modern techniques in molecular biology. Genetic transformation of sugar-cane varieties has been achieved by the CTC laboratories and various transgenic sugar-cane varieties are currently being field-tested.

The sugar-cane borer (*Diatraea saccharalis*) is a major pest in Southern Brazil. It is currently subject to biological control, as the sugar mills produce large quantities of parasites to be released in the field to control the borer.

Other technological changes experienced in sugar-cane agriculture were more related to the introduction of machinery for soil preparation and conservation, particularly in the Southern states of Brazil in the last forty years. Some intensification of machinery use has been observed in the last two decades in various crop operations, from soil tillage to harvesting, and particularly in cane loading. Lesser significant advances were introduced in planting the cane. All the cane is still planted by hand, although a cane planting machine has recently been developed by COPERSUCAR and licensed to DMB. Two other planting machines are now being marketed by agricultural machinery manufacturers such as Sermag and Brastoft.

Despite this progress, harvesting continues to be the least advanced operation in cane production. Cane fields are systematically burned to allow manual harvesting. This scenario is changing, however. It will probably take eight years until all land feasible for mechanised harvest will have green-cane management in the state of São Paulo, and fifteen to twenty years until the process can be considered concluded in Brazil. Environmental pressures, legislation enforcement (Gov. Est. SP 1988) and cost reduction are pushing towards mechanical harvesting of 'unburned cane' (Furlani Neto *et al.* 1996). In addition, there is the potential to generate revenues from cane residues, which is not being properly evaluated by the producers yet.

Particular attention is given in this chapter to the development of cane harvesting, where there is important potential for the reduction of sugar-cane production costs. Other possibilities for cost reduction can be achieved with the introduction of operation research techniques in agricultural management and the adoption of precision agriculture that allows a more rational use of resources and increases the agricultural yields.

Softwares based on geographical information systems (GISs), with embodied electronics, indicate the effect of productivity-related variables, such as soil fertility, diseases, insects, weeds, soil compaction, and harvesting methods. They compare information and provide differentiated application rates for herbicides and fertilisers, controlling thus the impacts on cane productivity. The use of satellites for variety identification in planting areas has been successful in 87 per cent of cases.

The use of the ARENA software in cane harvesting and transport management has helped to simulate and optimise the scheduling of loaders, harvesters and lorries which transport the sugar-cane to the mills. Reductions of 5–13 per cent in the cost of sugar-cane were brought about by the use of this software.

Another agricultural management software package developed by CTC makes it possible to take into account constraints such as cane variety, ratoon age, maturation curves and available equipment in deciding when and where to harvest. It also helps decisions concerning reform areas to be ploughed out after the cane has gone through its economical productive period.

### 5.1.4 Main sugar-cane residues: trash, bagasse and vinasse

The first main residue from sugar-cane is the trash, which is the sum of the plant's top and leaves. The trash represents around 25–30 per cent of the total energy in the plant, yielding up to 10 t/ha/year of dry matter. Three methods for cleaning cane stalks in the field coexist in Brazil at the present time: machine, labour, and fire.

A very low percentage of green sugar-cane is harvested manually in Brazil; this practice represents less than 1 per cent of the total harvest and is mostly used to produce seed cane. Although field burning represents an energy loss, it is still a current practice in Brazil. State laws restrict field burning, particularly around urban areas in Southern Brazil.

After hand-cutting, the whole stick cane is brought to the mill where it is washed and crushed. To avoid heavy sucrose losses, billet cane does not go through washing. In the crushing process, the fibre is separated from the cane juice. The milled fibre, soaked with some of the imbibition water, is called bagasse. The bagasse is sent straight to the boilers, to provide, through co-generation units, electricity, mechanical power, and thermal requirements to the entire sugar/alcohol making process. The bagasse contains approximately 30–40 per cent of the total energy available in the sugar-cane plant.

After the juice has been extracted, it has to be cleaned before crystallisation takes place. In the cleaning process, the impurities are flocculated and settled by decantation and separated using vacuum filter press technology. This last equipment produces a hot solid material called filter cake, which usually returns to the field in the form of fertiliser, mainly as a K source.

If the process is directed to produce sugar, the mill also produces molasses, a by-product used in many countries as a raw material in the candy and animal feed industries. The molasses are also used in some countries as raw material for the production of ethanol, as for example in Cuba, where it is employed in the production of rum. In Brazil, molasses are blended with clear cane juice for making fuel ethanol. In this process, a residue from the distillation is produced: the vinasse. This highly pollutant liquid is generated at a rate of 10–15 l of vinasse/l of ethanol and represented a major concern at the beginning of PROALCOOL, when there was a fear about what could eventually happen with improper vinasse disposal.

Although other residues can be listed, the most important are certainly trash, bagasse, and vinasse, because of the large amounts produced and their economic potential as valuable by-products.

## 5.2 From cane burning to a sustainable crop

Burning, planting, and moving was the first type of agriculture practised in Brazil. To open new areas for agriculture, the first farmers had to clear the forest and fire was used for that end. Today, however, in Southern Brazil, this practice is no longer acceptable, because large quantities of cultivated land surround urban areas and pre-harvest cleaning fires are a nuisance to the population of those areas.

However, changing from burning to a full green cane cropping requires careful planning. The Brazilian landscape is often hilly and thus, appropriate, low cost, planting and harvesting methods are needed to overcome not only cane harvesting difficulties, but also trash recovery, bailing, and transport.

### 5.2.1 The present status of sugar-cane harvesting in Brazil

In spite of the large production of sugar-cane, mechanical harvesting is still hardly employed in Brazil. Although there is not any precise figure about the number of harvesting machines in operation in Brazil today, this number will not be greater than 600 machines, which harvest approximately 50,000 tonnes/machine per season – approximately 30 million tonnes, or 10 per cent of the total 300 million tonnes produced in the 1997/8 harvesting season. Frequent reference is made to the existence of quite high mechanisation rates, but that relates to isolated cases, such as Usina São Martinho, where 89 per cent of the cane was mechanically harvested in the 1993/4 season.

There was a strong interest in mechanical harvesting at the beginning of the 1970s, as a result of labour shortage forecasts (Stupiello and Fernandes 1984). Implementation at that time, however, was unsuccessful and interest faded in the 1980s, in part due to the deteriorating performance of the Brazilian economy. Mechanical harvesting has regained interest in the mid-1990s (Furlani Neto *et al.* 1996). The main difference this second time is the economic plan implemented by the Brazilian Government, which makes more evident the need for cost reductions and the fact that mechanical harvesting does contribute in this direction. Today's cost for manual harvesting and loading of 'burned cane' exceeds US$4.00/t (Coletti 1997), while mechanical harvesting hardly reaches US$2.00/t (Lima 1998). In the case of green cane harvesting, data are still unreliable, but there are indications that manual cutting exceeds US$6.00/t, while mechanical harvesting does not exceed US$3.00/t.

The burning of sugar-cane prior to harvest is common practice in Brazil, as it increases the throughput in both manual and mechanical harvesting. Existing green-cane harvesters, in fact, present 30 to 40 per cent lower daily tonnage with unburned cane (Ripoli *et al.* 1990). Consequently, in the state of São Paulo, green-cane harvesting is practised only in a 1 km radius around the cities, as enforced by the local law (Gov. Est. SP 1988).

The reason for preventing the burning of sugar-cane fields is to avoid the emission of pollutants, e.g. particulate matter and $CO_2$ to the environment, that have impacts on human health (i.e. respiratory illnesses) and on human amenity. However, field burning also results in sucrose losses by exudation; Ripoli *et al.* (1996) show ethanol losses in the range of 59 to 135 l/ha. Work done by Fernandes and Irvine (1986) on the commercial sugar-cane fields of several companies indicated that the actual sugar yield was below the fields' potential, both for manual cut (–17 per cent) and chopper harvester (–21 per cent). These losses were due harvest failure, loss of cane harvested, deterioration after burning, and the impurities included by the chopper harvester or the grab-loader of whole stick cane.

Furthermore, cane residues, which could otherwise have an economic value, are lost in the burning process. Long-term trash blanketing can reduce N fertiliser applications by 40 kg N/ha as a result mainly of reduced nitrate leaching (Vallis *et al.* 1996).

Both from the ergonometric and economic points of view, green-cane harvesting moves in the direction of mechanical harvesting, but any significant expansion in Brazil will depend on some improvements to the available mechanical harvesting technology, such as:

1   The machine throughput and harvesting costs must be little affected by the amount of trash.
2   The technology should allow the trash to be partially removed from the field, because of the important energy potential of trash and the fact that, at the moment, there are not adequate cane varieties and agronomic experience to manage trash blanketed cane fields (Sizuo and Arizono 1987).
3   The machine should leave part of the straw in the field, for weed control and moisture conservation, in cases where agronomic management techniques are well established.
4   The technology should keep the present whole cane processing system, to avoid unnecessary investments associated with the change to chopped cane, as well as to avoid the sucrose losses associated with the chopping and cleaning processes, which represent an unacceptable step back.

Present field experience with whole cane harvesters, together with recent developments on whole cane mechanical cleaning (Tanaka 1996), as well as on machine right-angle turning and pilling (Braunbeck and Magalhães 1996), added to the known potential of computerised engineering resources applied to machine design, allows us to anticipate that it is realistically feasible to generate a whole set of cane harvesting equipment with the above-mentioned features.

### 5.2.2  Mechanical cane harvesting throughout the world

There is no single mechanical harvesting system available at the present time that is able to handle the wide range of field conditions existing over the world. Field conditions can vary from hilly land and presence of rocks or stumps to dry or wet soils, requiring planting either at the bottom of the furrow or at the top of the ridge. Sugar-cane yields can vary from 60 to over 200 t/ha, the stalks being erect or recumbent with length in the range of 2–5 m. Tropical field conditions differ markedly from those in the countries where the leading machinery manufacturers are established, making the available equipment always an adaptation instead of machines developed specifically for local conditions.

The Australian chopped cane principle and the Louisiana whole cane system are the two main harvesting technologies available in the world today. Other cane producers, who make use of mechanical harvesting, use derivations of these systems. Such is the case of Cuba, which utilises 'KTP' cane harvesters, designed on the basis of the Australian principle (Gómez Ruiz 1992).

The Australian technology, illustrated in Figure 5.1, was developed because of a lack of sufficient labour to harvest the fields. The prototypes were developed by farmers in

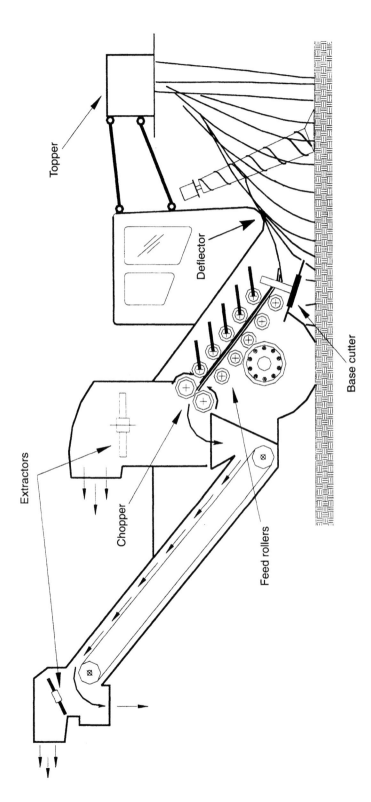

*Figure 5.1* Schematic view of a chopper harvester

the 1950s, and introduced chopping as a way to mechanically transfer the harvested product by free fall to the transport vehicle running side by side with the harvester. Chopping eliminates the whole cane loading operation. However, the cost of an adequately managed whole cane system can be inferior to that of a chopped one if cane losses and harvester idle time are accounted for in the cost analysis (Braunbeck and Nuñez 1986).

Louisiana's 'soldair' is the most cost-effective mechanical harvesting system for whole cane available at present (Richard *et al.* 1995). This harvesting system was developed for erect cane, which is usually a result of short growing periods of about 7 months; consequently it is not satisfactory to cut the Brazilian cane crops, mainly the first cut ones, which fall easily at the harvesting period.

In the US, there are four different sugar-cane regions: Florida, Hawaii, Louisiana, and Texas. In Texas the chopped cane system is used (Rozeff 1980). In Hawaii a locally adapted 'push-rake' system is used, which combines higher costs and lower cane quality. In Florida different systems are encountered: chopped cane, whole cane and manual cut.

In Colombia, where the government has set the year 2005 as the deadline to eliminate pre-harvest burning, work is underway to develop a harvester for collecting and chopping cane and cane residues. A major challenge in the development of the machine is represented by the high yields characteristic of Colombia's Valle del Cauca sugar-cane producing region (Ripoli *et al.* 1992).

The systems mentioned above are not well suited for Brazilian green cane from the standpoint of harvesting cost, cane quality, and losses. Technical developments are required to meet topographic, agronomic and sugar-cane processing needs typical of the Brazilian context.

### 5.2.3 Towards green-cane harvesting in Brazil

Mechanical harvesting expansion faces financial and technical constraints in Brazil, which are mainly due to:

1 shortage of skilled labour;
2 bad field lay-outs and poorly performing harvester technology for the existing fields;
3 lack of capital;
4 existing whole cane transportation and reception at the factory different from the emerging chopped cane system;
5 soil slopes must be under 12 per cent for present harvester design, which limits its use to about 45 per cent of the sugar-cane producing areas.

Mechanical harvesting is growing faster in the state of São Paulo, not only because of better topography, but also due to the better road network and the higher availability of skilled labour to operate a mechanised system. When cane producers of other states, such as Goiás and Mato Grosso, implemented mechanical harvesting they faced serious difficulties in hiring skilled labour to operate and maintain the machinery. It takes several years until the harvester fleet can reach a production of 400 t/machine-day, as a season average in 24 hour/day operation. Usina São Martinho, in the state of São Paulo, has exceeded production of 600 t/machine-day, largely due to skilled labour and adequate infrastructure, while Usina Sta. Helena, in the state of Goiás, after six years' operation, achieved an average harvester throughput of 400 t/machine-day.

A lower quality cane results from inadequate technology at the machine base cutter, which consists of two flat discs with approximately 900 mm diameter each, as shown in Figure 5.2. This defines a 1,800 mm wide plane which requires a perfectly levelled soil for the discs to operate very close to the surface without cutting the soil. Recommendations for efficient operation of base cutters indicate land flat and levelled. Brazilian undulated cane areas promote feeding of large quantities of soil into the harvester, which is removed inside the machine but about 0.5 per cent soil still remains in the cane. It has an impact in the cane process at the factory, especially because washing is not feasible on chopped cane. The solution to this problem will arise from existing and future developments on alternative cutting and feeding mechanisms more than from insisting on further level cane fields, which creates adverse agronomic conditions for ratoon longevity and moisture conservation. It is important to point out that the soil content of hand-cut cane does not exceed 0.15 per cent when loaded with properly operated rotary push-pilers.

The technical innovation and investment capacity of the Brazilian agricultural machinery industry is still rather small. It essentially limits itself to promoting the chopped cane technology developed abroad, predominantly in Australia and Germany.

There is also a technical barrier related to the potential use of machinery in areas with unfavourable topography (Braunbeck et al. 1999). Conventional two-wheel steering results in severe manoeuvre drawbacks as land slope increases. Techniques such as four-wheel steering and four-wheel traction, as well as steel or rubber tracks, would allow mechanisation to be extended to up to 90 per cent of the areas which are currently occupied by sugar-cane. Today's harvesters are mainly one-row machines with a high

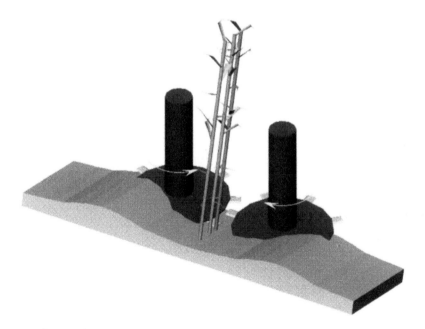

*Figure 5.2* Chopper harvester cutting into the soil to reach the bottom of the cane stalk

centre of gravity; they use two-wheel traction in the rear and two-wheel steering in the front. This driving and traction configuration is acceptable for agricultural tractors whose alignment with the line of motion is not so important. In the case of harvesters the lack of machine alignment with the crop lines leads to cane losses and frequent stops due to clogging (Ridge and Dick 1984).

Chopper harvesters present several critical points in terms of cane losses, such as the double disc basecutters, the feed rollers, the chopper and the cleaning extractors. These components are responsible for losses of 7–15 per cent (Ridge et al. 1984; Fernandes and Irvine 1986).

The successful implementation of mechanical harvesting in Brazil needs to address a series of technical issues, based on topographic, agronomic and sugar-cane processing conditions typical of Brazil.

### 5.2.4 Trash recovery, bailing and transport

Very little information is available in Brazil concerning trash recovery and use. Some experiments were conducted on trash recovery by Usina Sta. Elisa, from Sertãozinho, SP, in cooperation with Dupont and Class. Also, COPERSUCAR is conducting experiments on trash recovery and use in boilers, bailing trash with Class and Case machines; however, no conclusive reports are available yet.

After harvesting the green cane, trash may be left on the soil to dry for a few days. When the trash is nearly dry, with a 30 per cent moisture content, it can be recovered. If left on the soil it may cause fire or slow down ratoon sprouting. So, it is generally accepted that at least part of the trash should be recovered. Specialists' recommendations vary from 50 up to 90 per cent of trash recovery. They say that leaving a little organic material may bring some agronomic benefits, helping to control weeds and increasing long-term soil fertility. An experiment, reported by Molina et al. (1995), using a roller-type trash bailer, processed 5.7 t/h, recovered 83 per cent of trash with 30 per cent moisture content and obtained low density bales with 120 kg/m$^3$.

For the recovery a series of operations is required, starting with raking the trash into continuous windrows after natural drying in the field. The trash should be baled to make transportation economically viable. The commercial balers compact up to 150 or even 200 kg/m$^3$ density.

Two types of field baler are in the market today and they are classified by the bale geometry they produce: cylindrical and rectangular. A round-bale machine has a press chamber into which the trash is conducted directly after pick-up. The continuous rotation of the bale helps to compress the trash as it is fed into the chamber. The material is then steadily and increasingly pressed until the bale reaches the final diameter. After tying, the baler rear door opens hydraulically and the bale is pushed out onto the ground. The technical parameters supplied by one baler maker are the following: power required to operate the baler, 51 kW; collecting width, 1,600 mm; bale size, 1,200 × 1,200 mm; approximate productivity, 30 bales/h. The rectangular balers supplied by Class produce a bale with 150–200 kg/m$^3$ density. Usually the rectangular balers produce higher density bales than the cylindrical ones.

The operating costs may be determinant to make trash recovery feasible. The costs reported by Molina et al. (1995) were US$7–25.00/t, depending on local conditions, such as topography, infrastructure and available technology.

## 5.3 The use of trash and bagasse

Although trash and bagasse are obtained in different parts of sugar-cane production and processing, they are, somehow, similar in nature, because of their composition.

### 5.3.1 Trash and bagasse characteristics as fuels

From the point of view of energy consumption, the most important characteristics of a fuel are its composition, heating value and other properties related to the energy conversion technology adopted. An analysis conducted at the State University of Campinas' Alternative Fuels Laboratory obtained the results presented in Table 5.1 for eucalyptus, bagasse and trash from sugar-cane.

Other important characteristics such as the ash melting point, critical when a fuel is used in boilers, are not properly documented. Experiments on trash combustion are also not reported.

*Table 5.1* Composition (%) and high heating value (HHV) of eucalyptus, trash (kJ/kg) and bagasse from sugar-cane

| Sample | Moisture content | Volatiles | Fixed carbon | Ash | C | H | HHV |
|---|---|---|---|---|---|---|---|
| Eucalyptus | 11.9 | 80.2 | 19.8 | 0.0 | 49.6 | 6.0 | 18,494 |
| Cane trash | 10.5 | 74.7 | 15.0 | 10.3 | 43.2 | 5.6 | 15,203 |
| Cane bagasse | 9.9 | 75.4 | 13.7 | 10.8 | 43.6 | 6.2 | 17,876 |

### 5.3.2 Trash and bagasse use for electricity generation

In some cases the leafy residues, trash, may be separated at the cane cleaning centres, as in Cuba. In any case, these residues can provide an excellent source of animal feed and a valuable source of energy. Alternative uses can be found in the production of paper and board, furfural, fertilisers and even ethanol production, through a previous hydrolysis stage. Cane residues can represent about 23 per cent of the green matter weight, a proportion similar to that of bagasse (Ripoli *et al.* 1990; Payne 1991). However, in the case of green harvesting, not all the cane residues are likely to be used, because it is desirable to leave a portion in the field. Up to now there has been no conclusive study on the optimum share to be left in the fields. It is believed that this share depends on the soil type, cane variety, etc. The knowledge of the amount of leafy residues to be left in the field for 'mulching' is highly important because it determines how much of it can be commercialised, generating extra revenue.

The value added to the sugar and alcohol industry from the exploitation of by-products can provide a substantial contribution to the industry's financial sustainability and its diversification of revenue sources. Furthermore, the valorisation of the energy content of the sugar-cane industry wastes contributes to meeting electricity and heat demands, thereby reducing the dependence on fossil fuels and helping to decrease environmental pressures.

If it is assumed an average cane stalk yield of about 75 t/ha for Brazil, then the corresponding above-ground green sugar-cane yield is likely to be 105 t/ha. About 30 t/ha of the above-ground biomass consists of cane residues, with an average moisture content of about 50 per cent; and bagasse with a moisture content of about 50 per cent represents approximately another 30 t/ha. Given nearly 4 million ha of sugar-cane plantations in Brazil, the cane residues amount to about 120 million tonnes. By no means all these residues are recoverable. Under a conservative estimate that only 50 per cent of the fields are suitable for mechanical harvesting systems (which would render the recovery of crop waste viable), and that only 50 per cent of the waste from these fields is recoverable because of agronomic reasons and losses at various stages of the process, then about 30 million tonnes could be used for energy production purposes. Crop waste, combining trash and bagasse, possesses a gross calorific value of about 17 GJ/t dry matter. Hence, the energy content of the recoverable residues would be about 510 PJ, or 88 million barrels of oil equivalent per year, or about 42,000 GWh of electricity, assuming a 30 per cent conversion efficiency (Bauen *et al.* 1998).

So far bagasse has been used at the mill site to generate steam and in some cases to co-generate steam and electricity to satisfy the mill's needs, which are typically 350–500 kg of steam and 15–25 kWh of electricity per tonne of cane milled, and to dispose of the bagasse. As the latter is available in excess of the needs, efficiency in generation has not often been an issue. However, in many countries the industry is showing growing interest in generating surplus electricity which can be sold to the public grid and thus provide additional revenue.

Modern condensing-extraction steam turbine (CEST) co-generation systems and biomass integrated gasification gas turbine (BIG-GT) co-generation systems can meet the plant requirements and produce substantial amounts of surplus power. In the case of BIG-GT systems, where electricity is produced more efficiently than in CEST systems, efficiency gains are likely to be required in the utilisation of process steam. Because of the higher efficiency and the likely lower capital costs at the capacities typical of sugar and alcohol plants, BIG-GT systems should produce lower electricity costs. To minimise such costs, high capacity factors are required for the generating plants, and the use of cane residues to complement bagasse, particularly outside the milling season, thus becomes a necessity. Williams and Larson (1993) indicate that complementing alcohol production in Brazil with electricity generation could lead to alcohol and electricity costs that are competitive with conventional alternatives.

Countries other than Brazil are considering converting sugar-cane residues into energy sources, for the reasons discussed above. Mauritius has a 'Bagasse Energy Development Programme', which aims to capture the bagasse potential within an overall least-cost power station capacity expansion plan, and to promote technology development for the efficient utilisation of sugar-cane biomass for energy production, with emphasis on the exploitation of sugar-cane residues (Beeharry 1996). Excess bagasse in some mills is used to generate electricity, which is sold to the public grid. Bagasse-derived electricity meets 8 per cent (75 GWh) of the annual electricity consumption in Mauritius. Plans for expanding this capacity are underway, with negotiations to construct a 32 MWe plant. The use of cane residues would double the potential electricity supply from sugar-cane and would be accompanied by economic and environmental benefits.

The Colombian sugar-cane industry is also showing interest in co-generation surplus electricity sales. There are electricity supply deficits in many Colombian regions and notably in those where the sugar industry is concentrated. Electricity sales to the grid could provide an additional source of income and income diversification. Cane residues are a valuable source of energy, especially in view of the ban on pre-harvest burning of sugar-cane fields, which will be enforced in the year 2005. Furthermore, the use of cane residues for energy generation may allow for the continued use of bagasse in the paper and board industry, whose use is likely to be otherwise displaced towards a more profitable electricity generation.

The Brazilian sugar-cane industry already practises co-generation and sells some surplus electricity. About 15 mills have signed contracts with local power supply utilities. Imminent blackouts, caused by increasing electricity demand in Southern Brazil, are forcing authorities to find rapid solutions. However, the local authorities and mill-owners still do not realise that the sugar and alcohol industry can supply part of future electricity demand with relatively moderate investments. First, the authorities should enforce existing legislation to ban fire in cane fields; second, they should create conditions for a reasonable economic return on the required investment. The mill-owners, on the other hand, should not spoil 30 t of biomass/ha just to make the harvest easier.

### 5.3.3 Use of bagasse for energy production in other industries

Wood from native forests has been historically used to provide energy through its combustion. Also in the sugar mills in Brazil, wood was and still is used, to a limited extent, to complement bagasse, which is the main source of energy in this industry. Bagasse can provide enough heat, through its combustion in steam boilers, to produce all the process heat required in the sugar/alcohol production and also to generate enough electricity by turbo-generators to drive all electric motors and other electricity requirements, such as lighting.

After meeting the plants' internal needs there still exists a bagasse surplus in Brazil. Many mills simply adjust their boilers to burn all excess bagasse, because, otherwise, the residue will deteriorate without any benefit. However, in the state of São Paulo, where most sugar-cane is produced, firewood is expensive and transported from distant locations using diesel trucks. The wood is usually employed in the manufacture of bricks, in some food-processing plants, bakeries and restaurants. However, this firewood is becoming scarce and costly and is being substituted, in many cases, by natural gas and fuel oil.

There is, however, significant potential for bagasse to substitute for firewood, in the short term. A limiting factor apparently is a lack of entrepreneurship to disseminate a 'bagasse culture'. One example of a successful substitution is given by Destilaria Rosa, located in Boituva, SP. Their owners have installed a small brick industry just beside their distillery. There, the excess bagasse, nearly 30 per cent of the total available, is used to produce low-cost bricks. Tests were conducted in their furnace to evaluate both the efficiency and quality of the bricks (Aradas *et al.* 1998). In spite of the good results obtained, the example has not been followed by other mills.

The use of bagasse as a fuel outside the sugar and alcohol mills has been more intensive in the orange juice industry, where bagasse boilers, similar to those employed in the mills, have been installed. Also, there are some uses of bagasse as a fuel in, for example, the vegetable oil industry. The market price of bagasse varies, depending on the local availability and delivery distance. In the state of São Paulo usually the prices are in the range US$5–12/t.

There are some small companies in Brazil that produce equipment especially designed to burn vegetable residues, including cane bagasse: Andrade & Andrade, from Leme, SP, which manufactures the biomass burner 'Fire 100', Máquinas Walter Siegel, from Agrolândia, SC, which manufactures the 'Biochamm' equipment, and Irmãos Lippel, located in Agrolândia, SC. This equipment is simple in conception and is sold at accessible prices, around US$1,600. Today, just one manufacturer has more than 2,000 biomass burners installed in Brazil, mainly in the bricks industry.

### 5.3.4 Alternative uses of bagasse and trash as fuels

#### 5.3.4.1 Production of pellets and briquettes

There is no experience with the pelletisation and briquetting of sugar-cane trash. Although it is not practised commercially, experiments on bagasse pelletization have already been carried out (Bezzon 1994 and Cortez and Silva 1997).

Despite the fact that sugar-cane trash and bagasse are fibres with the same origin, their physical and chemical characteristics may differ significantly. The sugar-cane bagasse has a smaller particle size because it is milled in the juice extraction process. This results in finer particle sizes (Olivares Gómez *et al.* 1998) if compared to the unprocessed trash.

Some technical difficulties are associated with the bagasse briquette's long-term integrity. The bagasse high moisture content, nearly 50 per cent w.b., is considered the most negative factor when briquetting is considered. Bezzon (1994) conducted experiments heating up the bagasse up to 200–300°C before briquetting (1 cm diameter by 2 cm in length). The applied pressure ranged from 20 to 25 MPa and yielded briquettes with densities from 1,000 to 1,240 kg/m$^3$. The results were promising but experiments with larger briquettes are still required. Heating the briquette can melt lignin, resulting in a fibre-binding material.

#### 5.3.4.2 Small-scale gasification and charcoal production

Biomass gasification is defined as a thermal conversion process producing a fuel gas by partial oxidation under high temperatures. This conversion can be conducted in different types of reactor, depending on the biomass characteristics. Because bagasse is a finely ground material, the fluidised-bed reactor is the recommended technology.

The gas produced in small-scale gasifiers, usually in the capacity range from 100 kW to 1 MW, is often consumed in internal combustion engines coupled with electric generators.

A fluidised-bed prototype with 280 kW thermal capacity was designed and built at the State University of Campinas, with the help of Termoquip, a local gasifier manufacturer, to process sugar-cane bagasse and trash. The first tests allowed the identification of difficulties in handling wet bagasse in the feeding system. Experiments were then conducted with bagasse pellets, with which the reactor presented reasonable yield for air ratios (0.17–0.22). The fuel gas average heating value was 4 MJ/Nm$^3$, which can be considered satisfactory when air is used. The higher efficiency values obtained were 29 per cent and 33 per cent, respectively, for cold and hot fuel gas, obtained with a gas to air ratio of 0.22. Limitations in the gasifier operation range did not allow higher values for the gas to air ratio to be tested as far as efficiency is concerned.

The economical use of trash will depend on investigating more economical technologies to handle, transport and use the material, transforming it into a more valuable commercial product. Further research is needed regarding the conversion of trash not only to electricity, ethanol and fuel gas, but also to charcoal, a product with a large market in Brazil. A large part of charcoal production in Brazil still employs low efficiency furnaces, wasting wood (Rosillo-Calle et al. 1996). Charcoal production from trash and bagasse could benefit both the sugar and the steel industries in Brazil; no technologies, however, have yet been developed to convert these by-products.

### 5.3.5 Bagasse for animal feed and weed control

Although the production of energy may be the natural destination of bagasse, other possible uses have been widely investigated (Paturau 1982). In Brazil bagasse has been used as an ingredient of animal feed and in weed control.

Due to its high fibre content, cane bagasse may supplement animal feed if its fibres are pre-digested. This can be carried out in a reactor where the cellulose cells are broken down and hydrolysed. The hydrolysed bagasse is then compressed, making pellets which are easily handled and eaten by the animals. This practice has been welcomed by cattle farmers, especially during the dry winter months when naturally-grown grass is less available.

Some small farms in the surrounding areas of the sugar and alcohol mills are using bagasse for weed control, but this is still on a small scale.

## 5.4 Vinasse disposal technologies employed in Brazil

Vinasse is produced from cane juice and/or molasses fermentation. Chemically, vinasse composition varies according to the soil type, sugar-cane variety, harvest method, and industrial process used in the production of ethanol. Its colour, total solids content, and acidity are all parameters which may vary according to the type of vinasse and processes employed.

Generally, vinasse presents a light brown colour and a low total solid content (of 2–4 per cent) when it is obtained from sugar-cane juice, and a dark reddish colour, with a total solids content ranging from 5 to 10 per cent, when it is produced from cane molasses. The hazardous substances present in the vinasse generate a very high Biological Oxygen Demand (BOD), ranging from 30,000 to 40,000 mg/l (Bhandari et al. 1979), and a low pH of 4–5, because of the organic acids present in it, which are corrosive, requiring stainless steel or fibreglass containers to resist them. Vinasse contains unconverted sugars, non-fermented carbohydrates, dead yeast, and a variety of inorganic compounds, all of which contribute to the BOD.

Disposal of vinasse by spreading on river basins is not a recommended method due to its high BOD, which can cause serious damage to aquatic life, especially when dumped in large volumes, as was often the case with most ethanol distilleries. This practice, which was utilised when production was much smaller, had to give way to other methods.

The better-known technologies for vinasse disposal may be grouped as:

1   **Ferti-irrigation**. This is the most common disposal method utilised in Brazil; it is described below. It has a low initial cost and takes advantage of vinasse's fertiliser potential;

2   **Biodigestion**. Vinasse biodigestion has the advantage of on-site disposal, while a good fertiliser is produced and energy is recovered through the generation of biogas. This disposal method is particularly interesting from an energy substitution point of view, because the sugar-cane industry produces very large amounts of vinasse and still depends on diesel oil for sugar-cane transportation; biogas can substitute for the latter in the lorries' diesel engines.

3   **Animal feed**. This disposal method is used by some distilleries in the USA, e.g. at Shepherd Oil Distillery, Mermentau, Louisiana, the same practice is used with high-grade protein vinasse obtained from distilleries producing spirits. Vinasse obtained from molasses or high-test molasses fermentation is given straight to animals, usually beef cattle. Although good results have apparently been obtained, studies on nutrition and other effects have not been reported so far. Some research has also been conducted on the use of dry vinasse as fodder feed (Kujala *et al.* 1976; Kujala 1979);

4   **Production of fungi**. Research has been carried out at the Brazilian Institute of Technology (INT) on fungi production from vinasse (Araújo *et al.* 1977). Although the technology has not evolved from laboratory level, it remains a promising alternative when high-value products are sought;

5   **A construction material**. Studies conducted by Rolim and Freire (1996) on brick-making show that, for some applications, vinasse can be a useful building material. However, further research is required as vinasse is a highly hygroscopic material and misuse of bricks made with such material, if exposed to water or humid weather, could cause the bricks to collapse.

6   **Incineration**. Basic research has been carried out by Nilsson (1981), Spruytenburg (1982) and Cortez and Brossard (1997), who have experimented with the incineration of pure vinasse and its emulsions with heavy oils. Although vinasse incineration technology was presented as being commercially available in the early 1980s, by some companies, e.g. Alfa Laval and HCG, in practice, it failed to live up to expectations.

There are other alternative disposal methods, not yet tested on a large scale, e.g. ultra filtration and/or reverse osmosis, centrifugation, and production of single-cell proteins.

As vinasse is produced in large volumes, the possibility exists to reduce the amount to be disposed of by recycling it in the fermentation process. Vinasse may be partly used to dilute the sugar-cane juice or molasses in the fermentation step. The juice or molasses need to have their Brix adjusted to allow proper yeast growth, a process that normally requires water for dilution. Alfa Laval developed a process called 'biostil', which uses vinasse to dilute the molasses prior to the fermentation step (Cook 1983). This feedback system saves energy and water, producing an effluent with a higher solids content (approximately 16 per cent).

Research has also been conducted into the development of more resistant yeast strains, which can also lead to reductions in the production of vinasse, making its disposal a simpler problem.

From the vinasse disposal methods discussed above, only two are practised in the Brazilian sugar and alcohol industry: ferti-irrigation and biodigestion.

### 5.4.1 Ferti-irrigation of vinasse

In agriculture, the introduction of ferti-irrigation (Orlando *et al.* 1995) was probably the most important technological improvement in the last twenty-five years. In Brazil, most of the stillage is currently applied to the soil as a fertiliser because of its high potassium content. For this reason, it is necessary to have a good understanding of the soil composition to ensure that the amount applied corresponds to the specific soil needs.

The disposal problem is aggravated by the high transportation costs associated with the large volume of stillage. Disposal of surplus vinasse in ponds is often used when transportation distances (costs) become prohibitive. The method consists of pumping the vinasse or transporting it by trucks to the sugar-cane fields, where it is applied using either open channels or conventional irrigation systems.

Vinasse application may increase the productivity of sugar-cane by 5–10 per cent (COPERSUCAR 1989, 1997).

### 5.4.2 Biodigestion of vinasse

The vinasse disposal through biodigestion is used in just a few distilleries in Brazil. The two best known examples are the São Martinho and São João distilleries, both in the state of São Paulo.

At São Martinho, the vinasse is treated using a thermophilic continuous process in a UASB reactor with a controlled temperature of 55°C. Vinasse comes to the process at 85°C from the ethanol distillation. The biodigestion reactor has 5,000 $m^3$ capacity and the residence time is 1 day. About 50 per cent of the biogas produced is used in a yeast dryer, which has a capacity of 40 t; the rest is burnt in the bagasse boilers. The biogas composition is about 60 per cent methane and 40 per cent carbon dioxide, with a heating value of 21,500 kJ/$Nm^3$. The resulting effluent is used as a fertiliser.

At the São João distillery the large-scale continuous reactor installed in 1986 also degrades the vinasse producing biogas, reducing by 85 per cent the COD load. A mesophilic process (temperature of 35–37°C) is used in this case. The whole biogas produced fuels a fleet of lorries, which transports the sugar-cane from the fields to the mill, and some utility cars. The distillery produces 300 $m^3$ ethanol/day, generating about 3,000 $m^3$ of vinasse/day. It employs a METHAX-BIOPAQ system for biodigestion, which treats one-third of the available vinasse, some 1,000 $m^3$/ day.

The São João biodigester has been operating for ten years. The reactor has a capacity of 1,500 $m^3$ and the associated gasometer 600 $Nm^3$. The generated biogas is 70 per cent methane and 30 per cent carbon dioxide, with traces of impurities. The biogas is purified and compressed up to 220 atm in reservoirs of 400 $Nm^3$ capacity. The daily production is 6,500 $Nm^3$ (96 per cent methane). Therefore, the conversion factor biogas/vinasse is 11 $Nm^3$/$m^3$. The compressed methane is used to fuel 41 vehicles (29 lorries and 12 utility cars). This represents 50 per cent of the mill's lorries and 40 per cent of its utility cars.

In the 1997/98 harvest the use of biogas as an automotive fuel in São João distillery was discontinued and the generated biogas was used for yeast drying only. There were two main reasons: 1) the low prices of Diesel oil, which is heavily subsidised in the country; and 2) difficulties in obtaining spare parts for the modified diesel engines used by the distillery's lorries. Mercedes Benz, which manufactures lorries in São Paulo, could eventually supply specially built gas engines and lorries for that purpose, but the prices would be high unless the technology is adopted by several other distilleries.

### 5.4.3 Economic and environmental aspects of vinasse disposal

The economics of vinasse disposal using ferti-irrigation and biodigestion depends, to different degrees, on the market price of diesel. Ferti-irrigation is highly affected by fuel costs. Moved by high vinasse transportation costs, some plants have installed pumping systems, which operate together with road transportation. To dump excess vinasse, some factories use sacrifice areas, which probably cause pollution problems. Environmental studies have not proved harm on applying small doses of low solid vinasse to the field, although there is speculation about possible long-term underground water contamination. Some environmentalists express concern about a possible insect disequilibrium caused when vinasse is spread over the fields, particularly in the forty-eight hours following the application.

Biodigestion economics depends on several technical parameters that affect the residue residence time, as well as the end use and value of its main by-products. A preliminary evaluation was conducted by Silveira (1986), who studied the benefits of a complete plant coupled with a 75,000 l of ethanol/day distillery. The biodigestion system treated 1,170 $m^3$ of vinasse/day and produced 6,200 $Nm^3$ of biogas (95 per cent methane) in the same period. The input was 225 kWh of electricity and 72 $m^3$ of water every 24 hours. The labour requirement was three shift operators and one shift supervisor. The annual benefits were calculated to be equivalent to 17 per cent of the total investment when the methane was compressed and used to fuel a converted diesel lorry fleet.

A similar analysis was conducted by Ribeiro (1986) for a 150,000 l ethanol/day distillery, treating around 1,305 $m^3$ of vinasse/day and producing 4,760 $Nm^3$ (72 per cent methane) for automotive use and 16,033 $Nm^3$ for fuelling boilers. Total investment was calculated to be around US$687,000, including the biodigesters, filling pumps, vinasse cooler, secondary materials and installation. The biogas purification and compression unit cost was US$320,000 and the methane distribution system US$171,000. To allow the methane to be used as a fuel the diesel lorries have to be converted, resulting in an additional cost of US$260,000. Therefore the total investment was US$1.17 million. The annual operation and maintenance costs added a further US$97,000. The calculated payback period was 6.6 years, representing annual savings of US$203,000. The plant produced sufficient fuel to power sixty-five lorries, forty-two tractors, sixteen loaders, and thirteen other utility vehicles.

Gomes (1986) provides a typical investment analysis of biodigestion from vinasse from different substrates, based on different Brazilian conditions. His figures lead to a payback period of approximately two years.

Another important source of revenue that can be derived from biogas is the production of dried yeast. São Martinho's figures show that operating an industrial unit of 9,000 t of dried yeast/season costs only US$111.6/t of dried yeast, compared to US$170/t when steam is used. The internal rate of return is calculated to be 36 per cent, with a payback of 3.6 years.

In the METHAX biodigestion plant built at São João distillery, average figures for effluent composition indicate the presence of 328 kg of N/day, 100 kg of P/day and 2246 kg of K/day.

## 5.5 Conclusions

Cultivation of sugar-cane in Brazil has experienced an important growth after the implementation of PROALCOOL, when the planted area has expanded dramatically. Together with recent developments regarding Brazilian environmental legislation, the crop is about to go through an important benchmark when whole sugar-cane harvesting will have to be practised. The green-cane management, which will reduce $CO_2$ emissions, is new to the country. However, there are still lessons to be learned in managing the excess biomass to be left in the fields, together with changes in pest control caused by the absence of burning. Also, the biomass to be removed from the fields will represent an important amount of raw material that should be used in the best possible economic way, improving the mills' competitiveness.

The intensification of biomass use in Brazil depends on three key factors: competitive costs, regular supply, and commercially available and dependable technologies. Currently, bagasse has difficulties to compete with fuelwood (US$3–7/t) and cheaper residues, such as sawdust and cotton, coffee and peanut husks. There is a need for a constant supply, to assure long-term dependence; the new bagasse and trash clients cannot be left without supply for a period as long as six months. The packing, bailing and transporting of equipment, and the technology for using the trash are not fully developed and tested for the market yet. There is still need for improvements, notably to reduce the most expensive operations. Equipment with higher capacities and lower costs should have to be developed particularly for trash recovery.

The more intensive use of sugar-cane by-products is considered essential to improve the sugar-cane agro-industry sustainability. Current market price levels discourage investments on their use, either to generate electricity or to produce oil substitutes. Government action is required to boost such activities, in terms of enacting favourable legislation and providing economic incentives, as is happening in some other countries.

## 5.6 Bibliography

Aradas, M.E.C., Cortez, L.A.B. and Silva, E.E.L. (1998) 'Bagasse as a brick kiln fuel', *International Sugar Journal*, **100** (1189) 16–25.

Araújo, N.Q., Visconti, A.S., de Castro, H.F., da Silva, H.G.B., Ferraz, M.H.A. and Salles Filho, M. (1977) 'Produção de Biomassa Fúngica do Vinhoto', *Informativo do INT*, Instituto Nacional de Tecnologia, Rio de Janeiro, Jan, Feb, Mar, 12–19.

Bauen, A., Cortez, L., Rosillo-Calle, F. and Bajay, S. (1998) 'Electricity from sugar-cane in Brazil', in Kopetz, H. *et al.* (eds) *Biomass for Energy and Industry, Proceedings of the 10th European Biomass Conference*, CARMEN, June 8–11, Würzburg, Germany, 341–4.

Beeharry, R.P. (1996) 'Extended sugar-cane biomass utilisation for exportable electricity production in Mauritius', *Biomass and Bioenergy*, **11** (6) 441–9.

Bezzon, G. (1994) 'Síntese de Novos Combustíveis Sólidos a Partir de Resíduos Agroflorestais e Possíveis Contribuições no Cenário Energético Brasileiro', MSc dissertation, School of Mechanical Engineering, UNICAMP, Campinas, SP.

Bhandari, H.C., Mitra, A.K. and Malik, V.K. (1979) 'Treatment of distillery effluent',

*Proceedings of the 43rd Annual Convention of The Sugar Technologists Association of India*, Kanpur, India, G65–72.

Braunbeck, O.A. and Magalhães, P.S.G. (1996) 'Dispositivo Virador de Cana Inteira, Instituto Nacional de Propriedade Industrial', INPI-UM-7501b02, SP, Brazil.

Braunbeck, O.A. and Nuñez, G.J.S. (1986) 'Corte, Carregamento e Transporte-Análise de Custos e Desempenhos', *Boletim Técnico Copersucar*, No. 35/86, 44–53, Centro de Tecnologia Copersucar, Piracicaba, SP.

Braunbeck, O., Bauen, A., Rosillo-Calle, F. and Cortez, L. (1999) 'Prospects for green cane harvesting and cane residue use in Brazil', *Biomass and Bioenergy*, (in press).

Coletti, J.T. (1997) 'Principais Procedimentos para Redução de Custos', *JornalCana*, April, 18–19, Brazil.

Cook, R. (1983) 'BIOSTIL – A Breakthrough in Distillery Design' (advertisement), *Sugar y Azucar Yearbook*, 96–99.

COPERSUCAR (1989) *PROÁLCOOL, Fundamentos e Perspectivas', Cooperativa de Produtores de Cana, Açúcar e Álcool do Estado de São Paulo Ltda.* São Paulo, SP: COPERSUCAR.

COPERSUCAR (1997) *Company information from Cooperativa de Produtores de Cana, Açúcar e Álcool do Estado de São Paulo Ltda.* São Paulo, SP: COPERSUCAR.

Cortez, L.A.B. and Brossard Perez, L.E. (1997) 'Experiences on vinasse disposal part III: combustion of vinasse-#6 fuel oil emulsions', *Brazilian Journal of Chemical Engineering*, **14** (1) 9–18. Campinas, SP.

Cortez, L.A.B. and Silva-Lora, E. (eds) (1997) *Tecnologias de Conversão Energética da Biomassa*, Editora da Universidade do Amazonas, Manaus, AM, July.

Fernandes, A.C. and Irvine, J.E. (1986) 'A comparison of the productivity of the chop-load system of harvesting sugar-cane with the hand-cut, grab-load system', *STAB*, **4** (6) 105–10, July–August, Piracicaba, SP.

Furlani Neto, V.L., Ripoli, T.C. and Villa Nova, N.A. (1996) 'Avaliação de Desempenho Operacional de Colhedora em Canaviais com e sem Queima Prévia', *STAB*, **15** (2) 18–23, nov/dez, Piracicaba, SP.

Gomes, I. C. (1986) 'Avaliação do Vinhoto como Substituto do Óleo Diesel e outros Usos. Coleção', SOPRAL No. 10, Cap. VI, 85, São Paulo, SP.

Gómez Ruiz, A. (1992) 'Sistema Cubano de Cosecha en Verde',Seminário ECNOCANA/92, Araras, SP.

Gov. Est. SP (Governo do Estado de São Paulo) (1988) 'Dispõe sobre a Proibição das Queimadas, Decreto No. 28.848', p.1, 30/08/88, São Paulo, SP.

Kujala, P. (1979) 'Distillery fuel savings by efficient molasses processing and stillage utilization', *Sugar y Azucar*, (October), 13–16.

Kujala, P., Hull, R., Engstrom, F. and Jackman, E. (1976) 'Alcohol from molasses as a possible fuel and the economics of distillery effluent treatment', *Sugar y Azucar*, **71**, 28–39.

Lima, L.O. (1998) 'Custos de Colheita Mecanizada', *STAB*, **16** (4) 29–30, Piracicaba, SP.

Molina, Jr., W.F., Ripoli, T.C., Geraldi, R.N. and do Amaral, Jr., J. (1995) 'Aspectos Econômicos e Operacionais do Enfardamento de Resíduo de Colheita de Cana-de-Açúcar para Aproveitamento Energético', *STAB*, May–June, **13** (5) 28–31, Piracicaba, SP.

Nilsson, M. (1981) 'Energy recovery from distillery wastes', *International Sugar Journal*, (September), **83** (993) 259–61.

Olivares Gómez, E., Brossard Perez, L.E., Cortez, L.A.B., Bauen, A. and Larson, D.L. (1998) 'Considerations about proximate analysis and particle size of sugar-cane bagasse and trash', *The 1998 ASAE Annual International Meeting*, paper 98–6011, Orlando, FL, July 11–16.

Orlando, F.J., Bittencourt, V.C. and Alves, M.C. (1995) 'Vinasse application in a Brazilian sandy soil and nitrogen watertable pollution', *STAB*, **13** (6) 9–13, (July–August), Piracicaba, SP.

Payne, J.H. (1991) 'Cogeneration in the cane sugar industry', *Sugar Series*, 12, Amsterdam: Elsevier.

Paturau, J.M. (1982) *By-products of the Cane Sugar Industry*, Sugar Series, 3, New York: Elsevier.

Ribeiro, I. de S. (1986) Conference presented in: 'A Avaliação do Vinhoto como Substituto do Óleo Diesel e Outros Usos', published as *Coleção SOPRAL No. 10*, Sociedade de Produtores de Açúcar e de Álcool – SOPRAL, São Paulo, SP, 45–66.

Richard C., Jackson W. and Waguespack Jr., H. (1995) 'Improving the efficiency of the Louisiana cane harvesting system', *International Sugar Journal*, 98 (1168) 158–62.

Ridge, D.R. and Dick, R.G. (1988) 'Current research on green cane harvesting and dirt rejection by harvesters', *Proceedings of Australian Society of Sugar-cane Technologists*, 19–25.

Ridge, D.R., Hurney, A.P. and Dick, R.G. (1984) 'Cane harvester efficiency', *Proceedings of Conference on Agricultural Engineering*, 118–22, Bundaberg, Australia, 27–30 August.

Ripoli, T.C., Mialhe, L.G. and Brito, J.O. (1990) 'Queima de Canavial – O Despedício Não Mais Admissível', *Álcool & Açúcar*, 10 (54) 18–23, jul/ago.

Ripoli, T.C., Stupiello, J.P. and Martinho, W.L.R. (1992) 'A Cana-de-Açúcar no Valle del Cauca', *STAB*, 10 (3) 37–40, (January–February), Piracicaba, SP.

Ripoli, T.C., Stupiello J.P., Caruso J.G.B., Zotelli H. and Amaral J.R. (1996) 'Efeito da Queima na Exudação dos Colmos: Resultados Preliminares', *Anais do Congresso Nacional STAB*, 63–70, Maceió, AL.

Rolim, M.M. and Freire, W.J. (1996) 'Solo-Vinhaça Concentrada: Aplicação na Fabricação de Tijolos (Soil-concentrated vinasse: application in brick production)', In: *Workshop Reciclagem e Re-Utilização de Resíduos como Materiais de Construção Civil*, São Paulo, SP, ANTAC – Associação Nacional de Tecnologia de Ambiente Construído, Proceedings for Discussion, 118–23.

Rosillo-Calle, F., de Rezende, M.A.A., Furtado, P. and Hall, D. (1996) 'The charcoal dilemma: finding sustainable solutions for the Brazilian economy', *Intermediate Technology Publications*, London.

Rozeff, N. (1980) 'An investigation of sugar-cane scrap in the Rio Grande Valley', 22–7, ISSCT, V.1.

Silveira, A.M. (1986) Conference presented in: 'A Avaliação do Vinhoto como Substituto do Óleo Diesel e Outros Usos', published as *Coleção SOPRAL No. 10*, Sociedade de Produtores de Açúcar e de Álcool – SOPRAL, São Paulo, SP, 29–44.

Sizuo, M. and Arizono, H. (1987) 'Avaliação de Variedades pela Capacidade de Produção de Biomassa e pelo Valor Energético', *STAB*, 6 (2) 39–46, (November–December), Piracicaba, SP.

Spruytenburg, G.P. (1982) 'Vinasse pollution elimination and energy recovery', *International Sugar Journal*, (March) 73–4.

Stupiello, J.P. and Fernandes, A.C. (1984) 'Qualidade da Matéria-Prima Proveniente das Colhedoras de Cana Picada e seus Efeitos na Fabricação de Álcool e Açúcar', 45–9, *STAB*, (March–April), Piracicaba, SP.

Tanaka, F. O. (1996) 'Despalhe de Colmos de Cana-de-Açúcar Inteira (*Saccharum spp.*) por Rolos Oscilantes com Diferencial de Velocidades', MSc dissertation, School of Agricultural Engineering, UNICAMP, Campinas, SP.

Vallis, I., Parton W.J., Keating B.A. and Wood A.W. (1996) 'Simulation of the effects of trash and N fertilizer management on soil organic matter levels and yields of sugar-cane', *Soil and Tillage Research*, 38 (18) 115–32.

Williams, R.H. and Larson, E.D. (1993) 'Advanced gasification-based biomass power generation', in Johansson, T.B., Kelly, H., Reddy, A.K.N. and Williams, R.H. (eds), *Renewable Energy Sources for Fuels and Electricity*, Washington, DC: Island Press.

# 6

# SUGAR-CANE INDUSTRIAL PROCESSING IN BRAZIL

*Isaías de Carvalho Macedo*
*and Luís A.B. Cortez*

## 6.1 The sugar-cane industry as an energy producer

Sugar-cane production in Brazil was 224 million tonnes in 1989, 241 million tonnes in 1994, and 302 million tonnes in 1998. The proportion of sugar-cane used for ethanol production was approximately 65 per cent in 1989 and close to 50 per cent in 1998. Although the ethanol/sugar ratio is decreasing, the industry is at least 50 per cent an 'energy' industry. The renewable nature of the fuel produced can be seen in Table 6.1.

The average energy balance (output/fossil fuel input ratio) calculated with the data from Table 6.1 is 9.2, an exceptional figure even for biomass-based systems. The figures in Table 6.1 were obtained from the current sugar-cane mills situation in Brazil: low efficiency steam-based co-generation systems based on bagasse, nearly 95 per cent autonomous mills in electric power and providing all the needs in thermal energy (Goldemberg *et al*. 1993).

For 1998, the contributions of the sugar-cane industry to the Brazilian energy system are the supply of nearly 200,000 barrels (oil equivalent)/day of ethanol (for a Brazilian domestic production of nearly 1 million barrels oil/day) and the bagasse-based co-

*Table 6.1* Average input–output energy flows for burned cane, sugar mill with ethanol distiller, 1996, São Paulo (MJ/t cane)[a]

|  | Fossil energy input | Energy output |
|---|---|---|
| Sugar-cane production and delivery (agriculture) | 190 | |
| Cane processing (industry) | 46 | |
| Ethanol produced | | 1,996 |
| Bagasse surplus | | 175 |
| Total (external flows) | 236 | 2,171 |

*Source*: Macedo (1998)
*Note*
[a] Fuel only, no sugar

generation of electric, mechanical and thermal power. Bagasse is used at a rate of 0.14 t (DM)/t cane, leading to an annual production of 40 million t (dry matter). Nearly all the power from bagasse is used in-house, at the mills; estimates indicate a use of 3,600 GWh for electric power plus 4,500 GWh for mechanical drives.

Brazilian cane production represents nearly 25 per cent of world cane production, corresponding to 13.5 per cent of sugar and 55 per cent of the world's ethanol. The Brazilian sugar-cane uses about 5 million ha, spread over several regions (1.5 per cent of 'arable' land). The number of industrial units is nearly 350, all privately owned. The ethanol industry provides fuel for the largest biomass to liquid fuel programme in the world (about 3.8 million cars using 100 per cent ethanol; 24 per cent ethanol blend in all 'gasoline' cars).

The trends today show that hydrated ethanol consumption is decreasing, while anhydrous ethanol (gasoline blends) is increasing, together with the possibility of blends with diesel oil. Overall the production may level for the next years. As presented in Chapter 5, the recent legislation limiting sugar-cane burning in the largest production areas (São Paulo state) is leading to a much larger biomass availability in the next twelve years; also, new technologies are being investigated to provide more efficient energy transformation systems.

It is expected that at least 50 per cent of the cane in the major producing areas will be harvested without burning in the next twelve years. The amount of biomass available for energy is being evaluated, taking into account the following variables:

1   average trash in sugar-cane (tops and leaves);
2   main agronomic routes to harvest green-cane with trash recovery;
3   soil properties with trash left on field;
4   advantages of trash left in field for herbicide elimination;
5   trash properties as fuel recovery costs;
6   utilisation as boiler fuel or for gasification;
7   environmental impacts of trash recovery/utilisation.

The results to date are shown in Table 6.2. They were obtained assuming an average trash availability of 10 t (DM)/ha and that harvesting cane without burning is feasible only on 55 per cent of the total planted area, where harvesters can operate and recover either 50 per cent or 100 per cent of the trash available (depending on agronomic considerations).

*Table 6.2* Biomass for energy: estimates of bagasse and trash availability

| Trash recovery (%) | Trash recovered (t (DM)/t cane) | Total bagasse (t (DM)/t cane) | Total biomass available (t (DM)/t cane) | Fuel oil equivalent (kg/t cane) |
|---|---|---|---|---|
| 100 | 0.075 | 0.14 | 0.215 | 77 |
| 50 | 0.0375 | 0.14 | 0.177 | 61 |

*Source*: COPERSUCAR (1998)
Note
Equivalencies include fuel-processing losses, storage losses, combustion-expected efficiencies

For an annual yield of 300 million tonnes of cane, with a trash recovery of only 50 per cent in 55 per cent of the planted area the annual amount of new biomass available will be 11 million tonnes (DM). The uses proposed for this biomass are mainly in the energy sector. Conventional technologies can be employed, such as traditional biomass-fired steam boilers and furnaces. Two other technologies, both under development, are being seriously considered: gasification and power generation with gas turbines (discussed in Chapter 9) and hydrolysis followed by fermentation to produce ethanol. Pilot plant developments in both areas are under way.

The prospects, even with conventional technologies, are very interesting. The development of the co-generation energy market in Brazil is overdue, and it will be important as soon as the economy starts a new growth period. The co-generation industry has grown significantly. It is flexible today, being able to process variable amounts of ethanol or sugar. Process synergistic gains have been made with respect to molasses fermentation and in the production of very low colour raw white sugar, when ethanol is processed simultaneously.

In Brazil, most of the industrial units do not refine sugar, this is done, in many cases, in specific refineries. An overview of fabrication processes, some technology advances and energy/environmental considerations are presented in the following section.

## 6.2 Process in the sugar–ethanol factories

### 6.2.1 Raw material

In Brazil, all sugar is produced from sugar-cane. Today, around 50 per cent of the grown cane is used to produce fuel alcohol (ethanol). The sucrose is produced and stored in the crop, from where it is extracted and purified.

The adequate planning of the interface between the field (cutting and adequate varieties selection) and industry (rapid and efficient processing), to avoid deterioration and sugar loss, is of great importance. The average sugar-cane composition for industrial processing at the industry gate is: 8–14 per cent fibres; 12–23 per cent soluble solids; and 65–75 per cent water.

The processing flow diagram is presented in Figure 6.1. It shows all the basic processing stages briefly described in this chapter. Brazilian sugar-cane processing does not differ in essence to that throughout the world, except for the parallel processing to ethanol. For a more complete review on sugar processing see Hugot (1972), Meade and Chen (1985), Delden (1981), and Payne (1991).

### 6.2.2 Transport, weighing, unloading and stocking cane stalks

The transportation of cane from the fields in Brazil is most commonly by road with trucks hauling compositions. They carry whole stick cane from manually cut cane, or chopped cane cut into 20–25-cm pieces from mechanical harvesting. The loads are weighed at the factory. Some loads are selected and sampled for later laboratory determination of sucrose content in the raw material. This evaluation allows for agricultural control, the transport payment, the grinding control, the calculation of the industrial yield, and the determination of the values for the cane payment.

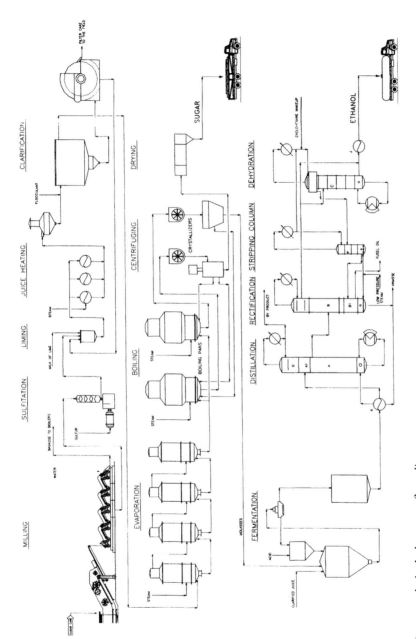

*Figure 6.1* Sugar and alcohol process flow diagram

The cane may be unloaded for temporary stocking or directly fed into the factory mills.

In the first case, the unloading is done though rolling bridges equipped with hydraulic grabs or cranes of the Hillo type, in a cane or in a covered space. The cane stocked in the cane yard is usually unloaded on to feeding tables (an inclined feeder of cane to the carrier, with a leveller and transporting chains) by loaders equipped with rakes. In the case of eventual failures in the transporting system and a corresponding interruption during the night shift, a certain amount of cane is kept stocked in the cane yards or in closed spaces. The stocked cane should be replaced at frequent intervals. Chopped cane, which should not be stocked, is directly unloaded for the milling process. The direct unloading may be done using the rolling bridges, cranes of the Hillo type, and, in the case of chopped cane, through hydraulic side dumping of truck–wagon compositions.

### 6.2.3 Cane juice extraction

The feeding table receives the cane stocked in the cane yards or directly discharged by the trucks and transfers it to the mill tandem (the complete mill complex, consisting of crusher or shredder and all mills, together with drives and reducing gears). The tables may have a conventional inclination of 5–7°, or up to 45°; the latter uses a higher operating speed and allows more uniform feeding for the cane carrier (conveyors used to transport the cane to the milling tandem, which consist of chains and slats).

The cane washing, performed on the feeding tables, is to remove foreign material, such as soil, rock, gravel or sand, to obtain a better quality juice and to reduce wear and tear on the equipment. There is a strong tendency to eliminate the cane cleaning system using water. Water, once a plentiful natural resource, is becoming a concern in some regions in São Paulo state and limitations on use are being discussed among politicians and associations representing different sectors of society. For example, the Piracicaba River basin, where many sugar and alcohol factories are located, is a region where water supply is becoming a serious problem, leading to discussions on different water use. The alternative is the introduction of dry cleaning systems, which, besides eliminating water, also allow the removal of some of the plant impurities, such as soil. Use of this method has substantially increased with the adoption of mechanical green-cane harvesting systems.

### 6.2.3.1 Cane preparation

The cane is prepared to increase its density (and consequently the milling capacity), and also to force the opening of cane cells to a maximum, in order to free the juices and allow higher extraction yields. The system used for cane preparation is made up of one or two sets of knives, the first preparing the cane to be sent to the shredder. The latter is formed by a drum with a concentric shaft, on which sets of oscillating hammers are mounted on arms. As it turns, it forces the cane to pass by small openings (of 1 cm) through a shredding plate. The tangential speed of the shredder, of 60–90 m/s, allows preparation indexes up to 80–92 per cent. This index measures the relation between the sugar in the cells opened by the shredder and the original sugar in the cane stalks.

### 6.2.3.2 *Feeding the mills*

After preparation, a tramp iron separator, using an electromagnetic field, removes nearly 90 per cent of the iron materials from the cane, to protect the mill tandem. Then the mill tandem is fed by forcing the cane into the carrier. Inside the cane carrier, the prepared cane forms a column with higher density, which improves the feeding and the milling capacity. The cane level inside the carrier is used to control the conveyor's speed and, consequently, the feeding of the mills.

### 6.2.3.3 *The cane milling*

The cane basically consists of the juice and the fibre. Since the sugar is dissolved in the juice, the objective of the milling process is to extract the largest possible amount of the sugar in the cane. On an industrial scale, two basic extraction processes are used: milling and diffusion. In Brazil, a large majority of the factories operate with mills. Milling is strictly a volumetric process and consists of 'displacing' the juice in the cane. This 'displacing' is obtained through the cane passing between two rollers submitted to a determined pressure and rotation. The displaced volume corresponds to extracted juice. A secondary objective of the milling process, although very important, is the production of the final bagasse in conditions of moisture that may allow proper burning in the boilers.

Each group of rollers constitutes a milling unit. The number of units utilised in the milling process varies from four to seven, and each one of them is formed by three main rollers: the 'front roller', the 'top roller', and the 'back roller'. Sometimes the mills have a fourth roller, named the 'pressure roller', which improves the feeding and the extraction efficiencies. The load that is applied on the bagasse blanket is transmitted by a hydraulic system that acts on the top roller.

### 6.2.3.4 *Imbibition*

The cane, passing successively through the various milling units, will have its juice removed gradually. The process of adding water to the bagasse is denominated imbibition and seeks to dilute the juice remaining in the bagasse, increasing the extraction of sucrose. There are several possible schemes to conduct the imbibition; the most frequently used is the composed imbibition, which consists of adding water between the last two milling units and making the extracted juice from the last milling set return to the previous one, and so on until reaching the second milling unit. Usually, the juices originated in the last two units are blended and will constitute the 'mixed juice'. With this system it is possible to obtain an extraction yield of 92–97 per cent, and a final moisture content in the bagasse of approximately 50 per cent wet basis.

### 6.2.4 *Generation of energy*

After the juice extraction, the sugar-cane bagasse is made up by 46 per cent fibre, 50 per cent) water, and 4 per cent dissolved solids. From 240 to 280 kg of bagasse can result per tonne of cane, and the sugar in it represents one of the process losses. Bagasse is used as a fuel in the boilers and ensures the sugar industry's electricity self-sufficiency and, in some cases, significant surpluses.

The steam generated in the boilers, at an average pressure of 18–21 kgf/cm$^2$ (1.76–2.06 MPa), is used to drive the steam turbines, where thermal energy is transformed into mechanical energy. Some of the turbines are responsible for driving the crushers, shredder, mills and turbo-pumps, while others drive the plant. The low-pressure steam, 1.3–1.7 kgf/cm$^2$, from the turbines is a source of thermal energy during the production of sugar and alcohol.

### 6.2.5 Production of sugar

#### 6.2.5.1 Preliminary juice treatment

The cane juice obtained from the extraction process presents several types of impurity. The preliminary treatment of the juice is envisaged to eliminate as much as possible insoluble impurities (sand, clay, bagacillo, etc.), whose contents vary from 0.1 to 1 per cent. Fixed and mobile screens are the basic equipment employed in their treatment. With openings between the bars of 0.5–2 mm, and placed very close by the mills, the fixed screens eliminate the coarse material in suspension (bagacillo). The retained material, constituted mainly by cane juice and bagacillo, returns to the first and second milling unit. The juice screening is also conducted by several types of mobile screen (Dutch State Mines (DSM), rotary, vibratory), and openings (0.2–0.7 mm), with efficiencies from 60 to 80 per cent. The retained material also returns to the mill.

#### 6.2.5.2 Juice weighing

After the preliminary treatment, the juice mass flow rate is quantified through flow rate measuring devices, allowing better control of the chemical process.

#### 6.2.5.3 Juice chemical treatment

After the preliminary treatment, the cane juice still contains minor impurities that may be soluble, colloidal, or insoluble. The physico-chemical treatment comprises coagulation, flocculation, and precipitation of the impurities, which should then be eliminated by sedimentation. It is also necessary to correct the pH value to avoid inversion and decomposition of the sucrose during the production of sugar. The treated juice may then be sent for the production of sugar and alcohol.

JUICE SULFITATION

This consists of the absorption of SO$_2$ by the juice, lowering its original pH to 4.0–4.4. The sulfitation is usually conducted in an absorption column. Due to the high solubility of SO$_2$ in water, absorption levels of up to 99.5 per cent may be obtained with this equipment. The gaseous SO$_2$ is produced in the sugar mills through combustion of sulphur in the presence of air, in special furnaces. The objective of the juice sulfitation processes is:

1   to inhibit reactions that could give colour to the juice;
2   to cause the coagulation of soluble solids;

3   to form a $CaSO_3$ precipitate;
4   to decrease the juice viscosity, making easier the evaporation and crystallisation operations;

The average sulphur consumption is about in 3 g/kg of sugar.

### LIMING

In this process hydrated lime $(Ca(OH)_2)$ is added to the treated cane juice to increase its pH value to between 6.8 and 7.2. The liming can be carried out in batches, or in a continuous process. Lime is produced in the sugar and alcohol mills through the 'burning' of limestone $(CaCO_3)$. Limestone gives off $CO_2$ when heated in a kiln. The burned CaO has some impurities (which are unburned and burned particles), some silica and magnesium (Delden 1981).

This treatment eliminates colourants in the juice, neutralises organic acids and forms sulphite and phosphate calcium that, as it sediments, draws with it the impurities present in the liquid. Lime consumption varies from 500 to 1000 g/TC, according to how rigorous the treatment is.

### HEATING AND SEDIMENTATION

Calcium is heated up to about 105°C in order to speed up the coagulation and flocculation of non-sugar protein colloids, and emulsify the fats, gum and waxes, besides promoting the removal of gases.

The sedimentation of impurities is conducted in a continuous process in a clarifier (a large vertical cylindrical tank where juice is given time for decantation). The decanted juice is sent to the factory's evaporation sector for further concentration. The sediment impurities, with a solid concentration of approximately 10° Bé (degree Beaumé is a scale used in sugar mills and refers to the percentage composition of sugar solutions), constitute the mud, which is removed from the clarifier bottom and sent to filtration, to recover the sugar that is present in it. The addition of polymers (about 2 g/tonne cane) increases the decanting speed and improves the juice quality.

The residence time of the juice in the clarifier, depending on the equipment used, varies from 45 minutes to 4 hours, and the mud quantity removed from the juice ranges from 15 to 20 per cent of the juice weight that enters the clarifier.

### 6.2.5.4 *Filtration*

Before it is sent to the rotary vacuum filters, the mud removed from the clarifier receives the addition of approximately 5 kg of bagacillo/TC, which helps the filtration. The filtrated juice returns to the process. The filtercake is recycled in the cane fields as a fertiliser. The sugar loss in the filtercake should not be higher than 1 per cent.

### 6.2.5.5 *Evaporation*

The first step to concentrate the juice is conducted in continuous evaporators. These are multiple effect systems, to save energy. The juice presents, initially, a concentration of 14–16° Brix, reaching, at the end, 50–58° Brix, when it is called syrup.

### 6.2.5.6 Sugar crystallisation

After leaving the evaporators, the syrup is sent to another concentration stage, where the formation of crystals of sugar occurs, as a result of the precipitation of the sucrose dissolved in water. There are two types of crystallisation: boiling crystallisation and cooling crystallisation.

The boiling process employs vacuum pans, which work individually, in batch or continuously. The water evaporation gives rise to a mixture of 50 per cent crystals surrounded by molasses (a sugar solution with remaining impurities) that receives the name of massecuite (a mixture of crystals and molasses before separation in the centrifugal). The massecuite concentration is around 91–93° Brix (percentage by weight of total soluble solids, as determined by a refractometer), and its temperature, at the discharge, is approximately 65–75°C. Depending on needs, it is possible to work with systems of one, two, or three massecuites.

In the crystallisation by cooling, the massecuite is discharged from the vacuum pans into the crystallisers, where a slow cooling takes place, generally helped by water or air.

### 6.2.5.7 Centrifugation of sugar

The cooled massecuite is sent to the centrifugation sector and is discharged into the centrifugal. Inside, the molasses are separated from the sucrose crystals. The process is then completed by washing the crystals of sugar with water and steam, still in the centrifugal's interior. The removed molasses are collected in a tank and returned to the vacuum pans to recover the dissolved sugar that they still contain, until a higher degree of exhaustion is achieved. The remaining material is also so-called final molasses, and is sent for the production of ethanol. The sugar discharged from the centrifugals presents a high moisture content (0.5–2 per cent), as well as a high temperature (65–85°C), due to the washing with steam.

### 6.2.5.8 Drying of sugar

The cooling and drying of sugar is conducted in a dryer, a metallic drum through which passes, countercurrent, an airflow blown by a ventilator. As it leaves the dryer, with a temperature 35–40°C and moisture content ranging from 0.03 to 0.04 per cent, the sugar is sent for packing, weighing, and storage. The overall efficiency of the production of sugar and ethanol is, on average, around 88 per cent.

### 6.2.6 Production of alcohol

Alcohol is obtained through fermentation of the cane juice or a mixture of molasses and the cane juice; therefore, through a biochemical process. Before it is sent to the fermentation process, the juice receives a purifying treatment.

### 6.2.6.1 Juice treatment for the distillery

After passing through the primary treatment, as described before, the juice undergoes pasteurisation: rapid heating and cooling. A more complete treatment of the juice includes the addition of lime, heating, and further decanting, analogous to the procedure employed in the production of sugar. The cooling of the juice is usually conducted in two steps:

1   the hot juice is passed through a regenerative-type heat exchanger; in countercurrent with the cool mixed juice, the hot juice in this equipment is cooled to approximately 60°C;
2   a final cooling to about 30°C is carried out in plate-type heat exchangers, when the water, the cooling fluid, uses a counter-current to the juice flow.

Free of impurities (sand, bagacillo, etc.) and sterilised, the juice is ready to be sent to the distillery.

### 6.2.6.2  Preparation of mash and yeast for the fermentation process

The mash is a mixture of molasses and juice, with a solids concentration of 17–22° Brix. In case it is required, water can be used for Brix adjustment, in order to facilitate its fermentation.

The fermentation method most commonly used in the Brazilian distilleries is the Melle-Boinot, whose main characteristic is the recovery of yeasts through the 'wine' centrifugation.

The recovered yeasts, before returning to the fermentation process, receive a rigorous treatment, consisting of diluting with water and adding sulphuric acid until the pH reaches 2.5, or lower (2.0), in case there is a bacterial infection. This suspension of diluted and acidified yeasts, named 'yeast inoculum', is stirred for 1–3 hours, prior to its return to the fermentation tanks.

### 6.2.6.3  Fermentation

It is in this phase that the sugars are transformed into alcohol. The reactions occur in tanks where the mash and the acidified yeasts are mixed in a 2:1 proportion, respectively; the mash is fed in 3–6 hours. The sugars (sucrose) are transformed into alcohol, according to the simplified Gay Lussac reaction:

$$C_{12}H_{22}O_{11} + H_2O \rightarrow 2C_6H_{12}O_6$$
$$C_6H_{12}O_6 \rightarrow 2CH_3CH_2OH + 2CO_2 + 98.2 \ kJ$$

During the reaction, an intense liberation of $CO_2$ occurs, the solution is heated and some secondary products are formed, such as higher alcohols, esters, glycerol, aldehydes, etc. The fermentation time varies from 4 to 12 hours. At the end of this period, practically all the sugars have been consumed.

As the fermentation ends, the average alcohol content in the fermentation vessels is 7–10 per cent v/v, and the mixture is called 'fermented wine'. Due to the large quantity of heat released during the process and the need to maintain the temperature at relatively low levels (34°C), the wine should be cooled with water. This can be carried out internally in the fermentation vessels through coils, or with external heat exchangers.

The fermentation process can be conducted in batches or continuously, using open or closed fermentation tanks. In the latter the evaporated alcohol is recovered by absorbing it in water.

### 6.2.6.4 Wine centrifugation

After fermentation, the wine is sent to continuous centrifuges, for yeast recovery. The recovered concentrated yeasts return to the tanks for treatment. The lighter phase in the centrifugation, the wine without yeasts, is sent to the distillation columns.

### 6.2.6.5 Distillation

The wine coming from fermentation has in its composition 7–10° GL (per cent in volume) of alcohol, besides other liquid, solid, and gaseous components. Among the liquids components, besides the alcohol, water is found in contents of 89–93 per cent; glycerin, higher alcohols, furfural, acetic aldehydes, succinic acid, and acetic acids make up the rest. The main solid components are bagacillo, yeasts and bacteria, non-fermentable sugars, mineral salts, and albuminoid materials, $CO_2$ and $SO_2$ are the main gaseous components.

The alcohol present in the wine is recovered by distillation, which utilises the different boiling points of the several volatile substances present in the wine to separate them through this heating. The operation is conducted with the help of seven columns, distributed in four stages: the distillation itself, the rectifying, the dehydration, and the cyclohexane recovery.

### 6.2.6.6 Dehydration

The hydrated alcohol, the final product of the distillation processes, is an alcohol–water mixture with an alcoholic content of about 96° GL. Distillation cannot increase this content further.

The hydrated alcohol can be commercialised as it is produced from simple distillation, or it can go through a dehydration process, which utilises an additional product (today cyclohexane is employed) to form a tertiary azeotropic mixture with water and alcohol. This addition results in ethanol, after another distillation, with a 99.7° GL content.

### 6.2.6.7 Vinasse or stillage

One important by-product from alcohol production is vinasse, also known as stillage, or spent wash. Its composition presents basically all substances introduced in the sugar-cane production and transformation, except sugar and alcohol. Its production rate is around 10–15 times that of alcohol and presents a significantly high BOD (Biological Oxygen Demand), which was considered an environmental threat at the beginning of PROALCOOL, because of its high pollution potential. (See also Chapter 5.)

## 6.3 Technological advances in the past fifteen years

Technological advances in the sugar and alcohol industry were strongly conditioned by the rapid growth of ethanol production. The latter grew steadily from about $0.5 \times 10^6$ m$^3$ in 1974 to $12 \times 10^6$ m$^3$ in 1985; from then to 1995 the growth was much smaller, up to $15 \times 10^6$ m$^3$. Processing of sugar-cane went through three distinct periods, as far as technology is concerned:

1 Large increases in equipment productivity (to cope with the growing alcohol demand from 1974 to 1985); this was done, eventually, with a lower regard to conversion efficiencies.
2 An increasing concern with conversion efficiencies, in all process levels, starting from 1980.
3 The search for better management of the processing units, starting with the planning at the agricultural level, and including instrumentation and automation at the factory, with suitable operational control systems, from 1985 (in the low-growth period).

Some examples of technological advances, in the period, are presented below. In the past few years, overall conversion efficiency in the industry rose, on average, by 1 per cent annually, with gains in productivity gains and cost reductions through better management.

### 6.3.1 Sugar-cane milling

Milling was mostly influenced by the need to increase productivity; in the period 1974–84 the so-called 'Brazilian four-roll milling system' was developed, with forced feeding through the Donnelley chute and a fourth roll, leading to very high feeding rates. Values evolved from 11.3 to 14.4 (tonne of cane/h)/m$^3$ roll. As expected, in the following years, the need for higher extraction rates led to the development and implantation of suitable imbibition systems. Today, the best Brazilian mills have the highest feeding rate (per weight) in the world, with an extraction yield up to 97.5 per cent.

### 6.3.2 Fermentation

This area has gone through a very large transformation, again led by the need for productivity and, latterly, efficiency increases. The main technological leads forward were:

- 1980 – better engineering: centrifugation, vat cooling;
- 1982 – microbiological control;
- 1985 – first large-scale continuous fermentation;
- 1986 – yeast treatment; operational control tools;
- 1989 – yeast population dynamics → DNA fingerprinting;
- 1992 – selected yeast;
- 1994 – second-generation continuous fermentation ($10^6$ l/day).

Some results of productivity gains obtained in fermentation are given in Table 6.3.

*Table 6.3* Productivity gains in fermentation

| Parameter | 1977 | 1995 | |
|---|---|---|---|
| | Average | Average | Best |
| Yield (%) | 83.0 | 91.2 | 93.0 |
| Production time (h) | 14.5 | 8.5 | 5.0 |
| Ethanol wine content (%) | 7.5 | 9.0 | 11.0 |
| Anti-foam utilisation (g/l) | 1.0 | 0.25 | 0.0 |

### 6.3.3 Instrumentation and automation

The first efforts in digital technology started in 1982, gradually substituting for the analog control systems in the sugar mills. In fact, the level of instrumentation and automation at that time was very low. The main concepts, the building up of modules for each factory sector, with allowance for future integration, were established at that time.

From 1984 on, the development of specific sensors and control strategies for sugar mills was undertaken (first in milling, followed by evaporation and fermentation). From 1986 on, areas such as co-generation (boilers) and distillation were included. In the 1990s most areas have been covered, and supervisory controls are already established in some sugar mills. The sectorial control strategies are being continuously updated. In 1996, the first linking between PLCs and field instrumentation, through supervisory systems, was made.

### 6.3.4 Energy production and utilisation

The evolution of the energy co-generation system at the sugar mills in Brazil progressed slowly, due to the fact that the state-owned electric power system exerted many barriers to energy sales by private producers, for most of the 1975–95 period. So, the evolution was only towards self-sufficiency, not even going up to high levels of efficiency. Today, the sample formed by sugar mills associated to COPERSUCAR, which process 65 million tonnes cane/year, shows that:

1  all the sugar mills in the sample have co-generation systems based on bagasse; most of them expanding steam from 22 to 2.5 bar (2,200 to 250 kPa);
2  self-sufficiency grew from 60 per cent (1980) to 95 per cent (1995);
3  only 4 per cent of the electricity produced was sold to the grid.

Several studies have been carried out to estimate the co-generation potential in the sugar and alcohol industry, considering conventional and 'future' technologies (essentially, bagasse/trash gasification and BIG-GT cycles). The state-of-the-art technology for biomass-based co-generation involves high-pressure steam cycles (9,000 kPa); trash collection/utilisation systems are being evaluated, as non-burning sugar-cane areas grow, following the current legislation.

152

It is expected that the partially privatised electric power system will evolve to allow (and encourage) the development of the industry's full co-generation potential; in this case, available technology could bring the mills' thermal energy requirements down to nearly 300 kg of steam/t of cane (from 450 kg, today). Bagasse surplus would then be much higher than the current values of 5–15 per cent.

## 6.4 Energy and environmental issues

The sugar and alcohol industry is well controlled for environmental impacts; its main effluents (stillage, filter cake, and boiler ashes) are recycled to the sugar-cane fields. Processes are usually easy to keep clean (no very high temperatures, no toxic exhausts, no releases of chemicals, and no sulphur in boiler exhausts).

The main advantages of adding ethanol to gasoline (replacing lead) are well known. Another large advantage is the reduction in $CO_2$ emissions for the whole life cycle of sugar-cane and its products. Values for Brazil, referring to 1996, are presented in Table 6.4.

The impressive result in Table 6.4 (some 20 per cent of all fossil fuel emissions, in Brazil) is probably going to increase with the growing utilisation of sugar-cane trash as a fuel. With the new (1997) legislation, some areas will have to stop burning sugar-cane before harvesting (as has been discussed in other chapters of this book), allowing for a large biomass surplus. Reaching (in some 8–12 years) 55 per cent of the total sugar-cane area, the available additional biomass could be $20 \times 10^6$ t (dry matter)/year. An extensive programme for development/evaluation of trash collection/preparation systems and processes for trash utilisation is under way.

*Table 6.4* Net $CO_2$ emissions of the Brazilian sugar-cane agro-industry

| Source | $10^6$ t Carbon (equiv.)/year |
|---|---|
| Fossil fuel utilisation in agriculture | + 1.28 |
| Methane/other GHG emissions in sugar-cane burning | + 0.06 |
| $N_2O$ emissions from soil | + 0.24 |
| Substitution: ethanol for gasoline | − 9.13 |
| Substitution: bagasse for fuel oil (food industry) | − 5.20 |
| Net result | − 12.74 |

*Source*: Macedo (1996)

## 6.5 Conclusions

The Brazilian sugar-cane industry, the largest in the world, utilises essentially, for sugar production, basic and consolidated technology that is currently available in many countries, including India, Australia, and South Africa, among others. In terms of technology, overall, this industry may be considered well positioned, as Brazil could well be regarded as the world leader in alcohol production technology, both on a small and on a large industrial scale.

In terms of energy efficiency, most of the Brazilian mills are self-sufficient, some presenting a surplus of electricity which they commercialise. There are good prospects for future commercialisation of electricity by the mills, where the estimated potential may reach several GWs, depending on the investments and adopted technology.

The industry overall energy efficiency should be improved through the implementation of green-ccane harvesting, allowing trash recovery. In economic terms, the potential for cost reduction appears to be more concentrated in the agricultural sector, which represents 60 per cent of the production costs. Also, the mills depend on the quality of sugar-cane stalks produced in the field; therefore, it is clear that closer cooperation between factory management and agricultural raw material supply is essential for overall successful sugar/alcohol production. That is also the case for co-generation of power with higher efficiencies.

New uses for the bagasse and the trash are also considered very important for creating new economic possibilities and increasing the industry flexibility. Among the most important opportunities are the conversion of fibre into alcohol, through hydrolysis, and animal feed production.

Environmentally, it seems that the most important change in the mills in the near future will be the increasing use of dry cane cleaning, with its implication of eliminating the use of large amounts of water. Another future benefit is the more efficient use of bagasse, which may lead to fossil fuel consumption reduction in other industrial sectors. More work is also being conducted to decrease the vinasse production, by developing new yeast strains capable of enduring higher alcohol concentrations. Also, other measures, such as vinasse evaporative cooling, could allow significant reduction in the vinasse volume and corresponding disposal costs.

Finally, the Brazilian sugar and alcohol industry has been living under a constant challenge, represented by the need for production cost reduction and sustainability. Its impressive size and unique characteristics have demonstrated that the sugar-cane business could produce not only sugar but also fuel alcohol in an environmentally-sustainable manner.

## 6.6 Bibliography

COPERSUCAR (1998) *Geração de Energia por Biomassa: Bagaço de Cana-de-açúcar e resíduos. Projeto. BRA/96/G31 (PNUD/MCT)*, Centro de Tecnologia Copersucar, *Relatório Março 1998* (RLP-04).

Delden, E. (1981) *Standard Fabrication Practices for Cane Sugar Mills*, Sugar Series 1, Amsterdam: Elsevier Scientific, 253.

Goldemberg, J., Monaco, L.C. and Macedo, I.C. (1993) 'The Brazilian fuel–alcohol program' in *Renewable Energy: Sources for Fuels and Electricity*, Island Press, 841–63.

Hugot, E. (1972) *Handbook of Cane Sugar Engineering*, revised 2nd edn, New York: Elsevier Press.

Macedo, I.C. (1996) 'Greenhouse gas emissions and energy balances in bio-ethanol production and utilization in Brazil (1996)', *Biomass and Bioenergy*, **14** (1) 77–81.

Macedo, I.C. (1998) 'Greenhouse gases and bio-ethanol in Brazil', *International Sugar Journal*, **100** (1189) 2–5.

Meade, G.P and Chen, J.C.P. (1985) *Cane Sugar Handbook*, 11th edn, John Wiley & Sons, 1134.

Payne, J.H. (1991) *Cogeneration in the Cane Sugar Industry*, Sugar Series 12, Amsterdam: Elsevier Scientific, 329.

7

# PULP AND PAPER

*Sergio Valdir Bajay, Mauro Donizeti Berni*
*and Carlos Roberto de Lima*

## 7.1 Introduction

The Brazilian pulp and paper industry comprises 220 companies that operate 255 industrial units, located in sixteen Brazilian States. The wood utilised by the industry is harvested exclusively from planted forests, which currently cover 1.4 million ha, mainly eucalyptus (62 per cent) and pine (36 per cent). Brazil is the world's seventh largest pulp producer and ranks twelfth in paper production. In 1997 the sector employed 67,000 people in industrial activities and 35,000 in forestry work, totalling 102,000 direct jobs (BRACELPA 1997a).

This chapter is divided into three main parts. The sections that make up the first part deal with the forestry activities directed to produce pulp wood. The manufacturing of pulp and paper in Brazil is evaluated in the second part, with emphasis on energy consumption and environmental impacts. Finally, paper recycling, which is a very important issue in Brazil and many other countries, is discussed in the third part.

## 7.2 Sustainable forestry

### 7.2.1 Types of sustainable forestry

Sustainable forestry means a set of activities that are environmentally and economically sustainable. These two types of sustainability are, to some extent, antagonistic: improvements in the economic sustainability of a project often are obtained at the expense of worsening its environmental sustainability and vice versa. How to obtain balanced solutions between these antagonistic objectives is further discussed in this chapter.

Sustainability can be achieved in the exploitation of both native and planted forests. The latter can be large and conventional planted forests, or short rotation coppices, as are increasingly being practised in the European Union, particularly in the 'set aside' land. Since pulp wood in Brazil is produced only from large planted forests, mostly of eucalyptus and pine, this is the only situation discussed in this chapter.

155

## 7.2.2 Wood from large, conventional planted forests

### 7.2.2.1 The production chain of wood from large planted forests for industrial consumption

The forestry activities aiming at producing wood from large planted forests for industrial consumption can be broadly grouped in four stages: implementation; management; harvesting; and transport.

The establishment of forest plantations commences by gathering all necessary inputs. Soil preparation should be properly carried out prior to planting. This first stage includes the planting of seedlings and maintaining them during the first year and, in some cases, during the second year, depending on the species.

The management stage aims to provide the timber during the growing phase of the forest, with the required quality to its future use: saw-logs, pulp wood, wood panels or firewood. During this stage, common steps are: manual, mechanical and chemical weed control, maintenance of roads and terraces, application of pesticides, whenever necessary, thinning and cleaning, etc. Resource inventories are also carried out to follow the forest growth, to detect any need for corrective measures and to provide the database for forecasts of the future woody biomass availability and productivity, usually measured either in $m^3$/ha/year or in t/ha/year.

During the harvesting stage the timber is collected and sent, as a raw material, to a given transforming process. The main activity of this stage is the logging of the trees, which should be preceded by the cleaning of the cutting plots and inner roads maintenance. After the felling, tree branches should be cut off and the logs are then gathered for transport.

The final stage consists of transporting the wood to the roads outside the forest plots, and then from such roads to the processing plants.

The four stages should be carefully planned to comply with the technical requirements and to minimise costs and negative environmental and social impacts from the forestry activities.

The use of biotechnology is common today in Brazil for obtaining better-adapted tree clone species. The 'minimum farming' approach has also achieved widespread use in soil preparation, both in new land and replanted areas; the soil compaction is minimised with this technique.

Forest management is the stage of longest duration and the one that demands the least involvement; regular check-ups only are required. During the harvesting and transport stages, activities are intensified and are carried out in as short a period of time as possible, often with the help of large machines, e.g. harvesters, skidders and heavy lorries.

### 7.2.2.2 Energy consumption

The four stages of the production-chain of wood extraction from planted forests for industrial consumption, referred to in the previous section, require energy inputs. The most energy-intensive stages are harvesting and transport. The higher the energy consumption, the higher the cost. Typically, one $m^3$ of wood at the processing plant backyard costs about three times as much as uncut trees.

Comitre (1995) calculated a full energy balance for planting eucalyptus in the Region of Ribeirão Preto, SP. According to his calculations, the indirect consumption of energy in fertilisers and pesticides was responsible for 61.2 per cent of the total energy input, the fossil fuels consumption came next, with 30.8 per cent of the total, and the energy equivalent of the employed manpower made up the remaining 8 per cent. He found an energy output/input ratio of 36.8.

### 7.2.2.3 Environmental and social impacts

TYPES OF IMPACT

Forestry activities may cause both negative and positive impacts on the environment. Companies engaged in forestry for industrial uses of wood have, in recent years, received considerable attention from the media and have been under pressure from both consumers and environmentalist groups. This is because of the intensive use of natural resources in their business. The magnitude of the impacts can vary substantially according to the practices and technologies adopted by each company and environmental sustainability is perfectly achievable, as discussed in Section 7.2.4.

Silva (1992) shows that the implementation of a planted forest is the stage with the highest capacity to change the environment. From a total of sixty-two environmental impacts he identified in planted forests, twenty-six were at this planting stage, followed by twenty-two in the management and harvesting stages, and a further fourteen in transport.

From the sixty-two impacts, only five were considered positive, four in the implementation stage and one in the management stage. The positive impacts identified by Silva are:

1   creation of jobs in rural areas;
2   increased activity in the commercial sector, through the hiring of production factors;
3   contribution to regional development, due to the creation of infrastructure, e.g. forest road network;
4   increase in food supply to vertebrate animals and birds, caused by ploughing; and
5   decrease in global warming, due to the capture of $CO_2$, by the photosynthetic process by the growing forest.

This last benefit is considered by many as the most important one nowadays.

Among the negative impacts Silva listed, the following were picked up as examples:

1   increase in the concentration of gases and particulate matter in the atmosphere, due to fires;
2   decrease in superficial and deep water quality, because of contact with biocides and ash from forest fires;
3   soil compaction and erosion, with the consequent increase in watershed turbidity and silting;
4   reduction of the areas with native vegetation;
5   less space for native habitats;

6    decrease in the global productivity of the aquatic biota, due to the increased watershed turbidity;
7    land ownership changes, with possible rural exodus; and
8    visual impact of deforestation after the felling of the trees.

EVALUATION OF THE IMPACTS

The National Council for the Environment (CONAMA Resolution No. 1, from 23 January 1986), defines, in general terms, the various types of environmental impacts and responsibilities and provides guidelines as to how to elaborate a study of environmental impact and the corresponding report. The latter is the basic document upon which the licensing procedure of every new economic activity (e.g. planting large forests) that can significantly influence the environment relies.

There are a relatively large number of methods to evaluate environmental impacts. The choice of the most appropriate method for a given application depends on data availability, the particular characteristics of the project(s) involved and the products aimed at.

According to Moreira (1985), Magrini (1989) and Lira (1993), the following methods are widely used in Brazil to evaluate environmental impacts of large projects, including new planted forests: 'ad hoc' procedure, check-lists, overlay mapping and simulation models.

The 'ad hoc' procedure simply requires the organisation of meetings among experts of several areas to gather in a short period of time the data and information required to reach the study conclusions.

Check-lists were, in fact, the first approach employed to evaluate environmental impacts. They are usually simple, can cope with poor and scarce data and can be combined with the 'ad hoc' procedure. There are four types of check-list: descriptive only, comparative, questionnaires and weighted lists.

In the third method referred to above, several maps are elaborated, based on a computerised geographical information system, containing information about soil types, terrain slopes, types of vegetation, etc. The maps are then overlaid, as expressed by the method's denomination. The elaboration of the maps is usually automatic.

Mathematical models can simulate the structure and operation of environmental systems, considering all the biophysical and antropic relations possible to be understood and modelled.

The best way to use any of these methods or their possible combinations is through a multi-objective approach, as discussed in Section 7.2.5.

### 7.2.3 Productivity, costs and economic sustainability

The continuous incorporation of new cultivation techniques and the widening use of biotechnology have elevated the productivity of the Brazilian forests. According to Table 7.1, in a period of thirty years the productivity of planted eucalyptus forests in the country increased by an average of 5.4 per cent per year, with very positive effects in terms of cost reductions.

In 1997 the average wood productivity in the planted forests owned by the Brazilian pulp industry was 44 steres with bark/ha/year (17.2 t/ha/year) for eucalyptus and 37 stwb/ha/year (14.4 t/ha/year) for pine.

*Table 7.1* Evolution of the average productivity of planted eucalyptus forests in Brazil

| Period | Improvements introduced in forestry practices | Average productivity (t/ha/year) |
|--------|-----------------------------------------------|----------------------------------|
| 1960–65 | Hybrid seeds | 3.3 |
| 1966–70 | Hybrid seeds and the use of fertilisers | 4.6 |
| 1971–75 | Imported selected seeds and the use of fertilisers | 5.9 |
| 1976–80 | Previous measures plus selected pruning | 9.3 |
| 1981–85 | Use of fertilisers, cloned seeds and vegetative propagation | 11.9 |
| 1986–90 | Previous measures and additional selection | 15.9 |

*Source*: Rodés *et al.* (1990)

Establishment costs of planted forests depend on many factors, like cost of land, soil type, level of farming, road building needs, types of selected seeds, planting techniques, fertilisers and pesticides employed, fire control measures, transport distances, etc. Each one of these factors can vary significantly, producing a relatively large range of variation for the total cost. Such range can be estimated according to forest productivity levels, as shown in Table 7.2.

*Table 7.2* Maximum and minimum establishment costs of planted forests, according to their productivity levels

| Productivity | | Establishment costs (US$/ha) | |
|--------------|---------------------------|------------------------------|---------|
| Qualification | Average value (t/ha/year) | Maximum | Minimum |
| High | 13.1 | 2,000 | 1,750 |
| High/medium | 10.1 | 1,620 | 1,400 |
| Medium | 7.3 | 1,270 | 1,060 |
| Medium/low | 4.7 | 940 | 750 |
| Low | 1.3 | 500 | 350 |

*Source*: Rodés *et al.* (1990)

In the United States, the cost of firewood per unit of energy content varies usually between US$3.5 and US$4.5 per GJ. With advanced cultivation techniques and biotechnology, this range of costs can be lowered to US$2–2.5 per GJ (Larson 1993).

Carpentieri *et al.* (1993) estimated a unit cost of US$1.47 per GJ for eucalyptus produced in large planted forests in the Brazilian Northeastern region. Table 7.3 splits this cost figure into its several components. Also in 1993, the state of São Paulo Agency for Energy Application reported an average cost of wood in the state of US$1.55 per GJ. In 1997, in the state of Minas Gerais, the average cost of wood from reforestation was US$2.51 per GJ (MME 1998). As can be observed in Table 7.3, planting, felling and transport represent 75 per cent of that cost.

There are huge differences in the cost of land among the Brazilian regions. In the Northern and Northeastern regions, one hectare of land costs US$12,000–50,000, while in the Southern, Southeastern and Central regions the range is from US$77,000 to 137,000. It is understandable, then, why all the new large planted forests in the current

*Table 7.3* Eucalyptus unit cost components in the Brazilian Northeast region

| Component | Unit cost, in US$/GJ |
|---|---|
| Production of seedlings in nurseries | 0.06 |
| Land rent | 0.15 |
| Planting | 0.36 |
| Management and R&D | 0.06 |
| Cultivation | 0.10 |
| Felling | 0.38 |
| Transport | 0.36 |
| Total | 1.47 |

*Source*: Carpentieri *et al.* (1993)

decade have been located in the former regions. In an effort to make feasible investments in new forests in the latter regions as well, most pulp producers there give technical and financial incentives to small local landowners to plant forests and sell the wood for them; the reliance on their own wood supply has thus been decreasing (Silva 1998).

The economic sustainability of planted forests for pulp wood depends not only on low establishment costs, but also on the strong and frequent fluctuations of international pulp prices, on the financial facilities for new projects, eventual tax relief, and the quantity and quality of the available general infrastructure (e.g. roads, railways, harbours, housing, basic school, health service, etc.) required by the related business activities.

In order to mitigate partially the swings in company finances, caused by the strong oscillations of pulp prices on international markets, several pulp wood producers have decided to diversify the products of their forests. This is a world-wide trend, which is also occurring in Brazil.

In recent years, the Brazilian pulp wood producers have been complaining to government that their competitors in Chile, Mexico, Paraguay, Uruguay, Malaysia and Indonesia receive tax relief benefits and very favourable credit facilities, and they have been demanding similar treatment. The government responded positively regarding tax relief to their exports and to imports of forestry machinery, and better credit conditions in BNDES, the Federal Government-owned development bank. Some of the entrepreneurs' demands, however, were not met, such as the government being guarantor in long-term low interest rate international loans, and the use of part of the companies' income tax to form a fund available for cheap loans to the companies themselves (MICT 1997).

In order to face more robustly the competition from pulp producers in countries with substantial government incentives (see above), the smaller Brazilian producers expect to form partnerships with foreign groups. This would allow them, at least, to get access to international cheap credit (Silva 1998). They also expect some government action to re-establish fair-trade conditions for them.

Several public services in the country have deteriorated sharply during the last fifteen years, and what has been called, more recently, 'Cost Brazil', represents the losses and surcharges the companies bear as a consequence of this degradation. The privatisation of roads, railways, harbours, water and power supply companies, etc., currently under way, should improve some of these services in the medium term, though at still uncertain costs to the customers. What is more concerning, for the longer term, is that the country is lagging more and more behind other nations in terms of income distribution, education, health service and R&D, factors which are not only important elements influencing the competitiveness of countries, regions and industries, but are at the very root of sustainable economic development.

### 7.2.4 Environmental sustainability

The manufacture of pulp, for paper making, is a capital-intensive activity that requires large-scale production to be competitive in a highly globalised industry. This fact, together with the homogeneity and quality requirements of pulp wood, rules out the use of native forests as a source for the paper and pulp industry.

Three factors relating to environmentally sustainable practices strongly concern the pulp wood forest plantation managers: the Brazilian Forestry Code, the stringent licensing requirement for new forest plantations, and voluntary environmental certifications, which are becoming important elements in opening the door to international markets.

During the 1960s, the Brazilian industry grew substantially, including pulp and paper, which depends on forestry activities. In order to give economic incentives and minimise negative environmental impacts, Government policies and environment surveillance regarding forests in the country were upgraded at that time.

The Brazilian Forestry Code was promulgated by Law No. 4,778 in 1965. It stimulates preservation, through the formation of permanent forests, and a rational exploitation of native and planted forests. It also conditions the industrial consumption of wood to reforestation. In 1967, the Brazilian Institute for Forestry Development (IBDF) was created, to apply the Code and the Government policies in this area.

Law No. 5,106, from 1966, provided tax relief (up to 25 per cent of income tax) to people and companies wanting to invest in forest plantation projects approved by IBDF. This incentive boosted such projects, particularly after Law No. 1,376 was enacted in 1974, giving priority to projects involved in specific programmes created by the Government at that time, like the National Programme of Pulp and Paper and the National Plan of Charcoal. These programmes had key roles in turning Brazil into a large exporter of pulp, pig-iron and steel. The incentives were phased out in 1989.

In the same year, the Brazilian Institute for the Environment and Renewable Natural Resources (IBAMA) was created by Law No. 7,735, to supersede IBDF. The actions of the latter were well below expectations during its whole existence. This can be illustrated by the fact that the Forestry Code was detailed, to be fully applicable for regulation purposes, by Federal Government Decree No. 97,628, only in 1989.

The application of the Code and the delivery of eventual corrective actions have improved with the creation of IBAMA. However, according to Medeiros (1995), the Institute's material and human resources are still below what would be required for a country-wide and product-wide efficient regulation of the forestry activities in Brazil.

In terms of environmental protection, the Brazilian Forestry Code received in the 1990s the help of two powerful allies, the environmental certificates and the 'green stamps', or 'ecolabels'. This is the case of forest-based products, like pulp and paper, that are exported to countries with strict environmental regulation and/or environmentally-minded consumers.

The environmental certification of a forestry business may or may not result in a document, provided by an independent and respected institution, employing well-known and accepted environmental protection auditing principles and methods, assuring the environmental sustainability of the undertaking. The granting of a 'green stamp' or 'ecolabel' for a product, like pulp or paper, usually requires several environmental certificates, since the whole life-cycle of the product should be considered. The objective is to show to consumers that the product has better net effects on environment than competitors without the label. The intention is to increase the 'environmental quality' of products through the mobilisation of market forces (Modl and Hermann 1995).

Environmental certifications and 'green stamps', together with other product quality certifications, are gradually substituting, in a growing number of countries, excise duties designed to protect local products. This seems to be a definite trend nowadays, in a world dominated by 'free trade' principles.

As a result, most of the Brazilian pulp wood producers have clear-cut environmental policies and management systems designed to apply those policies. Some of them recently received the ISO 14,000 certification.

Such environmental management systems often prescribe, among other measures, the use of minimum farming techniques, the strict control of storage, use and disposal, according to environmentally-optimised procedures, of fertilisers and biocides and the use of plantations in conjunction with native forests (mosaics), together with corridors of native vegetation, assuring good levels of biodiversity preservation (BRACELPA 1997a).

### 7.2.5 A multi-objective optimisation approach

As already mentioned, often the search to get improved economic sustainability for a planted forest requires the worsening of its environmental sustainability and vice versa. Trade-offs should be evaluated and a balanced solution looked for. Multi-objective optimisation techniques can be helpful in this task.

In a classical mathematical programming problem there is a single objective function to be minimised or maximised, and a set of mathematical constraints which define the space for feasible solutions. Both can be linear or not. In complex systems, like those easily found in energy supply and demand schemes, interactions between mathematical programming and simulation computer routines are common.

In a multi-objective optimisation problem there are several objective functions. Usually they are conflicting, so no solution exists that optimises all of them simultaneously. The computer routines designed to solve such a problem generate non-dominated solutions, where improvements in any objective function can only be obtained at the expense of some of the other functions (Steur 1986). The non-dominated solutions quantify the trade-offs among the objective functions considered. If the decision-maker has his preferences defined in advance regarding the objective functions, in a sort of utility function, the multi-objective problem can be transformed into a single objective one. Otherwise, either he is happy enough with the set of non-dominated solutions generated by the algorithm, to choose one of them, or he may prefer to build his utility function with the help of the multi-objective optimisation algorithm. To do that, the latter should be of an interactive type.

In the planning of planted forests, objective functions relevant to their economic sustainability can be the maximisation of net benefits along the planning horizon, or the minimisation of the differences between attainable values for selected financial indicators and targets set for them. The environmental sustainability objective functions may comprise the minimisation of the forest's negative environmental/social impacts and/or the maximisation of the positive ones.

Formal multi-objective optimisation exercises are not yet a common practice in the planning of planted forests in Brazil. The basic concerns contained in such exercises have been taken into account in some recent forestry ventures in the country, but not as systematic optimisation approaches.

## 7.3 The paper and pulp industry

### 7.3.1 Products, production and plant types

Paper is an aqueous deposit of vegetable fibres in a sheet form. The vegetable fibres consist primarily of cellulose fibre, originating mainly from wood. The fibres are separated by pulping and can then, if desired, be bleached, to reach a white colour. The multitude of tiny fibres produced by this process is known as pulp.

The pulp is then dispersed in water, at which stage, recycled paper may be added. The mixture is then cleaned and refined. The refined material passes to the paper-making machine, where it is spread evenly over a moving, continuous wire mesh band, the *foudrinier*. Water drains from the pulp leaving behind a wet layer of paper on top of the *foudrinier*. This wet paper web passes from the *foudrinier* onto a felt. Together these pass through a succession of roll-presses to remove more water from the web. The moist paper web is then dried by passing it over a number of steam-heated cylinders (CEC 1983). Sometimes infrared or radio-frequency electric heaters give a final touch in the drying process, improving the cross-section homogeneity of the paper sheet.

In the paper and pulp industry there are mills which produce pulp as an end product, known as 'market pulp', for shipment to paper-producing plants. There are also mills which produce paper and board from purchased pulp and, increasingly, from recycled waste paper. Finally, there are integrated pulp and paper mills. The latter can enjoy a considerable cost advantage over paper mills. Their continuous processing eliminates the need for additional drying of the pulp, prior to transportation to the paper mill. This is wasteful in energy terms, since the dried pulp must be redispersed in water prior to further processing. Pulp mills can be subdivided into categories according to their fibrous raw material, and according to the chemical, mechanical or combination of processes which they utilise (Ewing 1985).

Most pulp for paper-making is produced from wood. Almost any type of wood can be used for the production of chemical and semi-chemical pulp, but there are limitations on the wood species which can be used to make mechanical pulp. In Brazil, eucalyptus is the main raw material for chemical and semi-chemical pulp. It is a type of hardwood and has short fibres. Pine is the main softwood employed in Brazilian pulp mills, of all types; it has long fibres. The Brazilian Pulp and Paper Association (BRACELPA) also classifies the chemical and semi-chemical pulp into bleached and unbleached.

BRACELPA names the mechanical pulp as 'high yield pulp', because of the low mass losses incurred in the conversion of wood into pulp through this process.

Other fibrous materials that have been used for pulp production, on a much smaller scale, include sugar-cane bagasse, cotton waste, rice straw, jute and reeds.

BRACELPA (1997a) classifies the paper and board manufactured in Brazil into the following categories: newsprint; printing and writing; packaging and wrapping; board; sanitary paper; and industrial and special purpose papers.

Newsprint paper must be suitable for printing at high speed. It is derived from mechanical pulp and produced on very large high-speed machines. Printings and writings are, generally, white paper made from a mixture of chemical and mechanical pulps with mineral fillers to increase the paper opacity. These papers are used for books, stationery and magazines. Wrapping and packaging include papers for wrapping, bags and sacks and board materials for boxes and cases. Wrapping papers are produced from recycled waste paper and from chemical pulps. Case materials include corrugated board, in which three layers of paper are laminated together. Sanitary tissues consist mainly of toilet tissue, kitchen paper, facial tissue and handkerchiefs. They are usually made from chemical and mechanical pulps (CEC 1983). Recently developed cost-effective de-inking technologies allow the use of recycled paper even in the production of newsprint, printing and writing papers.

Table 7.4 shows the production of chemical/semi-chemical and mechanical (high yield) pulp and paper in Brazil from 1988 to 1997. As can be observed in Table 7.4, there has been a continual growth in the production of both pulp and paper in the period 1988–97, except in 1990 (pulp and paper) and 1992 (paper), and in the importance of chemical/semi-chemical pulp over high yield pulp in the production pattern found in the country. In fact, the production of the latter has been practically stagnant in this period.

Tables 7.5 and 7.6 present the production, export, import and apparent consumption of pulp and paper, respectively, in 1997, according to the categories adopted by BRACELPA.

The statistics of Table 7.5 show that Brazil relies little on imports of market pulp and that it is a major exporter of such product, essentially hardwood bleached

*Table 7.4* Evolution of pulp and paper production in Brazil, 1988–97 ($10^3$ tonne)

| Year | Chemical/ semi-chemical pulp | High yield pulp | Total production of pulp | Paper |
|------|------|------|------|------|
| 1988 | 3,793 | 398 | 4,191 | 4,684 |
| 1989 | 3,944 | 426 | 4,370 | 4,871 |
| 1990 | 3,915 | 436 | 4,351 | 4,716 |
| 1991 | 4,346 | 432 | 4,778 | 4,914 |
| 1992 | 4,871 | 432 | 5,303 | 4,901 |
| 1993 | 5,010 | 461 | 5,471 | 5,301 |
| 1994 | 5,376 | 453 | 5,829 | 5,654 |
| 1995 | 5,433 | 493 | 5,936 | 5,798 |
| 1996 | 5,736 | 465 | 6,201 | 6,176 |
| 1997 | 5,904 | 427 | 6,331 | 6,518 |

*Source*: BRACELPA (1997a)

*Table 7.5* Production, exports, imports and apparent consumption of the several types of pulp in Brazil, 1997 ($10^3$ tonne)

| Pulp type | Production | Exports | Imports | Apparent consumption |
|---|---|---|---|---|
| High-yield pulp | 427 | 3 | 2 | 426 |
| Chemical and semi-chemical | 5,904 | 2,381 | 277 | 3,800 |
| Softwood | 1,282 | 5 | 259 | 1,536 |
| bleached | 122 | 5 | 255 | 372 |
| unbleached | 1,160 | 0 | 4 | 1,164 |
| Hardwood | 4,622 | 2,376 | 18 | 2,264 |
| bleached | 4,333 | 2,346 | 18 | 2,005 |
| unbleached | 289 | 30 | 0 | 259 |
| Total | 6,201 | 2,384 | 279 | 4,226 |

*Source*: BRACELPA (1997a)

*Table 7.6* Production, exports, imports and apparent consumption of the several types of papers in Brazil, 1997 ($10^3$ tonne)

| Paper type | Production | Exports | Imports | Apparent consumption |
|---|---|---|---|---|
| Newsprint | 265 | 13 | 471 | 723 |
| Printing and writing | 1,983 | 837 | 231 | 1,377 |
| Packaging and wrapping | 2,911 | 286 | 33 | 2,658 |
| Board | 648 | 53 | 61 | 656 |
| Sanitary papers | 565 | 29 | 2 | 538 |
| Others | 146 | 111 | 180 | 215 |
| Total | 6,518 | 1,329 | 978 | 6,167 |

*Source*: BRACELPA (1997a)

chemical/semi-chemical pulp. In 1997 36.7 per cent of these exports were to Europe, 35.7 per cent were to Asia, 24.5 per cent to North America, and the rest to other regions (BRACELPA 1997a).

Table 7.6 reveals that:

1   packaging and wrapping, and printing and writing are the most produced and consumed types of paper in Brazil;
2   the country is highly dependent on imports of newsprint; and
3   it exports significant amounts of printing and writing paper and, to a lesser extent, packaging and wrapping paper. In decreasing order, the exports in 1997 went to Latin America (43 per cent), Asia (23 per cent), Europe (19 per cent), Africa (8 per cent) and North America (7 per cent).

### 7.3.2 *Economic indicators*

The Brazilian pulp and paper industry comprises both large, and technologically advanced, plants and smaller units, which generally still employ outdated technologies. The opening of the Brazilian economy has brought increased competition and the need to export, particularly during recession periods to make up for falls in local sales. Smaller companies have been forced either to modernise or to close down. The first option often requires mergers. As a result of this process, the production share of the larger companies has been increasing and the smaller units, if they are to survive, should concentrate their efforts in specialised market niches.

The economic performance of the Brazilian pulp and paper industry shows a strong cyclical characteristic, which results from the composition of two well-marked cyclical phenomena. The first one is the cyclical movement of pulp price, which reflects the world-wide alternating over- and under-investment states of this highly globalised industry. The other cause is the high frequency cycles of the Brazilian economy in recent years. The consumption of paper and board is strongly correlated with economic growth and income distribution. The Brazilian economy has not experienced sustained periods of growth, during either the 1980s or the 1990s, and income distribution in the country has improved just slightly along these decades.

The recent business cycles of the pulp and paper companies installed in Brazil can be observed in Table 7.7, which presents their sales, labour expenses, social security payments, taxes, profits, and the industry absolute and unit (per tonne) value added, all in constant US$, from 1990 to 1996.

The production and sales statistics, usually presented by the companies and by BRACELPA as indicators of the industry's economic performance, do not show the real magnitude of this cyclical process. The industry's value added and, particularly, its unit value added statistics capture this process much better. In using the latter indicator, however, care should be taken not to double count, i.e. only exported market pulp should be added to the production of paper to arrive at the total production figure employed to calculate this indicator. Table 7.7 shows clearly the Brazilian pulp and paper business recession worsening from 1990 to 1997, the spectacular recovery in 1994 and 1995, and the equally expressive fall-back in 1996.

Table 7.8 shows the investments in permanent assets, along the 1990–96 period, of the Brazilian pulp and paper industry, and the ratio between these investments and the corresponding value added generated. This ratio is one of the driving factors of the industry's energy consumption.

The year 1989 was a very good one for this industry. This is indicated, in Table 7.8, by a high (investment)/(value added) ratio in 1990. The ratio decreases from 1991 to 1993, reflecting the industry recession. It recuperates in 1994 and increases until 1996. The latter high value is a result of two succeeding high profit years for the industry previously.

### 7.3.3 *Production stages and equipment*

The production chain of paper, including pulp making, can be divided into the following stages: wood preparation; pulping and washing; bleaching; pulp drying; stock preparation for the paper machine; paper forming; paper pressing; paper drying; and paper finishing.

166

*Table 7.7* Sales, wages, social security payments, taxes, profits and value added (Dec/1980, 10⁶ US$), and unit value added (Dec/1980 US$/tonne), of the Brazilian pulp and paper industry, 1990–96

| Year | Sales | Wages | Social security | Taxes | Profits | Value added | Unit value added |
|------|-------|-------|-----------------|-------|---------|-------------|------------------|
| 1990 | 4,903 | 565 | 352 | 756 | -36 | 1,637 | 283 |
| 1991 | 5,249 | 601 | 401 | 659 | -304 | 1,357 | 270 |
| 1992 | 5,871 | 635 | 475 | 678 | -160 | 1,628 | 251 |
| 1993 | 7,244 | 771 | 494 | 771 | -568 | 1,468 | 201 |
| 1994 | 7,927 | 796 | 542 | 765 | 558 | 2,661 | 346 |
| 1995 | 7,815 | 689 | 430 | 940 | 639 | 2,698 | 353 |
| 1996 | 6,669 | 733 | 401 | 825 | -470 | 1,489 | 181 |

*Source:* Bajay *et al.* (1997)

*Table 7.8* Investments in permanent assets and the (investment/value added) ratio of the Brazilian pulp and paper industry, 1990–96

| Year | Investments in permanent assets, in Dec. 1980 $10^6$ US$ | (Investment)/(Value added) |
|------|------|------|
| 1990 | 1,238 | 0.76 |
| 1991 | 581 | 0.43 |
| 1992 | 511 | 0.31 |
| 1993 | 249 | 0.17 |
| 1994 | 885 | 0.33 |
| 1995 | 1,314 | 0.49 |
| 1996 | 1,529 | 1.03 |

*Source*: Bajay *et al.* (1997)

There are also some ancillary activities that are very important for the technical, economic and environmental performance of many pulp and integrated mills: chemical recovery; production of some chemical inputs on the mills' premises; and steam and electricity production, usually in co-generation units.

In the following subsections, each of the production stages and ancillary activities referred to above are briefly described, together with the corresponding main equipment. Comments are made about the thermal energy and/or electricity consumption intensity of some of this equipment.

### 7.3.3.1 Wood preparation

Usually, wood preparation includes debarking, chipping and screening, together with handling of both logs and ships. Sometimes logs are debarked by hand in the forest. Chipping and screening are not necessary for stone groundwood mills, where pulp is produced straight from logs.

The smaller chips and the bark feed the so-called 'biomass boilers', which produce part of the steam and, when co-generation units exist, part of the electricity requirements of the mill.

In general, this production stage requires only the consumption of electricity, for debarking, chipping, screening and materials handling.

### 7.3.3.2 Pulping and washing

As pointed out in Section 7.3.1, there are three basic types of pulping: mechanical, chemical and a combination of both.

In purely mechanical processes, the raw material is broken down by physical means into an aqueous suspension of fibrous particles. In the oldest and still the most widely used process, short cut logs are pressed against a large revolving grindstone and ground into a pulp. The cellulose fibres are to some extent damaged by this process and the pulp produced contains most of the non-cellulose components of the pulpwood raw material. For these reasons, groundwood pulp has a much lower strength than chemical pulp. However, the yield is much higher, typically 90–95 per cent on a bone-dry weight basis (Ewing 1985).

A more recent development in mechanical pulping is the use of refiners to break down wood particles into fibres by forcing them between rotating steel plates with a variety of surface configurations. Refiner groundwood pulp has superior strength characteristics to those of stone groundwood pulp, although the power consumed to make a tonne of pulp is significantly higher. This process uses chips and sawdust. By using steam to heat the chips, so-called 'thermal-mechanical' pulp can be produced, which has certain superior characteristics to ordinary refiner groundwood pulp (Ewing 1985).

The main feature of a chemical pulping process is a heated pressure vessel called a digester, into which wood chips are introduced, along with chemicals. The purpose is to dissolve non-cellulose components of the fibrous raw material – primarily lignin – into an aqueous phase. After the pulp leaves the digester it is washed to remove the dissolved chemicals. Digesters may operate on a continuous basis or in batches and their heating may be direct or indirect. Depending on the degree of cooking and on the fibrous raw material, the yield of chemical pulp can vary between 40 and 65 per cent of the bone-dry weight of input fibre. Chemical pulp has at most 10 per cent lignin and a low level of impurities (Ewing 1985; Bajay *et al.* 1996).

The most frequently used chemical pulping process world-wide, including Brazil, is named 'Kraft' and employs sodium- and sulphur-based chemicals, in an aqueous solution called 'white liquor', to disintegrate the wood chips into individual cellulose fibres. The cooking process in the digester can last 2–5 hours, at a temperature around 170°C and pressures between 7 and 9 atmospheres (Bajay *et al.* 1996). The pulp is formed after discharge of the digested material into a blow tank and separation of the fibres from the process residue, known as 'black liquor', which is an aqueous solution containing all the process chemical inputs and the non-cellulose components of the chips. After pulp washing, part of the black liquor is fed back to the digester, to help the circulation of the process inputs without the need to add more water.

The Kraft process has been the preferred among the several chemical pulping processes known because it allows chemicals recovery, a great deal of heat recovery and the combined production of steam and electricity in co-generation units that burn concentrated black liquor. The production of such units meets a large part of the mill energy demand. These three factors make the process highly cost-effective.

There is a wide spectrum of processes which use a combination of chemical and mechanical means to produce pulp. Nearest to the chemical end of the spectrum are the semi-chemical processes which use digesters, as for chemical pulping, but generally employ less chemicals and steam, and follow the digesting process with a heavy refining stage. The mixed chemical and mechanical processes offer some very specific advantages in terms of product characteristics for certain end-uses. They are generally characterised by higher yields and lower chemical costs than full chemical pulping, and better strength characteristics than full mechanical pulping (Ewing 1985). Their yields are usually between 65 and 85 per cent (Bajay *et al.* 1996).

Mechanical pulping is an electricity-intensive consumption activity with no need of thermal energy. In chemical pulping, electric energy is required to operate mechanical drives, and steam for the pulping process; to some extent, lower heat requirements may be accompanied by higher electric energy requirements to drive pumps and other equipment associated with measures to reduce thermal energy consumption (Ewing 1985).

### 7.3.3.3 Bleaching

The bleaching stage is intended to give the washed pulp a white colour, for the manufacture of white paper. Bleaching is accomplished in several stages by mixing chemicals with the pulp, and passing the mixture through retention towers at elevated temperatures. The pulp is washed between stages. Steam is used to produce hot water for washing and to achieve the desired temperature in each bleaching stage. Electric energy is used for mechanical drives for pumps, agitators, washers, etc. (Ewing 1985).

Bleaching removes traces of lignin, resins, metallic ions and other impurities that provoke the brownish appearance of unbleached pulp, allowing the attainment of the desired white colour. Until recently, the chemicals most employed in this stage were elemental chlorine or chlorine products. This has been changing due to pressures from environmentalists, concerned with the public health hazards of elemental chlorine. Oxygen and ozone have been replacing it.

### 7.3.3.4 Pulp drying

Non-integrated market pulp mills usually dry the pulp to facilitate handling and reduce freight costs, and then ship it in baled form to the paper mills. In integrated mills an aqueous solution of pulp is kept in storage tanks, from where it is pumped straight to the paper machines. Sometimes, even market pulp may be pumped in slush form to a nearby paper mill, or shipped in a partially dried condition.

Pulp may be flash-dried, but it is normally dried in a continuous web in a machine not unlike a paper machine, and subsequently cut into sheets and baled for shipping. Electric energy is used to drive the machinery, including vacuum pumps and presses which mechanically remove water. Steam is used to further dry the pulp web, either in drying cylinders over which the pulp is passed, or by heating air which is blown under and over the sheet to dry it and carry it through the pulp dryer (Ewing 1985).

### 7.3.3.5 Stock preparation for the paper machine

This stage of the process varies according to the form and type of pulp and the type of paper to be produced. If pulp is procured in dried, baled form the bales must be broken down and the pulp slushed into a suspension in water. If waste paper is used, it will normally require cleaning and de-inking as well as slushing. Stock preparation usually also includes refining to obtain the fibre characteristics for the grade of paper to be produced, screening to remove foreign material and cleaning to remove sand. Chemicals, dyes and fillers, such as clay, may be added during stock preparation to impart specific characteristics to the paper (Ewing 1985). The stage has a high consumption of electric power, particularly for driving the refiners.

### 7.3.3.6 Paper forming

The paper fibres must be formed into a sheet and this is normally accomplished by starting with a very dilute suspension – usually less than 0.5 per cent fibre – and either discharging it onto a travelling wire screen, where water is removed by gravity and suction, or picking it up on a travelling felt blanket to which suction is applied. After forming, the consistency of the sheet is typically in the range of 15–20 per cent fibre (Ewing 1985). Only electric power is required in this stage, for motors driving pumps, including the suction-type ones, and the travelling wire screen or felt blanket.

### 7.3.3.7 Paper pressing

More moisture is removed by pressing the paper sheet between pairs of rollers. The sheet is normally carried through the rollers on a felt to support it and to help carry away the water which is pressed out. Vacuum may be applied to further facilitate water removal. After pressing, the sheet dryness is usually in the range 35–45 per cent (Ewing 1985). Again, just electric power is demanded in this stage, for driving the press rollers, the travelling felt blanket and the vacuum pumps.

### 7.3.3.8 Paper drying

After pressing, the paper sheet is dried, usually by supporting it on a felt and carrying it around a number of steam-heated drums. The paper sheet cross-section characteristics are often improved with the help of infra-red or radio-frequency electric dryers. After drying, the sheet is usually wound into reels at the end of the paper machine. At this stage, electric power is employed in motors that drive the heating drums, the travelling felt blanket and the pumps, and the highest heating load in the paper machine should be met in the form of process steam for the drying drums.

### 7.3.3.9 Paper finishing

Paper machine reels are usually rewound into small reels of widths determined by the end use. Other finishing operations include coating, treatment with chemicals to obtain specific surface properties, and cutting sheets of specific sizes. Electric power is always required to run the drives of the paper-finishing machines; thermal energy is also necessary when coating and paper surface treatments with chemicals are involved.

### 7.3.3.10 Chemical recovery

As described in Section 7.3.3.2, the residue of the Kraft chemical pulping process – the black liquor – is a slush containing the non-cellulose components of the wood chips and all the input chemicals. The black liquor is concentrated up to 50 per cent solids in multiple effect evaporators and up to 65 per cent solids in direct contact evaporators (Bajay et al. 1996). The organic part of this concentrated black liquor is burnt in the chemicals recovery boiler. The sodium-based molten salts remaining in the boiler after the combustion of the organic matter are dissolved in water, forming what is known as 'green liquor'. The Kraft process chemicals are finally recovered from the green liquor with the help of lime-reburning kilns, which usually are fired by oil or natural gas. The main requirement for steam in the chemical recovery system is for liquor evaporation. Electric energy is used for mechanical drives such as pumps, fans and agitators.

### 7.3.3.11 Internal production of chemical inputs

In some plants chemicals inputs, particularly those employed in bleaching, are produced on-site, which can have a significant effect on energy consumption. This has been the case of the joint production of chlorine, sodium and hydrogen in electrochemical cells, and more recently, oxygen, and eventually it will be the case for other industrial gases. The electrochemical cells have a very high specific consumption of electricity.

### 7.3.3.12 Generation of steam and electric power

As can be concluded after the reading of the previous sections, large amounts of steam and electricity are required in the production chain of paper. On the other hand, pulp making provides significant amounts of residues, which can be burned in boilers. Thus, the installation of co-generation units to consume these residues is a natural and economic choice that also brings about environmental benefits, since this type of thermal recovery has been so far the most cost-effective way of disposing of such waste.

In 1996, the Brazilian pulp mills met 82.6 per cent of their electricity consumption needs through self-generation. The corresponding figure for the integrated mills is 54.4 per cent. From the total electricity self-generated by the Brazilian pulp and paper mills in 1996, 53.2 per cent was produced in co-generation units burning black liquor, 21.3 per cent in such units fuelled by bark and other wood-processing residues, 10.7 per cent in small hydro power stations, 9.5 per cent in co-generation plants fed by fuel oil, 5.3 per cent in such plants consuming coal and, finally, only 0.1 per cent in co-generation plants burning natural gas (BRACELPA 1997b).

### 7.3.4 Energy consumption patterns

The recent evolution of the Brazilian pulp and paper industry energy consumption figures, as taken from the Energy Balance of Brazil, is presented and discussed in Chapter 2. The inter-fuel substitution processes that took place in the industry in the last two decades are reviewed there.

This section addresses, quantitatively, the pulp and paper industry energy consumption patterns, in terms of energy consumption intensity of the various production chain stages, distribution of energy consumption by end-uses, and unit energy consumption differences among products and manufacturing plant types.

In Section 7.3.3 a qualitative overview of energy consumption intensity in the production stages of the pulp and paper industry was provided. Table 7.9 complements that information with ranges of unit energy consumption figures for those stages, compiled by Ewing (1985).

Some surveys have been carried out in Brazil about the distribution of energy consumption by end-uses. Studying the pulp and paper industry-related data of those surveys in the mid-1990s, Bajay et al. (1998) observed that:

1  88–97 per cent of the electricity consumption in the industry was for mechanical drive, 2–3 per cent for illumination, 0–3 per cent for steam generation, 0–4 per cent for direct heat, and 0–9 per cent for electrochemical processes;
2  87–100 per cent of fuel oil use was for generating steam and 0–13 per cent for direct heat (lime kilns);
3  all the fuelwood consumed by the industry generated steam; and
4  97–100 per cent of black liquor and wood preparation residues were burnt to generate steam and the rest produced direct heat.

*Table 7.9* Unit heat and electricity consumption of the pulp and paper industry production stages

| Production stage | Heat energy, in GJ/t | Electric energy |
|---|---|---|
| Wood preparation | – | 14–23 kWh/m³ |
| Debarking | – | 2–4 kWh/m³ |
| Chipping and screening | – | 10–15 kWh/m³ |
| Materials handling | – | 2–4 kWh/m³ |
| Kraft pulping | 2.5–5.0 | 100–140 kWh/t |
| Bleaching | 1.0–6.0 | 130–200 kWh/t |
| Pulp drying | 2.5–6.0 | 130–190 kWh/t |
| Chemical recovery | 3.0–7.0 | 55–85 kWh/t |
| Stock preparation – liner board | – | 230–240 kWh/t |
| Stock preparation – newsprint | – | 95–100 kWh/t |
| Paper machine | 3.5–7.0 | 170–300 kWh/t |

*Source*: Ewing (1985)

*Table 7.10* Unit heat and electricity consumption per type of paper

| Type of paper | Thermal energy, in Mcal/t | Electricity, in kWh/t |
|---|---|---|
| Board | 2,164 | 600 |
| Packaging and wrapping | 2,214 | 504 |
| Printing | 2,905 | 854 |
| Kraft paper | 3,035 | 1,174 |
| Special grades of paper | 3,618 | 1,371 |

*Source*: Bajay *et al.* (1998)

The unit energy consumption of paper making varies according to the type of paper or board, as shown in Table 7.10.

Unit energy consumption also varies with plant types. Bajay *et al.* (1998) calculated the electricity and total energy consumption per tonne of output produced in market pulp plants, integrated plants, sanitary paper makers and plants producing other types of paper. This is a plant classification adopted by BRACELPA, from which the data for the calculations were obtained. The exercise comprised the main Brazilian states that produce pulp and paper and spanned the period 1989–96. The national average figures for 1996 are in Table 7.11.

*Table 7.11* Average electricity and total energy consumption per unit of output of the Brazilian pulp and paper industry in 1996, according to plant type

| Type of plant | Electricity consumption, in MWh/t | Total energy consumption, in GJ/t |
|---|---|---|
| Market pulp makers | 1.03 | 17.1 |
| Sanitary paper makers | 1.12 | 13.7 |
| Producers of other types of paper | 0.76 | 13.4 |
| Integrated plants | 1.20 | 20.4 |

*Source*: Bajay *et al.* (1998)

The average consumption figure for pulp makers in Table 7.11 is relatively high because the two largest market pulp manufacturers in the country produce chlorine and sodium in electrochemical cells within their plants, which increases substantially their unit electricity consumption.

### 7.3.5 Environmental impacts

The main environmental impacts of the pulp and paper industry are briefly discussed in this section, together with some measures the industry is taking to mitigate them. The discussion is channelled to industrial plant impacts since the environmental issues concerning the production of pulp wood are dealt with in Sections 7.2.2.3 and 7.2.4.

Pulp and paper plants are important sources of gaseous, liquid and solid pollutants (Bajay *et al.* 1996). They are discussed in this section, in this order.

At the wood preparation stage, very small wood particles can be released from the chippers, handling equipment and chip piles into the air. To control this, most of the equipment parts in contact with the wood should be encapsulated and the chip piles should be well planned and used.

During the pulp cooking and washing, black liquor concentration in the multiple effect evaporators and black liquor burning in the chemical recovery boiler, pollutant gases rich in lignin compounds, mercaptans and $H_2S$ are formed. They should be collected and disposed of in a gas incinerator.

The plant boilers and the lime kilns emit air pollutants like $SO_2$, $CO$, $CO_2$, $NO_x$ and particulate matter, as does any similar industrial equipment. The solutions are also similar: better quality fuels (in the case of fuel oil, for instance), filters, electrostatic precipitators, gas scrubbers, low $NO_x$ burners, etc. It is important to have in mind, however, that some of this equipment simply transforms gaseous pollutants into liquid or solid ones, which should be disposed of properly.

At the bleaching stage, chlorine-related toxic gases are released. This has been a main target of environmentalist pressures on the industry. As a result, most bleached pulp makers are changing to 'elemental free chlorine processes', or even 'total free chlorine processes', where chlorine compounds are also phased out. Oxygen, ozone and hydrogen peroxide have been replacing chlorine and its compounds. Biological processes employing fungi and enzymes are also starting to make their way and apparently show good prospects for the medium and long terms.

Pulp making and, to a less extent, paper making use large amounts of water, particularly for washing operations. Black liquor, pulp and waste paper residues pollute this water. Thus, it should be treated before it can be released to the environment. Physical (filtration, flocculation, etc.), chemical and biological (aerobic and, sometimes, anaerobic) treatments have been applied for this purpose.

When de-inking of recycled paper is carried out in the plant, special care should be taken with dangerous solid wastes that may result, due to the presence of heavy metals in the inks.

The long-term goal of the industry world-wide is to achieve a 'closed system' of water treatment.

Solid wastes of many kinds are produced along the whole production chain, ancillary activities and even from the management tasks. These should be disposed of at well-designed and maintained industrial landfills and/or be incinerated, possibly generating steam and electricity, but, at least, providing some heat recovery. Incineration is the only way to deal with solid wastes. The monitoring and control of all gaseous, liquid and solid wastes produced by a plant is a basic part of its 'Environmental Quality Management System'. Several Brazilian pulp and paper producers have installed, or are considering installing, such a system, in order to qualify for environmental certificates, which are increasingly important to open (or not to close) the doors to several export markets.

### 7.3.6 Prospects for improved energy efficiency in the Brazilian pulp and paper industry

There are two ways to improve energy efficiency in a pulp and paper plant: to reduce unit energy consumption at the production line stages and to generate steam and electricity (when applicable) more efficiently.

Several studies identified large energy conservation potentials in the pulp and paper industry world-wide, with a not-too-large set of energy efficiency improvement measures. Some of these studies and measures are reviewed later in this section, starting with those carried out for other countries and ending with the Brazilian case.

A study carried out in 1990 by the Office of Industrial Technologies, US Department of Energy, concluded that the best technologies commercially available at that time could reduce the forecast energy demand of the American pulp and paper plants for the year 2010 by up to 30 per cent. Some new technologies, still at the development stage, could add another 25 per cent to the forecast referred to above (World Energy Council 1995). In the following paragraphs, some of these technologies and their expected energy efficiency improvements are discussed, according to the production stage involved.

Improvements in digesters that operate continuously, a larger use of heat recovery in thermal–mechanical pulping, pulping with anthraquinone and its compounds, and better chemical–mechanical pulping combinations can reduce energy consumption at the pulping stage by 26 per cent by the year 2010. Advanced technologies, like pulping with alcohol-based solvents, pulping with enzymes from rotten wood fungi, chemical pulping with fermentation, and chemical–mechanical pulping without sulphur can provide a further 10 per cent energy saving. Experiments conducted in Sweden have demonstrated that the use of enzymes in pulping can produce an energy saving of 28 per cent.

Optimised waste heat recovery systems for the lime kilns and Tampella's compact system, which, simultaneously, burns the black liquor and recovers the pulping chemicals, through chemical reactions, can provide a 37 per cent economy in the energy consumption of this activity by the year 2010. Another 33 per cent saving can be added to that figure if new technologies, still at the development stage, such as the direct recovery of alkaline products and black liquor gasification, are included.

Extended flexible nip paper pressing, which increases the role of the more cost-effective mechanical drying in the overall paper sheet drying process, and optimised coupling between steam-heated drum drying and infrared radiation electric drying can decrease by 32 per cent the energy consumption in paper making by the year 2010. Developing pressing and drying technologies can add a further 27 per cent saving.

In Canada, ten 'state-of-the-art' technologies, chosen by the local pulp and paper industry representatives, can provide, by the year 2010, an electricity consumption economy of 3.8 per cent and thermal energy savings of 32 per cent (Clayton 1995). The technologies are:

1  newsprint paper recycling, after de-inking;
2  increase in the fibres/water mixture consistency, from about 1 per cent to 8–15 per cent, thus reducing water pumping needs;
3  secondary treatment of the plants' effluents, through aerobic or anaerobic biological processes;
4  recycling of the bleaching stage residues, in pulping plants that operate with the Kraft process;
5  optimisation of the refining intensity distribution among the several stages of the thermal–mechanical pulping process;
6  development of presses to reduce the moisture level of wood bark from 55–58 per cent to 45–48 per cent, in order to improve the efficiency and the particulate matter emission of the 'biomass boilers', where they are burnt;
7  'suspension burning' of the wood preparation stage residues in the boiler, in order to increase the equipment efficiency and decrease its size and emission levels;
8  burning of the residues referred to above in fluidised bed combustors;
9  water removal and incineration, in the 'biomass boilers', of the waste paper de-inking process residues; and
10 pulp bleaching through the sequential process OZEDP, with oxygen, O, ozone, Z, alkaline extraction, E, chlorine dioxide, D, and hydrogen peroxide, P.

The strong environmental concern of the Canadian industrialists is very clear in their choice of the technologies; it is equally evident the little weight they gave, in their decision, to electricity conservation. Eight of these technologies reduce the plants' unit energy consumption; the exceptions are the secondary treatment of the plants' effluents and the pulp bleaching through the OZEDP process.

The participants of a European workshop about 'new technologies for a rational use of energy', promoted by the European Commission in 1992, concluded that if the European countries (except Sweden and Finland) would upgrade their pulp and paper industries in order to align them, in terms of energy efficiency, to the French industry (which was, at that time, the most efficient among them), an energy economy of 20–30 per cent could be achieved (Commission des Communautés Européenes 1992). Improvements in the boilers and in the co-generation units employed in the industry, including the increasing use of natural gas instead of fuel oil, could provide energy savings of about 10 per cent. The same magnitude of unit energy consumption economy could be obtained, on average, with the use of high-efficiency electric motors and variable-speed motor drives.

The full implementation of twenty-one state-of-the-art technologies in the Dutch pulp and paper industry, including the optimisation of process control, better heat recovery and the installation of expanded flexible nip presses, would represent an energy conservation potential of 33 per cent in the year 2000 (World Energy Council 1995).

In Finland, one of the world's largest producers of pulp and paper, energy conservation measures have decreased the local industry unit energy consumption since the early 1980s. Malinem and Helynen (1994) forecast reductions of 8 per cent and 12 per cent on the unit consumption of electric and thermal energy, respectively, in the year 2005, compared to the values observed in 1990.

Potential short-term energy savings of up to 20–25 per cent were detected for the Indian pulp and paper industry. In Indonesia, South Korea and Thailand these economies were estimated at 30, 15 and 17 per cent, respectively (World Energy Council 1995).

A study carried out by the Brazilian utility CEMIG, in the context of the National Programme for the Reduction of Electric Energy Losses (PROCEL), estimated, based on a sample of 52 pulp and paper plants, electricity consumption potential savings of 14 per cent in an optimistic scenario and 8 per cent in a pessimistic scenario, with relatively simple and not very expensive conservation measures. Two important limitations of this assessment were that it concentrated mostly on small- and medium-sized plants and no process changes were considered. Replacement of underloaded and low-efficiency motors, better balancing of loads at feeders and improved illumination schemes were the main types of measures evaluated.

The comparison of the average unit electricity consumption of the various types of pulp and paper plants in Brazil with corresponding efficient units in the same country allowed Bajay (1997) to compose the energy economy figures of Table 7.12. They should be seen as upper limits of the economic potential achievable in the medium to long term (years 2005–2010).

Today, the production of electricity from biomass in the Brazilian pulp and paper mills is based on all steam cycle co-generation plants, usually with back pressure turbines and less often with condensation extraction steam turbines, both running in thermal parity. As mentioned in Section 7.3.3.12 and Chapter 2, the biomass fuels consumed in the mills are their wood-processing residues, i.e. refuse wood chips, tree barks and black liquor.

The amount of electricity produced with a given input of biomass can be substantially increased with the use of Biomass Integrated Gasifier-Gas Turbine (BIG-GT) systems (Berni et al. 1996; Berni and Bajay 1997).

Forestry wastes have not been used as a fuel in the co-generation plants of the Brazilian pulp and paper mills, despite their high potential to increase the mills' electricity self-generation capacity. The increasing mechanisation of the forestry activities, allied with the better prospects for surplus electricity sales to the public grid, will very likely change this picture. Assuming an efficiency of 40 per cent for a BIG-GT plant with fluidised bed gasifiers, a national average productivity of approximately 30 m³/ha of eucalyptus, a specific mass of the wood feedstock of 390 kg/m³ and a conversion factor of 1.2 kWh/kg of forest waste, 24 per cent of the mills electricity

*Table 7.12* Electricity conservation potentials in the Brazilian pulp and paper industry, by plant type

| Plant type | Electricity conservation potential, in % |
|---|---|
| Market pulp makers | 33 |
| Sanitary paper makers | 40 |
| Producers of other types of paper | 47 |
| Integrated plants | 36 |

*Source*: Bajay (1997)

consumption could be met by burning only forest residues (Berni and Bajay 1997). If black liquor is added, using the same sort of equipment (i.e. BIG-GT system), from 50 to 100 per cent self-sufficiency on electricity consumption can be achieved, depending on the underlying assumptions regarding black liquor and forestry residues availability (Berni *et al.* 1996).

Some energy efficiency gains have been obtained in the Brazilian pulp and paper industry, particularly at the larger plants, in the last two decades. Modernisation was a basic requirement for progress and often even for survival in this highly competitive and globalised market. The main goals of the modernisation process were to increase production, improve quality and decrease costs. Some achievements, in terms of better use of the industry wastes to supply energy, should be acknowledged, but no concerted and sustained energy conservation effort (as observed in some of the countries mentioned in this section), was ever exerted.

The driving forces of modern industrial energy conservation programmes are the search for increased competitiveness and for lower negative environmental impacts. The Brazilian pulp and paper industry decision-makers should realise that increasing production is not the only way to cut costs and increase revenue and, eventually, market share (the *fuite en avant* paradigm) and that well-planned and managed energy conservation programmes can lead to substantial cost reductions (and/or additional revenue through sales of surplus electricity generated in market pulp and integrated plants to the public grid) and environmental gains. The Canadian programme, reviewed in this section, illustrates well this last point. If Brazilian pulp and paper makers ignore this, they will lag behind their main competitors and are likely to lose the important positions in international markets conquered by their hard work and competence since the 1970s.

## 7.4 Paper recycling

### 7.4.1 Evolution of paper recycling in Brazil

Paper recycling helps to reduce the intensity of use of natural resources, decreases the need for waste disposal, decreases the specific energy consumption in paper manufacturing and, usually, also provides reasonable levels of profits for those in the business. These profits, however, oscillate as much as the price of pulp in the international market.

The savings, in terms of uncut trees and energy consumption, provided by recycled paper vary according to the raw material quality and technologies employed. In average terms, however, it can be stated that 30 kg of waste paper avoids the felling of a eucalyptus tree and 20 kg of recycled paper a pine tree. Recycling of newspapers can save from 50 to 78 per cent of the energy consumed in the production of such paper, from pulp manufactured by mechanical processes (Correia *et al.* 1994).

Paper recycling in Brazil is as old as the local paper industry. The first paper manufacturers in the country employed essentially waste paper and paper trimmings as their raw materials. At that time, most of the local paper demand was met with imports. The next jump forward of the emerging industry was the manufacture of paper with imported pulp, made with softwood. Only in the 1970s this industrial branch grew substantially, employing, then, mostly pulp produced in the country, mainly from hardwood. The larger availability of recyclable paper trimmings also advanced the recycling activities. Economic interest has been the only factor to support such expansion of recycling in the country.

Paper recycling levels practised in Brazil are well below not just the technical potential of the country but also the economic one (Correia *et al.* 1994). The activity has been limited to board and lower quality papers. De-inking technologies that have spread quickly in several parts of the world in the last few years have not arrived in the country yet on any significant scale; they allow the manufacturing of high quality products from waste paper. As the prices of pulp and recycled paper oscillate with high magnitude and frequency, the local recycling activity is not organised and operated efficiently, constituting an important barrier to its growth.

Consumption of recycled papers in Brazil in 1997 was 2.2 million tonnes. Compared to the apparent paper consumption that year, the recovery rate was 36 per cent (BRACELPA 1997a). In the USA, Japan and Great Britain the paper recovery rates that year were about 38 per cent, 50 per cent and 62 per cent, respectively (Paper Federation of Great Britain 1997).

It is important to point out now the limits to paper recycling. Tissues and special papers, for instance, cannot be recovered. The refining of the paper machine stock, for any type of paper, reduces the fibre length. So, there is a technical limit to the number of times the paper can be recovered, in order to keep the required new paper strength. A generally accepted limit is five times. It can be extended through adequate combinations of pulp and several types of recycled paper; the proportions of each component depend on the new paper type and quality and on the feedstock properties.

Rising land costs in the Brazilian regions where the main pulp wood markets are located tend to increase the economic attractiveness of recycling, even for higher quality papers. The pressure of environmentalist groups, particularly in what concerns the urgent need for higher levels of recycling and some processing of industrial and municipal waste, should also increase the supply of waste paper.

### 7.4.2 A balanced solution between minimum cost and maximum use of recycled paper

In the manufacturing of paper it is common to use both pulp and recycled paper as feedstock.

When the sole objective of the manufacturer is to minimise the sum of production costs, $Z_c$, he wants to determine the optimum amount of pulp, $x_{ip}$, and of each type $j$ of recycled paper, $x_{ij}$, employed to produce new paper type $i$, so as to achieve this goal, as expressed mathematically below:

$$MinZ_c = \sum_{i=1}^{m} \left( c_{ip}x_{ip} + \sum_{j=1}^{n} c_{ij}x_{ij} \right) \tag{7.1}$$

where $c_{ip}$ and $c_{ij}$ are the unit costs of pulp and recycled paper type $j$, respectively, to produce new paper type $i$.

From the point of view of maximum gains for the environment, the ideal objective function is to maximise the use of recycled paper as feedstock, $Z_e$, ie:

$$MaxZ_e = \sum_{i=1}^{m} \sum_{j=1}^{n} x_{ij}$$

(7.2)

When the solutions to objective functions (7.1) and (7.2) do not coincide, a multi-objective approach can be adopted (Leach *et al.* 1997), where a balanced solution between minimum cost and maximum environmental benefit is sought. Varying the weights given to each objective function will generate the trade-off curve between them, involving non-dominated solutions. With the possible help of an interactive multi-objective computer algorithm, the decision-maker can choose the most adequate solution for him, among those represented in the trade-off curve.

One constraint that should be met by all feasible solutions to the problem is the one that considers the efficiency of conversion of waste paper type $j$ to produce new paper type $i$, $\eta_{ij}$, in the amount, $P_i$, of paper type $i$ to be manufactured:

$$\sum_{j=1}^{n} x_{ij} \eta_{ij} + x_{ip} = P_i \qquad i = 1,...,m$$

(7.3)

A paper type $i$ usually requires a minimum amount of pulp in its production. This is expressed mathematically by the inequalities

$$\sum_{j=1}^{n} x_{ij} \eta_{ij} \leq (1 - \beta) P_i$$

(7.4)

where $\beta$ is the minimum share of pulp required in the production of paper type $i$.

Possible supply constraints of waste paper type $j$, $S_j$, and pulp, $S_p$, are represented by expressions (7.5) and (7.6), respectively.

$$\sum_{i=1}^{m} x_{ij} \leq S_j \qquad j = 1,...,n$$

(7.5)

$$\sum_{i=1}^{m} x_{ip} \leq S_p$$

(7.6)

Equation (7.7) indicates the total mass balance for the production scheme considered, where $\alpha$ represents the scheme overall losses in the manufacturing of all $m$ types of new papers, from pulp and from $n$ types of waste paper:

$$(1 - \alpha) \sum_{i=1}^{m} \left[ \sum_{j=1}^{n} x_{ij} + x_{ip} \right] = \sum_{i=1}^{m} P_i$$

(7.7)

The final mathematical constraints express that the values of the several types of feedstock cannot be negative, i.e.:

$$x_{ij}, x_{ip} \geq 0 \qquad\qquad i = 1,...m; j = 1,...,n \qquad\qquad (7.8)$$

This type of modelling can be applied to a particular paper mill, or for a whole region. In the latter case, it can be employed to quantify possible financial incentives granted by government to paper manufacturers in order to increase the paper recovery rate in that region.

## 7.5 Bibliography

Bajay, S. V. (1997) A indústria de papel e celulose: Seu consumo energético, por usos finais e tipos de plantas, evolução tecnológica e perspectivas de conservação de energia, in *Anais do 3° Congresso Latino-Americano sobre Geração e Transmissão de Energia Elétrica*, vol. 2, UNESP, Guaratinguetá, SP, 699–704.

Bajay, S. V., Carvalho, E. B. and Ferreira, A. L. (1997) *Relatório Técnico Final – Parte I: Diagnóstico econômico, competitividade internacional, banco de dados econômicos e cenários alternativos de desenvolvimento econômico*, Contrato ELETROBRÁS (PROCEL) / UNICAMP / FUNCAMP, NIPE / PRDU / UNICAMP, Campinas, SP.

Bajay, S. V., Carvalho, E. B. and Ferreira, A. L. (1998) *Relatório Técnico Final – Parte II: Medidas de conservação de energia e modulação de carga, potenciais de conservação, banco de dados sobre consumo energético e cenário de desenvolvimento setorial com programas institucionais de conservação*, Contrato ELETROBRÁS (PROCEL) / UNICAMP / FUN-CAMP, NIPE / PRDU / UNICAMP, Campinas, SP.

Bajay, S. V., Walter, A. C. S., Ferreira, A. L., Carvalho, E. B. and Athayde, M. A. P. (1996) *Desenvolvimento de Programas de Conservação de Energia Elétrica e Modulação de Carga nos Segmentos Industriais de Papel e Celulose e Fundição, na Região Administrativa de Campinas. Relatório da Atividade I: Caracterização técnica, econômica e ambiental dos segmentos industriais energo-intensivos objeto de análise*, Contrato ELETROBRÁS (PRO-CEL)/UNICAMP / FUNCAMP No. 4065/95, NIPE/UNICAMP, Campinas, SP.

Berni, M. D. and Bajay, S. V. (1997) 'Gasification technologies and the Brazilian potential for the production of electricity from wood and forest wastes' in *Proceedings of the 4th European Conference on Industrial Furnaces and Boilers*, vol. 2, Porto, Portugal: INFUB, 125–35.

Berni, M. D., Bajay, S. V. and Athayde, M. A. P. (1996) 'The potential of the black liquor and forestry residues for the generation of electricity and the corresponding environmental impacts in the Brazilian pulp and paper industry' in *Proceedings of the World Renewable Energy Congress*, vol. 3, Oxford: Elsevier Science, 1970–3.

BRACELPA (1997a) *Annual Report*, Brazilian Pulp and Paper Association, São Paulo, SP.

BRACELPA (1997b) *Consumo de Energia Elétrica no Setor de Papel e Celulose – 1996*, Associação Brasileira de Celulose e Papel, São Paulo, SP.

BRACELPA (1998) 'Resultados de 97 superam expectativas do setor', *Celulose & Papel*, **14** (61) 12–15.

Carpentieri, A. E., Larson, E. D. and Woods, J. (1993) 'Future biomass-based electricity supply in Northeast Brazil', *Biomass and Bioenergy*, 4 (3) 149–73.

CEC (1983) *Energy Audit No 3: Pulp, Paper and Board Industry in the European Economic Community*, Commission of the European Communities, Brussels.

Clayton, D. W. (1995) 'Impact of selected technologies on the Canadian pulp and paper industry in 2010' in *Energy Efficiency Improvement Utilizing High Technology – An Assessment of Energy Use in Industry and Buildings*, World Energy Council, London, 8.1–8.16.

Comitre, V. (1995) *Informação Econômica*, vol. 25, No. 10, São Paulo.

Commission des Communautés Européens – DG XVII (1992) *Nouvelles Technologies pour l'Utilisation Rationelle de l'Énergie dans l'Industrie de Papier-Cartons*, Agence de l'Environnement et de la Maîtrise de l'energie, Paris.

Correia, P. B., Berni, M. D. and Athayde, M. (1994) 'Otimização energética e ambiental através da reciclagem', *O Papel*, **55** (8) 34–9.

Ewing, A. J. (1985) *Energy Efficiency in the Pulp and Paper Industry with Emphasis on Developing Countries*, World Bank Technical Paper No. 34, Washington, DC.

Larson, E. D. (1993) 'Technology for electricity and fuels from biomass', *Annual Review of Energy and Environment*, No. 18, Annual Reviews Inc, 567–630.

Leach, M., Bauen, A. and Lucas, N. J. D. (1997) 'A systems approach to materials flow in sustainable cities: a case study of paper', *Journal of Environmental Planning and Management*, **40** (6) 705–23.

Lira, J. A. F. (1993) Impactos Ambientais da Exploração de Lenha em uma Floresta Plantada Localizada em uma Região Montanhosa, Dissertação de Mestrado, Universidade Federal de Viçosa, Viçosa, MG, Brasil.

Magrini, A. (1989) *A Avaliação de Impactos Ambientais*, CENDEC, Brasilia, DF.

Malinen, H. O. and Helynen, S. A. (1994) 'Possibilities for increasing power production with biofuels in the Finnish forest industry – present and future technologies and their effects on emissions', in *Proceedings of the Conference on Biomass for Energy, Environment, Agriculture and Industry*, vol. 1, New York: Elsevier Science, 779–92.

Medeiros, J. X. (1995) *Energia renovável no setor siderúrgico: Uma análise social, econômica e ambiental da produção de carvão vegetal para os alto fornos de Minas Gerais*, Tese de Doutorado em Planejamento de Sistemas Energéticos, Universidade Estadual de Campinas, Campinas, SP.

MICT (1997) *Ações Setoriais para Tornar a Indústria Brasileira mais Competitiva: Papel e Celulose*, Secretaria de Política Industrial/Ministério da Indústria, Comércio e Turismo, Brasilia.

MME (1998) *National Energy Balance – 1998*, Ministry of Mines and Energy, Brasilia, DF.

Modl, A. and Hermann, F. (1995) 'International environmental labeling', *Annual Review Environmental*, **20** 233–64.

Moreira, I. V. D. (1985) *Avaliação de Impactos Ambientais*, Rio de Janeiro, RJ: FEEMA.

Paper Federation of Great Britain (1997) *Annual Report*, London.

PROCEL (1990) *Projeto Diagnóstico do Potencial para Auditoria Energética – Relatório de Avaliação*, Rio de Janeiro, RJ.

Rodés, L., Barrichelo, L. G. E. and Ferreira, M. (1990) Biodiversidade e o projeto FLORAM: produtividade e condições ambientais, *Estudos Avançados*, **4** (9) 175–200.

Silva, M. J. (1998) Fabricantes discutem alternativas para reerguer o setor florestal, *Celulose & Papel*, **14** (61) 26–7.

Silva, R. P. (1992) *Simulação e Avaliação Econômica de um Programa de Reflorestamento, no Planejamento de uma Empresa Florestal*, Dissertação de Mestrado, Universidade Federal de Viçosa, Viçosa, MG.

Steur, R. (1986) *Multiple Criteria Optimization: Theory, Computation and Application*, Dallas: Wiley & Sons.

World Energy Council (1995) *Energy Efficiency Improvement Utilising High Technology – An Assessment of Energy Use in Industry and Building*, London.

# 8

# PRODUCTION AND USE OF INDUSTRIAL CHARCOAL

*Frank Rosillo-Calle*
*and Guilherme Bezzon*

## 8.1 Introduction

Charcoal is an economic activity of increasing importance in many developing countries, and one which is expected to increase significantly in the future. In Brazil, industrial charcoal has been produced for four centuries, although it did not reach maturity until the 1960s. Charcoal production peaked in 1989 when some 44.8 million m³ were produced, compared to 26 million m³ in 1996. Historically the bulk of charcoal has been produced from native forests, but for the past two decades there has been a gradual phasing out in favour of plantations, e.g. in the late 1970s about 90 per cent of the charcoal originated from native forests compared to 30 per cent in 1996, while the rest came from plantations. Charcoal is used in Brazil mainly as a reductor and thermal agent in industrial applications, e.g. pig-iron, steel, cement, etc.

This chapter assesses the industrial production of charcoal in Brazil. It considers briefly the history of charcoal production from native forests and plantations, industrial uses; reforestation activities, charcoal-making technologies, the implications of the utilisation of charcoal versus coke; environmental implications, and socio-economic factors. Finally, it addresses some possible directions in the future.

## 8.2 General background

It is extremely difficult to estimate global charcoal production since, in most cases, it is an integral part of the informal economy of many developing countries, characterised by small-scale operations involving a very large number of small farmers, and rural poor people. Estimates vary from 26 to over 100 million tonnes of charcoal produced annually world-wide. Charcoal is produced from forestry residues resulting from the expansion of agriculture, pasture land, waste from wood processing, saw mills, forestry thinning, and more professionally, from biomass plantations.

In most developing countries charcoal is mainly used as a domestic fuel for cooking and heating and also in cottage industries. However, it is also an important reduction and thermal agent in various industries. It is used in numerous metallurgical industries, especially pig-iron, foundries and forges, cement factories, and for chemical applications, as is the case in Brazil. Contrary to the general view, charcoal consumption has increased in recent years and is becoming a major source of energy as many people from rural and urban areas of developing countries convert from wood to charcoal use (Rosillo-Calle *et al.* 1996). Table 8.1 summarises estimated charcoal production and forecasts in developing countries from 1995 to 2020, representing 22.3 and 58.3 Mtoe (million tonnes oil equivalent) respectively. In 1995 Asia was the largest producer with 9.1 Mtoe, followed by Africa with 6.8 Mtoe and Latin America with 6.4 Mtoe. These figures should be regarded as conservative. The main feature of this table is that the IEA foresees an almost threefold increase in charcoal production and use from 22.3 Mtoe in 1995 to 58.3 Mtoe in 2020. Another important feature is the large energy losses associated with charcoal production, and the need to increase charcoal production efficiency.

The origin of charcoal-making is lost in prehistory. In Egypt, around 3000 BC, charcoal by-products were used for embalming and preparing bodies for burial. Some archaeologists have suggested that deforestation resulting from charcoal production may have caused the abandonment of a major iron-making centre around Lake Victoria in central Africa, where people began smelting iron about 2,500 years ago (Nooten and Raymaekers 1988).

In Europe, for example, it is known that charcoal-making was an important industry about 1100 BC for the recovery of iron and other metals from their ores. Charcoal was

*Table 8.1* Estimated world charcoal production in developing countries, 1995–2020

| Region | 1995 | 2010 | 2020 |
|---|---|---|---|
| East Asia | | | |
|   Share of charcoal in final biomass use (%) | 5 | 7 | 8 |
|   Charcoal production/use (Mtoe) | 5.6 | 7.8 | 9.2 |
|   Wood input in charcoal production (Mtoe) | 16.5 | 21.7 | 25.1 |
|   Energy losses in cultural transformation (Mtoe) | 10.8 | 14.0 | 15.9 |
| South Asia | | | |
|   Share of charcoal in final biomass use (%) | 2 | 3 | 4 |
|   Charcoal production/use (Mtoe) | 3.5 | 7.9 | 11.1 |
|   Wood input in charcoal production (Mtoe) | 12.6 | 28.2 | 39.5 |
|   Energy losses in charcoal transformation (Mtoe) | 9.1 | 20.3 | 28.4 |
| Latin America | | | |
|   Share of charcoal in final biomass use (%) | 9 | 9 | 9 |
|   Charcoal production/use (Mtoe) | 6.4 | 7.0 | 7.2 |
|   Wood input in charcoal production (Mtoe) | 13.2 | 14.5 | 14.9 |
|   Energy losses in charcoal transformation (Mtoe) | 6.8 | 7.5 | 7.7 |
| Africa | | | |
|   Share of charcoal in final biomass use (%) | 3 | 6 | 8 |
|   Charcoal production/use (Mtoe) | 6.8 | 19.1 | 30.8 |
|   Wood input in charcoal production (Mtoe) | 27.0 | 72.1 | 112.1 |
|   Energy losses in charcoal transformation (Mtoe) | 20.3 | 53.0 | 81.3 |
| Total developing countries | | | |
|   Share of charcoal in final biomass use (%) | 3 | 4 | 5 |
|   Charcoal production/use (Mtoe) | 22.3 | 41.8 | 58.3 |
|   Wood input in charcoal production (Mtoe) | 69.3 | 136.5 | 191.6 |
|   Energy losses in charcoal transformation (Mtoe) | 47.0 | 94.7 | 133.3 |

*Source*: IEA (1998)

the predominant fuel in the iron industry before the Industrial Revolution. For example, in England large tracts of land were deforested, causing the devastation of forests in many parts of the country, so that by the mid-sixteenth century, the deforestation situation was so bad in some areas that a series of enactments intended to preserve the country's woods were issued (Schubert 1957).

During the first few decades of the twentieth century the growing demand for steel in many countries, together with the growing capacity of the chemical industry, brought about an unprecedented demand for charcoal and its liquid by-products. However, when refined bituminous coal, coke and lignite became competitive with charcoal, the decline of the charcoal industry began once again. The development of 'rapid pyrolysis' in the late 1950s opened up a new category of raw materials, such as industrial wastes and agricultural and forestry residues, which were until then untapped (Emrich 1985; Rosillo-Calle *et al.* 1996).

## 8.2.1 Brazil

The origins of charcoal production in the country date back to the late sixteenth century, when the Sardinha's family was reported to be using charcoal in their foundries to produce iron ore. The charcoal-based industry has always been heavily concentrated in the state of Minas Gerais (MG) because of the large iron ore deposits. It can be said that the history of charcoal production is very much the history of pig-iron and steel production in MG. The history of charcoal production and use in Brazil can be briefly summarised as follows (Rosillo-Calle *et al.* 1996).

### 1591 to 1812

This includes the period between the registration of the first and second charcoal foundry. The main characteristic was the very low efficiency of charcoal production and use in iron works. Total charcoal and pig-iron production was, however, very small and remained almost an artisan activity.

### From 1814 to late 19th century

During this period various types of foundry were operating with greater efficiency and more specialised uses. During this period the first serious criticism appeared against the irrational destruction of the native forests resulting from the activities of farmers, cattle ranchers and charcoal-makers.

### From the late 19th century to around the 1930s

This period was characterised by some improvements in charcoal-making techniques, better operating methods, the introduction of more advanced blast furnaces, etc.

### 1930s–1970s

This period saw some improvements, a rationalisation of the pig-iron and steel sector, an increased productivity and greater awareness of the implications for native forests.

During the 1940s the iron and steel industry finally adopted a policy of 'afforestation' as the best and cheapest way to guarantee the supply of raw material to produce charcoal. This resulted in a rapid expansion of afforestation activities, particularly with eucalyptus, and a rapid expansion of the charcoal industry, stimulated by the increase in oil prices in the early 1970s.

## 1980s–1990s

The main development and expansion of the charcoal-based iron works took place from the mid-1980s in the Carajas region, in the state of Para. In the initial projections of the Grande Carajas, the planned capacity in the first phase of pig-iron production was 1.5 M tonnes annually, and of charcoal about 1.5 M tonnes. These projections never materialised for a combination of reasons (see Rosillo-Calle *et al.* 1996). By the end of 1998 charcoal production stood at about 0.5 M tonnes (approx. 1.6 M m$^3$), mostly saw mill and forestry residues.

There have been various serious attempts to produce charcoal from sustainable forest management schemes of native forests and also from plantations. However, these schemes have not been very successful partly because the socio-economic conditions of this region do not facilitate this kind of project. For example, there are many uncontrolled illegal deforestation and timber extraction activities which produce large amounts of saw mill and forestry residues, available to the charcoal-maker either free or at very low cost. He is hence discouraged from finding other alternatives, as may be the case in other parts of Brazil.

From 1990, however, a new preoccupation began to appear among some industrialists and professionals of the industry who were concerned with the high cost of charcoal, possible supply problems, environmental pressures, and the long-term future of the charcoal-based industry. Current thinking is that the long-term future of this industry depends on a combination of factors including: 1)the bulk of charcoal production being from eucalyptus plantations; 2) a rapid modernisation drive of the most backward sectors; and 3) a rapid increase in efficiency to reduce costs.

It is striking that the traditional methods of charcoal-making have changed little from ancient times to the present. As a study by the Food and Agricultural Organization (FAO) illustrates well, 'the only new factors are that the simple methodologies have been rationalised and that science has verified the basic processes which take place during carbonisation and spelled out the quantitative and qualitative laws which govern the process' (FAO 1985: 3).

Currently Brazil is one of few countries which carries out R&D in charcoal production and use on a meaningful scale, and where efficiency has reached about 35 per cent (quite high compared to many other parts of the world, where efficiencies of 10–15 per cent are common). Most of the charcoal producers around the world lack the resources and skills to do so; it is important to remember that this is a marginal activity in most countries. The survival of this industry must surely be linked to a rapid modernisation process and an increase in efficiency.

A shared characteristic of traditional charcoal-making is low efficiency and many and varied techniques, ranging from very simple and cheap earth mouths to the more efficient round and rectangular brick kilns. However, commercial charcoal-making often includes sophisticated and expensive technologies, such as retorts, with a high conversion efficiency, mostly used in industrial countries.

## 8.3 Industrial uses of charcoal in Brazil

The industrial use of charcoal in Brazil increased substantially after 1960, especially to provide carbon as a reduction agent for pig-iron and steel manufacture. From the 1960s to 1980s, several small charcoal-based pig-iron works were set up while some of the large steel companies modified their blast furnaces from coke to charcoal usage. For example, the Mannesmann's steel pipe mill was originally designed to operate using German coke, but Brazilian monetary exchange restrictions in the 1960s forced a change of approach and the company decided to start using charcoal in its reduction units. These activities were centred in the state of MG because of its large iron ore deposits and extensive native forests.

In the late 1980s various large steel-making companies started to substitute charcoal with coke for a combination of reasons:

1  difficulties in obtaining charcoal;
2  high cost of charcoal mainly because of transportation costs, e.g. distances of 400 to over 1,000 km are common;
3  new regulations for native forest preservation and environmental pressures which increasingly inhibited charcoal production from native forests;
4  low cost of imported coke;
5  financial advantages offered by coke. For example, to produce charcoal from plantations requires long-term investment (i.e. from land preparation to planting, to charcoal-making), which is financially expensive in a country like Brazil where interest rates are very high. Table 8.2 shows industrial uses of charcoal in Brazil between 1990 and 1996. It shows a decline of industrial charcoal from about 29 million $m^3$ in 1992 to about 26 million $m^3$ in 1996.

Currently new data seem to indicate that the cost of pig-iron from coke is higher than charcoal-based pig-iron due to: 1) an increase in coal prices in the international market and 2) an increase in the ratio of pig-iron per tonne of charcoal. This is because of

*Table 8.2* Industrial use of charcoal, 1992–96 (unit: $10^6$ $m^3$)

| Industrial sector | 1992 | 1993 | 1994 | 1995 | 1996 |
| --- | --- | --- | --- | --- | --- |
| Steel integrated mills | 6.7 | 8.0 | 7.0 | 7.6 | 5.2 |
| Pig-iron (independent producers) | 14.0 | 15.3 | 17.3 | 15.1 | 13.0 |
| Steel alloys | 2.9 | 3.1 | 2.7 | 2.9 | 2.9 |
| Cement | 1.3 | 1.5 | 1.9 | 1.7 | 2.2 |
| Primary metals | 1.2 | 0.7 | 0.5 | 0.6 | 0.6 |
| Others | 3.0 | 3.1 | 2.7 | 2.9 | 2.1 |
| Total | 29.1 | 31.7 | 32.1 | 30.8 | 26.0 |

*Source*: ABRACAVE (various Statistical Yearbooks)

increased productivity of charcoal from plantations together with its better qualities compared to coke. For example, the volume of charcoal from native forests needed to produce 1 tonne of pig-iron averages 3 $m^3$, compared to 2.5 $m^3$ of eucalyptus charcoal (Sadi 1997). Preliminary data indicate that the cost of pig-iron from coke is currently about US$150/t, compared to US$130/t of charcoal from eucalyptus. In addition, charcoal is of a higher quality, because of its lower level of impurities.

Thus, the increasing use of charcoal from plantations could be a major force in stimulating the charcoal-based industrial sectors. It is important to bear in mind that, because of the nature of the technology, small pig-iron mills cannot modify the blast furnaces to use coke instead of charcoal. In addition, there are some special steels that require very low level of impurities which can only be produced using charcoal. The use of coke to produce special steels would require further purification treatment that would put up the cost significantly. Generally, however, to survive, the small pig-iron producers have to use charcoal efficiently to be able to compete with coke-based pig-iron production.

## 8.4 Charcoal from native and planted forests

The expansion of the agricultural and pasture lands has had a very serious effect on deforestation in the Brazilian Southeast region, of which charcoal production has also played a role, particularly in MG where deforestation now represents about 80 per cent of the total area. Deforestation has been a misunderstood issue, often associated with large-scale charcoal production from native forests, a simple explanation to a very complex issue. Growing environmental concerns have resulted in the introduction of new environmental regulations since the early 1990s, which is having a direct effect on the production of charcoal from native forests, particularly in MG. The law requires a progressive reduction of the use of native forests for charcoal production, so that by the turn of the century the pig-iron and steel industry should be self-sufficient in

*Table 8.3* Industrial charcoal production from native forests and plantations, 1987–96 (unit: $10^3$ $m^3$)

| Year | Native charcoal[a] | % | Planted charcoal[b] | % | Total |
|------|------|------|------|------|------|
| 1987 | 27,725 | 80.7 | 6,624 | 19.3 | 34,349 |
| 1988 | 28,563 | 78.0 | 8,056 | 22.0 | 36,619 |
| 1989 | 31,900 | 71.2 | 12,903 | 28.8 | 44,803 |
| 1990 | 24,355 | 66.0 | 12,547 | 34.0 | 36,902 |
| 1991 | 17,876 | 57.7 | 13,102 | 42.3 | 30,978 |
| 1992 | 17,826 | 61.1 | 11,351 | 38.9 | 29,177 |
| 1993 | 17,923 | 56.5 | 13,777 | 43.5 | 31,700 |
| 1994 | 15,180 | 46.0 | 17,820 | 54.0 | 33,000 |
| 1995 | 14,920 | 48.0 | 16,164 | 52.0 | 31,084 |
| 1996 | 78,000 | 30.0 | 18,200 | 70.0 | 26,000 |

*Source*: ABRACAVE (various Statistical Yearbooks)

*Notes*
[a] Charcoal produced from native forest, including forestry residues
[b] Charcoal produced from plantations, mainly eucalyptus plantations

charcoal, either from plantations or from sustainable forestry management projects. As a result, the charcoal production from native forests has decreased from about 81 per cent in 1987 to 30 per cent in 1996, while charcoal from plantations has increased from 19 per cent to 70 per cent, respectively (Table 8.3).

However, there are currently ongoing discussions on how to modify this law to allow greater flexibility. For example, a large part of charcoal production from native forests originates from forestry residues and this potential will be wasted if it is not used for charcoal. It is argued that the law should allow the production of at least 20 per cent of charcoal from native forest residues.

Environmental pressures are forcing a rethink of the charcoal-based industrial sector, with mixed reactions that will have significant implications for this industry's future direction. For example, some steel companies are replacing charcoal for coke, while others, like Mannesmann and many small independent pig-iron producers, are concentrating efforts to cut costs by: 1) an increase in the overall efficiency, and 2) by producing the maximum amount of charcoal from eucalyptus plantations to achieve self-sufficiency. This will have the additional benefit of reducing transportation costs significantly, as most of these plantations will be sited near the centres of consumption.

Currently the largest producer of charcoal from plantations is Mannesmann Florestal, followed by Acesita and Belgo-Mineira. Mannesmann is committed to using 100 per cent charcoal, while Acesita has only one large blast furnace left using charcoal. Belgo-Mineira's long-term approach seems to be the gradual replacement of charcoal by coke, if the present company policy is maintained.

## 8.5 Afforestation programmes for charcoal production in Minas Gerais

To remedy the charcoal shortages the state of Minas Gerais (MG) has embarked on various afforestation programmes, of which FLOREMINAS (Polo Florestal Minas Gerais) and the Fazendeiro Florestal are good examples. The objective of FLOREMINAS was to plant approximately 3 million ha in a ten-year period, mainly for charcoal production, in an area of about 210,000 km² around the steel production areas.

Fazendeiro Florestal has received good support from the charcoal producers. The project aims to assist small farmers who want to use part of their land to plant eucalyptus, usually in areas of poor soil quality, which tend to be unsuitable for agriculture. Alternative schemes began to be devised in the early 1960s, which were aimed at supporting the small farmers, to improve their socio-economic conditions, while guaranteeing the supply of raw material (wood, charcoal, etc.) to the local markets. Altogether, over 70,000 ha had been reforested under such schemes from 1988 to 1996, with varying degrees of success (see Figure 8.1).

These schemes already appear to have brought various socio-economic benefits to the local small farmer because they:

1   allowed the incorporation of new land into a productive system which was previously of little value;
2   resulted in good financial gains to the small farmers; and
3   reduced the pressure on the native forests.

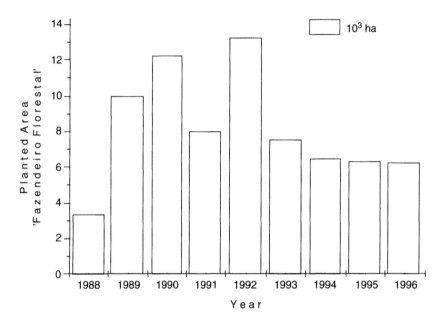

*Figure 8.1* Total afforested area according to the programme *Fazendeiro Florestal*
*Source*: ABRACAVE (1997)

Afforestation for charcoal production will have to be speeded up if the new law requirements are to be met. The large companies committed to charcoal-based steel-making, such as Mannesmann, are increasing their afforestation activities. For example, 'Mannesmann Florestal (MAFLA)' was created in 1969 with the chief objective of achieving self-sufficiency in charcoal for its steel mills (MAFLA 1997). The MAFLA research centre was responsible for introducing new and improved varieties, better management techniques, charcoal conversion technology, etc. For example, forest management has resulted in improved eucalyptus varieties with higher productivity and lower operational costs; pest and disease control; greater awareness of the environmental implications, etc. MAFLA harvests about 17,000 ha/yr and produces over 1.2 million m$^3$ of charcoal from eucalyptus plantations; it has over 125,000 ha of commercial forests located in the *Cerrado*, in the North and Northwest of MG.

## 8.6 Technological aspects of charcoal production

The technologies used for charcoal production depend, basically, on its final use and costs. For example, charcoal for steel- or pig-iron-making should have a fixed carbon content higher than 75 per cent (dry basis), while a charcoal for domestic cooking or heating may, on average, have 65–75 per cent of fixed carbon content, or even as low as 40 per cent in some cases.

Several factors affect the charcoal quality and yields, especially biomass composition, heating rates, final temperature and pressure. Woody biomass with higher density and lignin content results in a better quality charcoal and higher yields. Final temperature and heating rates are directly responsible for the calorific value of charcoal, its carbon content and yield. Pressure is also an important parameter, which is often not considered. A small increase in pyrolysis pressure results in considerable gains on charcoal yield (Antal *et al.* 1992).

Carbonisation can be conducted in several ways. Since oxygen concentration can be controlled it is not difficult to obtain charcoal, but the same cannot be said if the aim is to achieve high yields. Many techniques have been tried in the past, but three seem to be the most common today:

1  internal heating by controlled combustion of the raw material;
2  external heating by combustion of firewood, fuel oil or natural gas; and
3  heating with circulating gas (retort or converter gas).

Internal heating systems are the most commonly used in Brazil. External heating systems and heating with circulating gas are still little known. Two of the most common kilns in Brazil are the *rabo-quente* (hot tail) kiln and the *superficie* or *colmeia* (surface or beehive) brick kiln. Much larger rectangular kilns are gradually being introduced. The *rabo-quente* kiln is built with common bricks, usually without a chimney, and with one door, with an effective kiln volume holding between 4.5 t (the most commonly used) and 250 t of wood. The kiln diameter varies from 3 to 7 m and the kiln height from 3.5 to 3.7 m; the chopped wood length varies between 1.2 and 1.5 m.

The *superficie* or *colmeia* kiln is also built with common bricks, but with one to six chimneys and one or two doors. The effective kiln volume holds 17.5t–75 t of wood. Most of the charcoal produced from eucalyptus plantations is made in surface beehive kilns. Their advantages are symmetric carbonisation, very low construction cost, and the possibility of being built near the forests. The main disadvantage is lack of control of carbonization parameters (e.g. temperature, time, oxygen, concentration).

New technologies for charcoal production are being developed by various companies to increase charcoal yields but most of them are still comparatively more expensive than traditional surface round kilns. ACESITA had a charcoal modernisation programme, including a continuous carbonisation retort, which achieved an efficiency of about 35 per cent. The programme was, however, abandoned and to the best of our knowledge no new retorts are currently commercially operating in Brazil.

MAFLA is replacing the traditional round surface kilns by rectangular-shaped kilns with much larger capacity and a tar recovery system (see Figure 8.2). This rectangular kiln permits partial mechanisation, using trucks for loading wood and unloading charcoal. It also allows better temperature control which increases charcoal yield and utilisation of by-products.

ACESITA and Belgo-Mineira are also replacing their traditional round brick kilns by larger rectangular kilns of similar characteristics to the MAFLA ones. The charcoal costs for the ACESITA rectangular kiln is around US$4.18 per $m^3$ of charcoal, compared to US$4.93 per $m^3$ for the traditional brick kilns (Paranaiba 1998) (see Table 8.4). The wood loading and charcoal unloading are conducted using trucks, which can come inside the kiln, reducing the overall time for these operations. ACESITA is planning to build volatile and tar recovering devices to reduce atmospheric emissions.

In 1991 Belgo-Mineira built two rectangular kilns to compare their carbonisation process with that produced in the traditional brick kilns. The rectangular kilns'

*Figure 8.2* MAFLA's rectangular and round-shaped kilns
*Source*: MAFLA (1997)

performance was evaluated from 1991 to 1998; this has proved to have lower production costs and better operational conditions (e.g. mechanised loading and unloading) while producing the same quality charcoal as traditional brick kilns. The cost of the rectangular kiln in 1991 was US$17,000, 7.55 per cent lower than the round brick kiln with the same capacity. The kiln is 13 m long, 5 m wide and 6 m high, with a useful volume of 320 m³. Their charcoal conversion efficiencies are around 27 per cent (dry mass basis) with a complete carbonisation cycle of about twelve days (Naime 1998).

It is very difficult to calculate the theoretical value of charcoal conversion efficiency on yield because charcoal contains hydrogen and oxygen and because lignin contains more carbon than cellulose. Based on the average woody biomass composition, charcoal theoretical yields range from 44 per cent to 55 per cent (dry basis). The Hawaii Natural Energy Institute, in the University of Hawaii, has developed a process that increases charcoal yields. The process is based on the pyrolysis of biomass in a pressurised reactor under controlled temperature. The elevated pressures move the reaction kinetics towards charcoal production, decreasing the production of liquids and gases. The moisture and pyrolysis vapours act as catalysts at elevated pressures, helping to increase charcoal yields. This process produced yields of over 45 per cent for eucalyptus wood, with properties similar to commercial charcoals, e.g. 78 per cent fixed carbon content and 30 MJ/kg of high calorific value (Antal *et al.* 1996). (See also Chapter 9, Section 9.5.)

*Table 8.4* Main characteristics of ACESITA's rectangular kiln

| | |
|---|---|
| Kiln dimensions, l × w × h (m) | 13 × 4.6 × 4 |
| Wood capacity (m³) | 103.6 |
| Charcoal production (m³) | 60.9 |
| Carbonisation period (days) | 6 |
| Cooling period (days) | 6 |
| Charcoal production/month (m³) | 140.3 |
| Number of kilns | 20 |
| Cost of kiln (US$) | 20,000 |

*Source*: Paranaiba (1998)

Despite the lower overall efficiency of the conventional brick kilns compared to modern technologies, the bulk of charcoal in Brazil is still produced in these kilns because their lower installation and maintenance costs compensate for the lower productivity. It is expected that the modern technologies will replace the conventional kilns on a large scale only when their initial costs decrease or if wood prices increase significantly to compensate the initial higher investment costs. Nevertheless, the gradual replacement of the conventional brick kilns by the rectangular kilns by some of the leading charcoal-makers is very encouraging.

## 8.7 Charcoal versus coke: some environmental considerations

The Brazilian industry is increasingly being forced to compete in the international market with good quality products at low prices. Iron ore in Brazil is still very abundant and cheap, especially in Minas Gerais, but charcoal, which was the main carbon source in the past, has became more expensive and more difficult to obtain from native forests in the last decade, contributing to an increase in steel and pig-iron prices. This forced various steel companies to modify their blast furnaces from charcoal to use coke. Figures 8.3 and 8.4 show the pig-iron and raw steel production in Brazil, based on charcoal and coke as carbon source. They show an increase in the production of pig-iron and raw steel from coke in the last few years, while charcoal use has remained steady.

According to ABRACAVE, there are nine large steel companies operating sixteen blast furnaces and sixty small pig-iron companies operating 116 blast furnaces based on charcoal. Usually, the large steel companies have most of their blast furnaces operating with both coke and charcoal. In the case of the small pig-iron producers, all their blast furnaces operate only with charcoal.

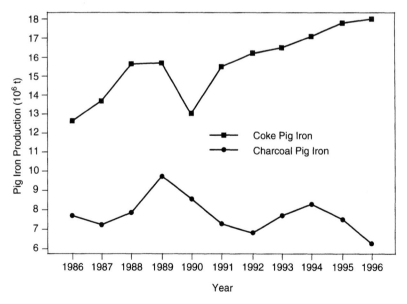

*Figure 8.3* Evolution of pig-iron production from charcoal and coke
*Source*: ABRACAVE Statistical Yearbook (1997)

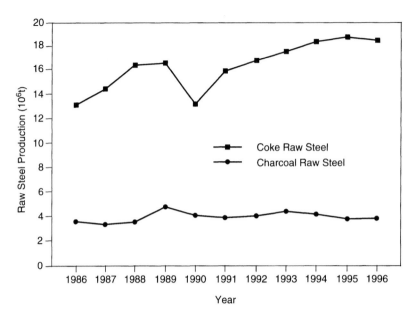

*Figure 8.4* Evolution of raw steel production from charcoal and coke
*Source*: ABRACAVE Statistical Yearbook (1997)

A characteristic of these furnaces is that they cannot be modified to operate with coke. Furthermore, these small producers have few alternatives except to improve the efficiency of the charcoal furnaces and to reduce charcoal production costs or to close down. In the past, most of the charcoal used by the small pig-iron producers originated from native forests, but in the future charcoal would have to be produced from plantations, particularly from small plantations or sustainable native forests, and this is forcing new thinking about this sector.

## 8.8 Environmental implications

Three major environmental concerns have arisen with regard to large-scale charcoal production in Brazil (Rosillo-Calle *et al.* 1996):

1   The first concerns the potential destruction of natural vegetation. However, the main responsibility for large-scale deforestation lies not with the charcoal industry, but with the expansion of agricultural and grazing land.
2   The second concern refers to the environmental impacts posed by large eucalyptus plantations. Four main objections have arisen in recent years: a) heavy concentration of plantations in some areas, particularly in the state of MG; b) concern that eucalyptus plantations use good agricultural land. However, many eucalyptus plantations are found mostly on poorer soils, which have been degraded and are often unsuitable for agriculture; c) concern about the allegations that eucalyptus contaminates the soils, drains it dry and thrives at the expense of other plants in the vicinity. However, there is little scientific evidence to support this view;

d) the lost of biological diversity compared to native forests. Plantations by their very nature will cause loss of diversity, but steps are being taken to improve local diversity. It is important to find a balance between sustainability, environment, and what is practically achievable.

3    The third major environmental concern refers to charcoal-making activities. Charcoal production undoubtedly has certain negative environmental impacts given the nature of the product. Today thousands of tonnes of chemicals derived from the charcoal-making processes are released into the atmosphere. This pollution could be prevented or largely reduced with better technology and better use of by-products. Generally, however, this activity is not a serious environmental hazard as it is a widely dispersed activity, away from population centres.

On the other hand there are clear environmental advantages to the charcoal-based industry, including low S and $NO_x$ and $CO_2$ neutral. With all its drawbacks charcoal is far more environmentally friendly than coke. For example, Table 8.5 shows that if tar recovery is included, for each tonne of charcoal-based pig-iron 890 kg of $CO_2$ are absorbed from the atmosphere, while 1,746 kg of $CO_2$ are released to the atmosphere when coke is used instead.

It should also be recognised that the charcoal-based industry has made considerable advances in recent years, notably in forest management, infrastructure, transport and mechanisation, although much remains to be done. Generally, the industry needs to be modernised and to have a more positive attitude towards the environment, to innovation and to investment in R&D, if it is to have a secure future. Particular attention should be paid to 'green products' which may have better commercialisation possibilities.

*Table 8.5* $CO_2$ and $O_2$ balance in the production of pig-iron from charcoal and coke

| *Charcoal-based pig-iron production* | | | *Coke-based pig-iron production* | | |
|---|---|---|---|---|---|
| *Process* | $O_2$ *kg/t of pig-iron* | $CO_2$ *kg/t of pig-iron* | *Process* | $O_2$ *kg/t of pig-iron* | $CO_2$ *kg/t of pig-iron* |
| Tree growing | +1,762 | −3,643 | Coal mining | 0 | 0 |
| Carbonisation with tar recovery | −792 | +963 | Coke production from coal | −306 | +160 |
| Blast furnace | −830 | +1,790 | Blast furnace | −684 | +1,586 |
| Total | 203 | −890 | Total | −960 | +1,746 |

*Source*: SINDIFER (1997)

## 8.9 Socio-economic factors

The charcoal industry has historically had a poor, if not negative, image because, rightly or wrongly, it has often been associated with poor working conditions, environmental damage, deforestation, backwardness, massive eucalyptus plantations, etc. In addition, large eucalyptus plantations have been associated in some areas with loss of agricultural land, soil erosion and environmental degradation (Rosillo-Calle *et al.* 1996).

The industry is looking for alternatives in favour of 'sustainable development' and is already actively involved in environmental and conservation programmes of various types. However, there are still many instances in which charcoal production is associated with destructive practices.

The gradual phase out of charcoal from native forests is having two important effects: i) the elimination of the most negative environmental impacts, together, perhaps, with some of the most abusive labour practices; and ii) serious employment losses, as can be seen in Table 8.6. The charcoal-based sector is a major source of employment, as can be appreciated from Table 8.6, e.g. in 1989 it directly employed over 267,000 people, compared to about 120,000 in 1996. Charcoal production from native forests in particular has been a major source of employment for many labourers and also an important secondary source of employment for many small *fazendeiros* (small farmers). Charcoal production from forest plantations is a more professional activity and is less labour intensive, although the socio-economic conditions are much better. Thus, it is important that alternative sources of employment are available for those more directly affected by these changes.

*Table 8.6* Employment in the charcoal-based industries, 1989–96

| Group | 1989 | 1992 | 1994 | 1995 | 1996 |
|---|---|---|---|---|---|
| 1. Afforestation – planting, harvesting, maintenance | 51,358 | 44,300 | 49,971 | 45,328 | 41,560 |
| 2. Charcoal from native forests – harvest, transport of charcoal | 134,600 | 75,200 | 64,050 | 62,953 | 32,911 |
| 3. Integrated steel production – direct employment by the steel industry | 46,772 | 33,800 | 28,838 | 23,260 | 20,934 |
| 4. Pig-iron – direct employment by the steel industry | 22,608 | 16,200 | 20,981 | 18,351 | 16,791 |
| 5. Iron/steel alloys – direct employment by the steel industry | 12,000 | 10,700 | 9,400 | 8,000 | 7,200 |
| Total | 267,438 | 180,200 | 173,240 | 157,892 | 119,396 |

*Source*: ABRACAVE (various Statistical Yearbooks)

Contrary to popular belief, however, many plantations have been set up in areas of low population density and poor and under-utilised land. This has stimulated development and increased employment in rural areas and has improved local socio-economic conditions.

Afforestation is labour intensive – about five men ha/yr – and costs per job created are much lower than in any other sector. For example, in 1992 the cost of creating a job in the metallurgical industry was the equivalent of US$419,400; in the agricultural sector US$12,980; in livestock US$11,180; and US$7,260 in afforestation activities (Abrahao and Furtado 1992).

## 8.10 Conclusions

In spite of being a centuries-old process, the charcoal-based industrial sector can be regarded as relatively new since its consolidation is almost a post-1945 phenomenon. An important characteristic of the industry, particularly the pig-iron sector, is the large number of small independent producers and its concentration in the state of Minas Gerais. The small mills found there are the backbone of the industry and are the most vulnerable to the international market fluctuations of pig-iron and steel products because of their generally poor performance and low efficiencies.

Historically, native forests have provided raw materials for the bulk of charcoal production. This situation is changing rapidly, as can be appreciated in Table 8.3: in 1987 almost 81 per cent of charcoal was produced from native forest, compared to 30 per cent in 1996.

This industry is a major source of employment generation with wide socio-economic ramifications. Charcoal-making and afforestation are labour-intensive activities and are often the main source of employment in many rural communities, particularly among unskilled and semi-skilled people, who are so numerous in Brazil. The progressive phasing out of native charcoal in favour of charcoal from plantations has addressed some of the most serious environmental concerns, but it is also having a serious negative employment impact on many labourers and small farmers, for whom charcoal-making was a major source of income. It is doubtful if afforestation activities, for example, will be able to absorb this loss of employment.

Considerable gains have been made in silvicultural and environmental matters. For example, different eucalyptus varieties and clones are now being planted in the same area, allowing greater biological diversity and biological control. Charcoal-making is becoming more professionalised, particularly among the major charcoal producers and consumers. More scientific principles are being applied, with careful monitoring of wood density, moisture content, and calorific value, factors hardly considered in the past (Rosillo-Calle *et al.* 1996).

Charcoal costs still constitute a major component of the final product price. However, prices are often determined by prevailing market conditions, such as demand for pig-iron and steel products, and do not necessarily reflect real production costs. Transport has been a major component of the overall cost, but as an increasing amount of charcoal is produced from plantations, this will have a lesser impact in the future. Efficiency, which is low, particularly for charcoal from native forests, and which also contributed to higher costs, is improving with the professionalisation of the industry.

The aggressive pursuit of by-products can also play a significant role in making the industry more competitive and modern, at least in the medium and long term, although market conditions and know-how may be major determinants. Furthermore, by-products offer a considerable potential and are an option which the industry should pursue more vigorously. Co-generation of electricity, together with production of some liquid fuels and chemical products, looks particularly promising.

Thus, in spite of many obstacles, the Brazilian charcoal-based pig-iron and steel industry is becoming increasingly competitive both in the domestic and international markets.

SINDIFER (Brazilian Pig Iron Association) has laid out a number of objectives to ensure the future of the industry (SINDIFER 1997), including:

1   To ensure the supply of charcoal, about 520,000 ha of eucalyptus should be planted in the next seven years.
2   To produce charcoal at prices lower than US$23/m³ by creating some kind of financial support mechanisms, e.g. low interest rates of approximately 6 per cent per year, and reducing the charcoal transportation distances to less than 150 km.
3   To increase charcoal quality and productivity to 2.5 m³/tonne pig-iron, by the introduction of new technologies, and mechanisation.
4   To stimulate the development of related activities, e.g. use of eucalyptus wood for building materials, better use of by-products, co-generation using blast furnace gas, etc.
5   To increase international marketing of pig-iron.
6   To develop cleaner carbonisation techniques.
7   To improve the socio-economic and working conditions of the labour force.

The survival of the charcoal-based pig-iron industry is strongly linked to the accomplishment of these goals if it is to have a long-term future. It is unrealistic to expect that this sector will survive in its present form, thus further changes will have to be implemented. There are many small producers who would not be able to survive a restructuring process because of high investment costs and competition from coke.

The phasing out of the production of charcoal from native forests presents new challenges and opportunities. The most significant advantages of producing the bulk of charcoal from plantations include: 1) lower cost, e.g. through lower transport cost, as the distances would be shorter, and higher efficiency; 2) better image, e.g. charcoal will be produced from environmentally sustainable forests and plantations; and 3) guarantee of supply of raw material, e.g. if plantations can be established successfully on a large scale.

## 8.11 Bibliography

ABRACAVE, Brazilian Charcoal Association, *Statistical Yearbooks*, Belo Horizonte, MG, various years.

Abrahao, J. and Furtado, D. B. (1992) Acesita Energetica no Alto Jequitinhonha, Acesita Energetica, Belo Horizonte, MG.

Antal, M. J., Mok, W. S. L., Varhegyi G. and Szekely, T. (1992) 'Review of methods for improving the yield of charcoal from biomass', *Energy and Fuels*, 4, 221–5.

Antal, M. J, Croiset, E., Dai, X., de Almeida, C., Mok, W. S. and Norberg, N. (1996) 'High yield biomass charcoal', *Energy and Fuels*, 10, 652–8.

Emrich W. (1985) *Handbook of Charcoal Making*, Series E: Energy from Biomass, vol. 7, D. Reidel Publishing.

FAO (1985) *Industrial Charcoal Making*, Food and Agriculture Organisation, FAO Forestry Paper 63, FAO, Rome.

IEA (1998) 'IEA biomass energy analysis and projections', in *Biomass Energy: Data, Analysis and Trends*, OECD/International Energy Agency, Paris, 151–66.

MAFLA (1997) 'Mannesmann Florestal', *Internal Report*, Belo Horizonte, MG.

Naime, M. A. (1998) 'Some considerations about the Belgo-Mineira rectangular kiln for charcoal production', in *Technologies for Improving Charcoal Production*, Belo Horizonte, MG: SINDIFER.

Nooten, F. and Raymaekers, V. (1988) 'Early iron smelting in Central Africa', *Scientific American*, 259 (1), 84.

Paranaiba, W. (1998) 'Evaluation of ACESITA's rectangular kiln', in *Technologies for Improving Charcoal Production*. Belo Horizonte, MG: SINDIFER.

Rosillo-Calle, F., de Rezende M. A. A., Furtado, P. and Hall, D. O. (1996) *The Charcoal Dilemma – Finding a Sustainable Solution for Brazilian Industry*, London: IT Press.

Sadi, P. (1997) personal communication, ACESITA Energy Division, Belo Horizonte, MG.

Schubert H. R. (1957) *History of the British Iron and Steel Industry*, London: Routledge & Kegan Paul.

SINDIFER (Brazilian Pig Iron Association) (1997) *Internal Report*, Belo Horizonte, MG.

# 9

# NEW TECHNOLOGIES FOR MODERN BIOMASS ENERGY CARRIERS

*Arnaldo Walter* et al.

## 9.1 Introduction

*Arnaldo Walter*

Biomass is an important energy source for many countries, but its traditional use, involving unsustainable biomass consumption and low conversion efficiencies, has a limited future. Increasing competition among energy companies and energy sources and growing social awareness about environment preservation and sustainable use of natural resources are driving forces for deep changes in the energy sector. In this sense, future prospects for biomass clearly point to modern energy carriers: electricity, liquid and gaseous fuels in substitution to direct use of solid feedstock.

Modern use of biomass will strongly depend on the technical and economic feasibility of new conversion processes, or even on the scale-up and on overcoming technical barriers of traditional ones. This chapter presents a review of the current possibilities of some of these technical options, discussing the basics of each technology, presenting information of the current status of R&D programmes, their economics, and future prospects.

The text is the product of multiple contributors. The thermal gasification of biomass is discussed by Richard Bain, Ralph Overend and Kevin Craig. The text about electricity production from biomass was written by Andre Faaij, Ausilio Bauen, and myself; while the contribution on ethanol production from cellulosic materials is by José Roberto Moreira, from Biomass Users Network and Centro National de Referência de Biomassa, from Brazil. Finally, the text about biomass pyrolysis is a contribution by Guilherme Bezzon and José Rocha.

## 9.2 Gasification for heat and power, methanol and hydrogen

*Richard L. Bain, Ralph P. Overend and Kevin R. Craig*

### 9.2.1 Introduction

In this section the thermal gasification of biomass is discussed. Principles of gasification, along with examples of gasifiers under development, are presented. Applications discussed include heat and power generation, and synthesis gas production for methanol and hydrogen. Technologies being commercialised today are primarily for thermal applications and electricity production. Development of commercial technologies for chemical synthesis is much further into the future.

Through the gasification of biomass, a very heterogeneous material can be converted into a consistent gaseous fuel intermediate that can be used reliably for heating, industrial process applications, electricity generation, and liquid fuels production. As received, biomass can range from very clean wood chips at 50 per cent moisture to urban wood residues that are dry but contaminated with ferrous and other materials, to agricultural residues, to animal residues, sludges, and the organic component of municipal solid waste (MSW). The process of gasification can convert these materials into carbon- and hydrogen-rich fuel gases that can be more easily utilised, often with a gain in efficiency and environmental performance compared to direct combustion of the biomass. Gasifier systems usually comprise the biomass fuel handling and feeding system, which is coupled by means of airlocks to the gasifier. The gasifier is usually a refractory-lined vessel and the gasification is carried out at temperatures of approximately 850°C at either atmospheric or elevated pressures.

The product gas has to be treated so it matches the end-use application. For close-coupled gasifier-combustor systems there is no clean-up of the gases. For gas turbine applications in a power system (see Section 9.3.3.1) the gas has to be free of particulates, tars, sulphur and chlorine compounds, and alkali metals to ensure the integrity of the turbine hot section. For internal combustion applications (see also Section 9.3.3.2) the gas must be cooled to ensure that a sufficient charge of energy can be put into each cylinder, and particular attention has to be given to tar and particulate contents to ensure that the valving and cylinders are protected. Fuel cell applications (Section 9.3.3.5) would require the gas to be mainly hydrogen without any significant sulphur or chloride contamination to protect the electrodes. For synthesis operations, such as methanol and hydrogen production, particulates and contaminants (e.g. $H_2S$) must be removed to prevent poisoning of downstream catalysts.

The process efficiency is really quite high: thermal gasification is typically 80–85 per cent efficient in converting the organic content of the feed into a fuel gas mixture. Ultimate biomass to electricity efficiencies higher than 45 per cent are forecast with the use of combined cycles to generate electricity. If the gases are converted to hydrogen, the limiting efficiency with fuel cells may be over 55 per cent (HHV basis). The environmental advantage is that the fuel gas is a much smaller volume to be processed than the combustion stream from a boiler. This and the generally lower treatment temperature of the biomass result in retention of metals (including alkali) in the ash and cyclone as salts that can be disposed of. The gas can easily be cleaned of acid gas components, including hydrogen chloride, before combustion and thus is environmentally superior to direct combustion.

### 9.2.2 Gasification

Gasification-based process design is affected by the gasifier medium, gasifier pressure and reactor type.

### 9.2.2.1 *Gasifier medium*

Gasification involves the devolatilising and converting of biomass in an atmosphere of steam or air (or both) to produce a medium- or low-calorific value gas. Air-blown or directly heated gasifiers use the exothermic reaction between oxygen and organics to provide the heat necessary to devolatilise biomass and to convert residual carbon-rich chars. In these directly heated gasifiers, the heat to drive the process is generated inside the gasifier. When air is used, the resulting product gas is diluted with nitrogen and typically has a dry-basis calorific value of about 5–6 MJ/Nm$^3$.

The dry-basis calorific value of the product gas can be increased to 13–14 MJ/Nm$^3$ by using oxygen instead of air. Oxygen production is expensive, however, and its use has been proposed only for direct heating gasification applications involving the production of synthesis gas where nitrogen is not permitted in downstream synthesis conversion operations. Oxygen typically costs US$40–60/tonne and typically is used at the rate of 0.25–0.3 tonne/tonne of biomass, a cost equivalent to US$10–20/tonne of biomass (Wyman *et al.* 1992).

An alternative to the air-blown process is indirectly heated gasification. Indirectly heated gasifiers heat and gasify biomass through heat transfer from a hot solid or through a heat transfer surface. Because air is not introduced into the gasifier, little nitrogen diluent is present and a medium-calorific gas is produced. Dry-basis values of 18–20 MJ/Nm$^3$ are typical.

### 9.2.2.2 *Pressure*

A second variable affecting gasification-based power systems performance is gasifier operating pressure. This aspect is especially important for the performance and economics of gasification-based power systems involving gas turbine-based cycles. Turbines typically operate at compression ratios of 10–20, giving turbine inlet pressures of 1.0–2.0 MPa. A pressurised gasifier will produce gas at a pressure suitable for direct turbine application and provide the highest overall process efficiency. To take full advantage of operating at pressure, however, a number of ancillary systems must be developed. Reliable, high-pressure feed systems have not been commercially proven.

At typical gasifier conditions (825°C and 2 MPa), tars, chars and volatile alkalis are generated. To maximise system efficiency, these materials must be removed from the hot product gas without lowering the temperature below the tar dew-point, typically about 540°C. Thus, hot-gas clean-up systems are required. Tars have relatively high heat contents and can be burned in combustors, but they may plug char filters and may form soot during combustion, which affects combustion stability. The removal of tars ensures an even and less luminous combustion process (to avoid radiative heat transfer problems at the turbine). Therefore, the first element of a hot-gas clean-up system will probably be a catalytic or thermal tar cracker. A catalytic tar cracker will operate at temperatures comparable to gasifier temperature (about 825°C); a thermal cracker will typically operate in a temperature range of 870–980°C.

After the tar cracker the product gas will be partially cooled to minimise the amount of alkali vapours, typically to 350–650°C. The product will then pass through a ceramic filter to remove solids, allowing also the removal of much of the alkali. Particulate removal is to protect the turbine blades from erosion. For gas-turbine applications the gas may still contain too much alkali, as the gas turbine limits for alkali are about

25 ppb in the turbine combustor exit gas. To ensure that the product gas meets alkali specifications, thus avoiding deposition and corrosion of the turbine blade materials, an alkali-getter bed may be added after the filter. Typical bed materials are emalthite or hectorite. Getter beds will probably be designed as parallel fixed beds, giving continuous operation while allowing for bed material replacement.

Alternatively, the gasifier can be operated at low pressure and the cleaned product gas compressed to the pressure required for gas-turbine application. In this case, a tar cracker will probably be used to minimise the amount of tar which must be handled during quenching. The product gas exiting the tar cracker will be conditioned to provide a suitable compressor feed. The water and tar content must be low enough to ensure no condensation during compression. Also, soluble tars which will affect lubrication oil properties must be eliminated. Usually a combination of heat exchange, to reduce the gas from tar cracker exit temperature to residual tar dew-point, and wet scrubbing is used. The water vapour content will be at saturation at scrubber exit temperature (about 90°C) and pressure.

### 9.2.2.3 Gasifier type

Four primary types of biomass-gasification reactor systems have been developed: fixed bed reactors, bubbling fluid-bed reactors, circulating fluid-bed reactors, and entrained flow reactors.

#### FIXED BED

Fixed bed gasifiers can be classified primarily as updraft and downdraft. Updraft represent the oldest and simplest gasifiers. The updraft gasifier is a counterflow reactor in which fuel is introduced into the top by means of a lockhopper or rotary valve, and flows downwards through the reactor to a grate where ash is removed. The gasifying medium, air or oxygen and possibly steam, is introduced below the grate and flows upwards through the reactor. Typical product exit temperatures are 80–100°C.

A wide range of condensable tars and oils, which can condense in product lines, is produced in the pyrolysis zone. For this reason updraft gasifiers are usually just operated in a close-coupled mode to a furnace or boiler to produce steam or hot water. Certain feeds with low-melting ash may have slagging on the combustion grate. In addition, feed particle size needs to be controlled to maintain a uniform bed.

In downdraft gasification, the air and product both flow in the same direction as the solid bed. Downdraft gasifiers are specifically designed to minimize tar and oil production. The fuel and pyrolytic gases/vapours move co-currently downwards through the bed. The exit gas temperature is typically around 700°C.

Downdraft gasifiers have the same general constraints on feed properties as updraft gasifiers. The feed needs to have a fairly uniform particle size distribution, with few fines to maintain bed physical properties and minimise channelling. The feed needs to have low ash with a high fusion temperature to prevent slagging. In addition, the feed moisture content needs to be less than about 20 per cent to maintain the high temperatures required for tar cracking. A variation on the downdraft gasifier is the crossflow gasifier, in which air is introduced tangentially at the bottom of the gasifier. The operating principle of the crossflow gasifier is the same as for the downdraft gasifier.

## BUBBLING FLUID BED

In a gas–solid fluidised bed a stream of gas passes upwards through a bed of free-flowing granular materials in which the gas velocity is large enough that the solid particles are widely separated and circulate freely throughout the bed. During overall circulation of the bed transient, streams of gas flow upwards in channels containing few solids, and clumps or masses of solids flow downwards (Perry and Chilton 1973). The fluidised bed looks like a boiling liquid and has the physical properties of a fluid. In fluidised-bed gasification of biomass the gas is air, oxygen or steam, and the bed is usually sand, limestone, dolomite or alumina.

A cyclone is used either to return fines to the bed or to remove ash-rich fines from the system. The bed is fluidised by a gas distribution manifold or series of sparge tubes (Hansen 1992). Biomass is introduced either through a feed chute to the top of the bed or through an auger into the bed. In-bed introduction provides residence time for fines that would otherwise be entrained in the fluidising gas and not converted in the bed.

The bed is usually preheated using an external burner fired by natural gas, propane or fuel oil. Fluidised-bed gasifiers have the advantage of extremely good mixing and high heat transfer, resulting in very uniform bed conditions. Gasification is very efficient, and 95–99 per cent carbon conversion is typical. Bubbling fluidised-bed gasifiers are normally designed for complete ash carry-over, necessitating the addition of cyclones for particulate control.

## CIRCULATING FLUID BED

If the gas flow of a bubbling fluid bed is increased, the gas bubbles become larger, forming large voids in the bed and entraining substantial amounts of solids. This type of bed is referred to as a turbulent fluid bed (Babcock and Wilcox 1992). In a circulating fluid bed, the turbulent bed solids are collected, separated from the gas, and returned to the bed, forming a solid circulation loop. A circulating fluid bed can be differentiated from a bubbling fluid bed in that there is no distinct separation between the dense solids zone and the dilute solids zone. Circulating fluid bed densities are about 560 kg/m$^3$ compared to a bubbling bed density of 720 kg/m$^3$ (Babcock and Wilcox 1992). To achieve the lower bed density, gas rates are increased from the 1.5–3.7 m/s of bubbling beds to about 9.1 m/s. The residence time of the solids in a circulating fluid bed is determined by the solids circulation rate, the attrition of the solids, and the collection efficiency of the solids separation device.

## ENTRAINED-FLOW

In entrained-flow gasifiers, pulverised feed is fed dry or in a slurry continuously into a pneumatic-flow reactor along with a relatively large amount of oxygen. The high temperatures caused by added oxygen almost completely destroy oils and tars. The high temperature (typically 1,300–1,400°C) also means that the ash is typically removed as a liquid slag. These gasifiers have been developed for coal, and only very limited testing with biomass has been performed. There are a number of reasons for the lack of application to biomass, but the high cost of feed preparation to reduce moisture content to low levels and reduce the particle size is the primary concern (Larson and Katofsky 1992). Entrained-flow reactors have been developed for coal by Shell, Texaco, and Koppers-Totzek.

### 9.2.3 Large-scale gasifiers

A number of large-scale gasifiers are being developed in the United States and Europe, primarily for application in the heat and power market. The Foster Wheeler low-pressure circulating fluid-bed gasifier has been in commercial operation in Finland and Sweden for a number of years, providing gas for lime kiln operation (Wilen and Kurkela 1997). The following discussion includes brief comments on a number of gasifiers in development.

#### 9.2.3.1 IGT/Carbona

The Institute of Gas Technology (IGT) has developed the RENUGAS® gasification technology, specifically for the conversion of biomass to low- or medium-heating-value gas (Lau *et al.* 1993). Biomass is fed to a single pressurised, bubbling, fluidised-bed-gasifier vessel. Inert alumina beads form the deep fluidised bed and provide stable fluidisation behaviour and needed heat capacity for efficient transfer of energy released by the combustion to endothermic devolatilisation and gasification reactions. The use of a deep single-stage bed of inert solids yields high carbon conversion. Feed is introduced into the fluidised bed by means of a high-speed screw. The process has been tested during more than 250 hours of steady-state operation at feed rates up to a 10.9 Mg/day and at pressures up to 3.45 MPa. Parameters studied have included gasification temperature, pressure, feed moisture, feedstock type, steam input, bed media, fluidised bed height, and gas superficial velocity. Biomass feedstocks gasified include maple wood chips, whole tree chips, California highway clippings, paddy rice straw, refuse-derived fuel, bark and paper mill sludge, bagasse from Hawaii, and alfalfa.

The RENUGAS® technology was scaled up to 91 Mg/day by the Pacific International Center for High Technology Research at the Hawaii Commercial and Sugar Company's sugar mill in Paia, Maui, Hawaii. Phase I of the Hawaii project was completed with the gasifier being operated longer than 100 hours at throughputs as high as 50 Mg/day and pressures as high as 1.14 MPa. A second phase of the Hawaii project was led by Westinghouse Electric Corporation, with the objective of testing a Westinghouse hot-gas clean-up filter system on a 10 per cent slip-stream basis. The system operated for 170 hours in 1997. The facility is now shut down.

#### 9.2.3.2 Carbona/Kvaerner

A variation of the IGT RENUGAS® (U-Gas) technology was developed for coal and then extended to various biomass feedstocks. The original developer was Tampella Power, Inc., through a joint venture with the Swedish utility Vattenfall, called Enviropower. In 1996 Tampella was purchased by Kvaerner. The employees of Tampella working in the gasification area formed a small company called Carbona Corporation. The Kvaerner/Carbona pressurised fluid-bed gasification pilot plant is based on IGT's U-Gas gasifier and includes all essential modules for research, component testing, and completing the development of the process for IGCC (Integrated Gasification-Combined Cycle) applications. The maximum thermal input is 15 MW. The gasifier, a single-stage pressurised fluid-bed gasifier that produces a low-calorific-value gas suitable for combustion in a gas turbine, can be operated at up to 3.04 MPa and 1,100°C. More than 3,000 Mg of biomass, primarily wood, have been gasified in more than 9,000 hours of operation.

The Kvaerner/Carbona technology has been selected as the gasifier manufacturer for the Minnesota Valley Alfalfa Producers (MNVAP) alfalfa gasification project in Granite Falls, Minnesota, USA. The Minnesota 'Alfagas' project is a cooperative research and development agreement between the US Department of Energy (DOE) and MNVAP, a farmer-owned cooperative, to produce protein (leaf meal) and electricity from alfalfa in a US$200 million multi-phase project. Other partners in the project include Enron, Carbona Corporation, Kvaerner Pulping, Westinghouse Electric Corporation, Stone and Webster, Great River Energy and the University of Minnesota. With the eventual construction of this new Agri-Power facility, the producers hope to produce electricity at a 75-MW scale from the stem portion of 640,000 tonnes per year of alfalfa, and about 320,000 tonnes per year of high-protein leaf pellets. A power purchase agreement has been signed with a local utility, Northern States Power, and detailed design is under way. Plant completion is projected to be in 2001.

### 9.2.3.3 *Foster Wheeler Energia OY (Ahlstrom)*

Foster Wheeler Energia OY (formerly A. Ahlstrom OY, Finland) has been developing a high-pressure circulating-fluid-bed gasifier for a number of years. This development has culminated in the construction and operation of a demonstration unit in Värnamo, Sweden. The capacity of the unit is approximately 82 tonne/day and uses ceramic filters for hot-gas clean up. In combined-cycle operation the generating capacity of the plant is about 4 $MW_e$ from a Alstom Gas Turbines Ltd (formerly European Gas Turbines) Typhoon series gas turbine and 2 $MW_e$ from a bottoming steam cycle (Ståhl and Neergaard 1998). Approximately 9 $MW_{th}$ of hot water is also produced for district heating. The gasifier started operating in 1993, and in the autumn of 1995 the first turbine operations commenced.

Foster Wheeler Energia OY (FW) has also supplied four commercial-scale atmospheric pressure circulating-fluid-bed gasifiers to the pulp and paper industry, with capacities of 17–35 $MW_{th}$ (Nieminen 1998). FW is presently operating a 43 $MW_{th}$ gasifier at Kymijärvi power plant in Lahti, Finland. The Kymijävi power plant is a 250 $MW_{th}$ pulverised coal boiler plant which typically co-fires 38 per cent natural gas by heat input. The gasifier operates as a low-pressure, air-blown, circulating fluid bed gasifier with air heating (from product gas) and no feed drying. Feed consists of mixtures of sawdust, wood residues, and recycled fuel (plastics, cardboard, paper, wood). With 50 per cent moisture-feed the product gas heating value is about 2.2 MJ/kg. The product contains 800–1,000 mg/$Nm^3$ $NH_3$ and 25–40 mg/$Nm^3$ HCN. Product gas is fed to the pulverised coal boiler at burners located below the existing coal burners. When biogas is substituted for coal at the 15 per cent heat input level, the boiler emissions change in the following manner: $NO_x$ decreases by about 10 mg/MJ, $SO_x$ decreases by about 20 mg/MJ, HCl increases by 10 mg/$Nm^3$, CO remains unchanged, and particulate matter decreases by 20–10 mg/$Nm^3$.

### 9.2.3.4 Thermiska Processer

The Thermiska Processor (TPS) gasifier process is a low-pressure, air-blown, circulating-fluid-bed design in which the primary gasification step is followed by the second circulating-fluid-bed cracker. In the cracker dolomite (mixed magnesium–calcium carbonate) is used to catalyse the conversion of tars to gases and lower molecular weight vapours. Tar cracking reduces the amount of organics removed from the product gas stream during water quenching operations, and thus maximises useful product recovery and minimises water-treatment costs. The TPS process has been in development for a number of years. In 1992, Ansaldo Aerimpianti SpA of Italy (Barducci *et al.* 1997) installed a commercial, two-bed unit in Greve-in-Chianti, Italy. The Greve units have a combined capacity of 30 $MW_{th}$, burning pelletised refuse-derived fuel. Limited operations have also been performed using hogged wood or agricultural wastes. Fuel gas generated at the plant is either burned in a boiler to produce steam for electricity production or used as fuel in a nearby lime kiln operation.

TPS has been selected to supply technology for a biofuel-based combined cycle power plant in Brazil. The project, co-funded by the World Bank Global Environment Facility and a number of Brazilian utilities, led by Eletrobras, will produce electricity from eucalyptus in a 32 $MW_e$ combined-cycle facility. TPS technology is also being used for an 8 $MW_e$ demonstration project in Yorkshire in the United Kingdom, called the Arbre Project. The clean syngas leaving will be compressed and injected into a gas turbine, an Alstom Typhoon. The hot gas turbine exhaust gas will be fed to a boiler where it will be combusted with the bypass syngas to raise steam for a steam turbine. The gross plant output is expected to be 10 $MW_e$, 4.75 $MW_e$ from the gas turbine and 5.25 $MW_e$ from the steam turbine. Parasitic loads will be 2 $MW_e$, giving 8 $MW_e$ for export to the local grid.

Similar low-pressure, air-blown gasifiers have also been developed by Lurgi (Merhling and Vierrath 1989). The Lurgi low-pressure circulating-fluid-bed gasifier is being used in the Bioelettrica, SpA Energy Farm Project in Cascina, Italy (southeast of Pisa), with funding provided by the European Union Thermie Programme. The unit size is a 41 $MW_{th}$ wood-fired gasifier, producing 12 $MW_e$ in combined cycle mode. Construction was due to start in May 1999, with commissioning planned for early 2001.

### 9.2.3.5 Battelle Memorial Laboratory

Since 1977 the Battelle Memorial Laboratory (Paisley and Overend 1994; Paisley *et al.* 1992), Columbus, Ohio, has been developing an indirectly heated biomass gasification process, called the Battelle High Throughput Gasification Process. A 2 $MW_{th}$ gasifier has been operating in West Jefferson, Ohio, since 1980. The Battelle gasification process produces a medium-calorific-value gas without using air or oxygen in the gasification step. The process uses two physically separate reactors, a gasification reactor in which biomass is converted into a medium-calorific-value gas and residual char, and a combustion reactor that burns the residual char to heat sand, which is circulated back to the gasifier to provide heat for gasification. The Battelle process utilises circulating-fluid-bed reactors to take advantage of the inherently high reactivity of biomass feedstocks and has demonstrated throughputs in excess of 3.90 kg/s-m². 

The Battelle process is being scaled to 42 $MW_t$ (182 dry tonne/day) as part of a major US DOE initiative to demonstrate gasification of renewable biomass for electricity

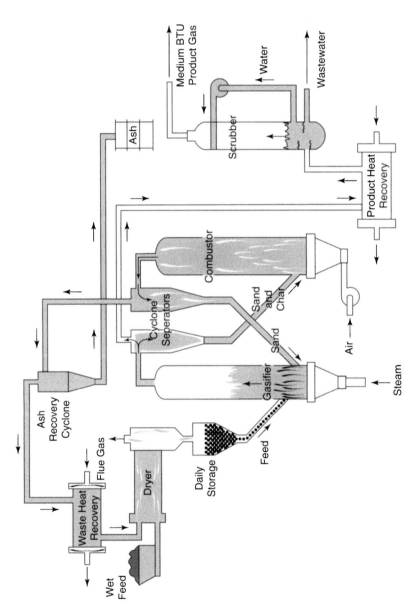

*Figure 9.1* Schematic representation of the Battelle/FERCO gasification process

production. The demonstration and validation of this gasification/gas turbine system is being undertaken at the existing 50 MW$_e$ wood-fired McNeil Power Generating Station in Burlington, Vermont. The objective of the Vermont project is to achieve a significant scale-up of Battelle's indirect gasification technology from a process development unit of 12 tonne/day throughput to a scale that would be large enough to handle commercial-size components (200 tonne/day or 45 MW$_{th}$). The economic analysis of the system in combined cycle at the 50–70 MW$_e$ scale showed attractive returns (Paisley *et al.* 1997).

In the process development unit, the scrubber is a once-through liquid design. Because of the site environmental constraints at McNeil, the scrubber is self-contained and consists of an atomiser quench, followed by a venturi scrubber – water balance is obtained by evaporating the excess water into the product gas stream and then to the McNeil boiler. There are no aqueous effluents from the gasifier. Solids (char fines, elutriated bed materials and other solids, and tars) that accumulate in the scrubber circuit are pumped as a slurry to the combustor for evaporation and incineration.

### 9.2.3.6 Producer Rice Mill Energy Systems

Producer Rice Mill Energy Systems (PRM) has developed a gasification system primarily targeted at processing rice hulls. The PRM system is a fully commercial gasification system, with thirteen operating systems in the United States, Australia, Costa Rica, and Malaysia, ranging in size from 30 to 300 tonne/day. Because of the size range, the PRM gasifier can be classified as either large or small. PRM gasifiers use an agitated bed with three air zones under a perforated grate, and are normally close-coupled to a boiler. Other feedstock, including sawdust, bark, and MSW, has been investigated.

### 9.2.4 Small-scale gasifiers

Many small gasification systems – for purposes of this discussion, systems less than 5 MW$_e$ size – have been developed since the beginning of World War II, with varying degrees of success. Because of the lack of access to fossil fuels, many of the European countries in World War II relied on small-scale gasifiers for civilian transportation. More than one million gasifiers were in use at that time (Reed and Gaur 1998). After World War II small-scale biomass gasification fell out of favour because of the convenience and cost of using petroleum products. With the large increase in oil prices in the mid-1970s, interest in small-scale gasification re-emerged, primarily for application in developing countries and for district heating applications in Europe.

Development of gasifiers for power applications in the late 1970s–early 1980s was supported by the World Bank, but the majority of projects were not successful, for various technical and commercial reasons. During the 1990s interest has again grown, with the primary driver being concerns over the use of fossil fuels and their impact on global warming. Units have been developed and tested in Brazil, China, India, Europe and the United States. Most systems are fixed-bed gasifiers, mostly downdraft gasifiers. A number of reviews of small-scale gasification systems have recently been published, including a survey of biomass gasification (Reed and Gaur 1998); a review of the World Bank projects in Brazil, Burundi, Indonesia, Mali, the Philippines, Seychelles, and Vanuatu (Knoef and Stassen 1997); the status of small-scale gasification in Europe (Novem 1996); and a review of prime movers for small-scale systems (Jakobsen *et al.*

1998). A summary of gasifiers/projects is given in Table 9.1. This table is by no means complete. A brief discussion of a few example systems follows below.

### 9.2.4.1 Bioneer gasifier

The Bioneer gasifier was developed in Finland in the late 1970s by VTT to replace imported fuels with low-cost indigenous fuels, such as peat and wood. Research and development were carried out through the mid-1980s, and in 1985–86 a number of commercial Bioneer plants were commissioned. One additional plant was constructed in 1996 by Foster Wheeler Energia OY. The Bioneer gasifier is a fixed-bed updraft gasifier that produces high levels of tar. The Bioneer gasifier systems do not incorporate a gas quench and clean-up system and are unsuitable for use in gas engines or gas distribution. In all the systems, hot product gas is combusted immediately in a gas boiler to generate steam or hot water for district heating.

### 9.2.4.2 Rural Generation Limited

Rural Generation Ltd, Londonderry, Northern Ireland, has commercialised the downdraft gasifier (originally developed at Enniskillen, Northern Ireland), as a combined heat and power system to produce electricity plus heat for drying grain and wood and for home heating. The gas produced from the gasifier is used as fuel for an IVECO diesel engine operated in a dual fuel mode (diesel plus biogas). Gas clean-up is accomplished with a cyclone, water bubbler and dry filter in series. The feed is 80 kg/hr of wood residues, switching to willow in the winter of 1999. Power production is 115 kWh/hr, with 80 kWh/hr coming from biogas; the percentage biogas is limited by allowable engine inlet temperature and some gas is flared. The unit has about 1,500 hours of operation to date.

### 9.2.4.3 Free University of Brussels/Dinamec

The Free University of Brussels and DINAMEC have teamed up to build and operate a prototype gasification system to power an indirectly-heated gas turbine. The scale of the project is 500 kW$_e$ in a co-generation mode using wood waste as fuel. Biogas is produced in the gasifier and then is fed to a close-coupled combustor. Hot flue gas from the combustor is fed to a heat exchanger to heat exhaust air from the compressor section of a gas turbine. The heated compressed air is then fed to the expansion turbine to produce power. The flue gas exiting the heat exchanger is then used to heat water for district heating application. The exhaust air from the turbine is used as combustion air in the combustor. The advantage of such a system is that there is no turbine contamination. A potential disadvantage is lower efficiency than comparable producer gas-fired direct turbine systems.

### 9.2.4.4 Indian gasifiers

A number of gasifiers have been developed in India in the 1980s and 1990s, with the prime driving force being a field demonstration programme of the Indian Ministry of Non-Conventional Energy Sources. Hundreds of small downdraft gasifier systems in

Table 9.1 Small-scale gasifier listing (Knoef and Stassen 1997; NOVEM 1996; Reed and Gaur 1998)

| Company/site | System capacity | Reactor type | Gas clean-up | Prime mover | Fuel | Comments |
|---|---|---|---|---|---|---|
| American High Temperature, Germany | up to 1,000 kW | downdraft | NI | NI | wood | demonstration. CHP |
| Ankur Scientific, India | to 500 kW | downdraft | | | wood, rice hulls, stacks, shells | |
| Balong, Indonesia | 20 kW | downdraft | cyclones, impingement cooler, filters | diesel | wood, rubber | community electricity |
| Bioneer, Finland | 1–15 MW | updraft | none | steam | wood | district heating |
| B9 Energy Biomass, N. Ireland | 200 kW (2–100 kW in parallel) | downdraft | proprietary | diesel | wood | electricity, district heating |
| Bolo, Philippines | 38 kW$_{me}$ | downdraft | cyclone, scrubber, filter | diesel | charcoal | irrigation |
| Cemig, Brazil | 15 kW, 1–2 MW | charcoal blast furnace gas | NI | NI | charcoal | |
| Chacra, Brazil | 23 kW | crossdraft | cyclone, cooler, paper filter | otto | charcoal | irrigation |
| Chiptec Wood Energy Systems, USA | <10 MMBtu/hr | updraft | cyclone | tested with Stirling engine | wood | institutional heating applications |
| Community Power Corporation, USA | 35 kW | downdraft | scrubber | tested with Stirling engine | wood pellets | research unit |
| Cratech, USA | 1.2 MW$_e$ | pressurized fluid bed | hot gas filter | gas turbine (?) | cotton gin trash | research unit |
| DASAG, Switzerland | 20–2,500 kW$_t$ | stratified downdraft | NI | NI | wood, agricultural wastes | |
| DML Dieselmotorenwork Leipzig | up to 2,000 | downdraft | NI | NI | various | commercial |

Table 9.1 continued

| Company/site | System capacity | Reactor type | Gas clean-up | Prime mover | Fuel | Comments |
|---|---|---|---|---|---|---|
| Dogofiri, Mali | 160 kW$_e$ | downdraft | ash flushing, scrubber, filters | otto | rice husks | industrial electricity |
| ENEA, Policoro, Italy | 15 kW$_t$ | downdraft | heat exchanger, filter, strippper | gas turbine | hazelnut shells, wood chips, wood pellets | research unit 280 kg/hr unit under construction for project in China |
| EPA Research Triangle Park (Mechem), USA | 1 MW | downdraft | cyclone, coolers, separators, filters | 800 kW Waukesha engine | wood | test unit at Camp LeJeune, NC |
| Fluidyne, New Zealand | 35 kW | downdraft | cyclone, cooler, filter | diesel | wood blocks and chips | |
| Free University of Brussels | 500 kW | fluid bed | cyclone | indirectly heated turbine | wood | demonstration |
| India Institute of Science, Bangalore | 3.7 kW to 5 MW | downdraft | NI | NI | wood chips, straw | |
| India Institute of Technology, Bombay | 3.7 kW | downdraft | NI | NI | dung briquettes | research unit |
| Itamarandiba, Brazil | 40 kW | crossdraft | cyclone, cooler, filter | otto | charcoal | industrial electricity |
| Kara, Netherlands | 35–150 kW | downdraft | NI | diesel (?) | wood | electricity |
| PRM Energy Systems (also Prime Energy), USA | 10–1,000 tpd | updraft | NI | NI | rice hulls | 13 commercial systems |
| Reflective Energies, USA | 100–1,000 kW | downdraft | NI | microturbine | TBD | research unit |
| Rural Generation Ltd, N. Ireland | 100 kW | downdraft | cyclone, scrubber, filter | diesel engine | wood, willow | electricity and district heating |
| Shandong Energy Research Institute, China | 500 Nm³/h gas | downdraft | NI | NI | ag residues | gas for villages |

Table 9.1 continued

| Company/site | System capacity | Reactor type | Gas clean-up | Prime mover | Fuel | Comments |
|---|---|---|---|---|---|---|
| Shawton Engineering, UK | 100 kW | downdraft | NI | otto | waste wood | research unit |
| Societe Martezo | 100–600 kW | downdraft | NI | NI | wood | electricity |
| Stwalley Engineering, USA | 40 kW | downdraft | NI | NI | wood chips, corn cobs | |
| Tata Energy Research Institute, India | 40 kW | NI | NI | NI | wood | |
| Technical University of Denmark | 400 kW | downdraft | NI | NI | straw | electricity |
| Tora, Burundi | 36 kW | downdraft | cyclone, coolers, scrubber, oil filter | diesel | peat | industrial electricity |
| Thermogenics | 20 TPD | inverted downdraft | mechanical cleaner, ESP | diesel or SI engine | wood, RDF, tyres | |
| Volund R&D | 4–6 MW | updraft | cyclone. condenser | steam | wood | includes district heating |
| Zaragoza, University of, Spain | 1,050 kg/h | fluid bed | NI | NI | wood chips, energy crops | research unit |
| Zaragoza, University of, Spain | <300 kg/h | downdraft | NI | NI | wood chips | research unit |

Note
NI = No information available

the 4–500 kW (shaft power) range have been constructed and operated, primarily to fuel compression engines for thermal energy, shaft power or electricity. The majority of the applications are in the agricultural sector, for things such as irrigation water pumping. Organisations involved in gasification in India include the India Institute of Technology (Bombay), the Indian Institute of Science (Bangalore), and Ankur Scientific (Baroda). A good overview of Indian biomass gasification systems is given by Professor Parikh of IIT (Parikh *et al.* 1994). Other references on Indian gasification development include Mukunda *et al.* (1993), Parikh *et al.* (1989), and Gaur *et al.* (1985).

As an example, Ankur Scientific has built more than 250 downdraft gasifiers of varying sizes. The largest is a 500 kW downdraft gasifier, designed for agricultural and small industrial applications. The major system components include the biomass feeding system, the gasifier, cyclone, scrubber, and filter (for engine applications).

### 9.2.4.5 Chinese gasifiers

Small-scale gasification development is led by the Shandong Academy of Sciences (Overend 1998). Biomass (wood, wood wastes, wheat and rice straw, and manure) is converted into a low-calorific-value gas using downdraft gasifiers, contaminant particulates and tars removed, and delivered to households through centralised gas-supply systems for small villages. Fourteen gasification systems are operating in Shandong Province, and a demonstration system for 100–200 households exists in the Beijing area. The XFF-2000 downdraft gasifier produces 500 $Nm^3/hr$ of gas at a reported efficiency of 72–75 per cent. The Chinese are also investigating the production of electricity using compression engine and Stirling engine gensets (see also Section 9.3.3.4).

### 9.2.5 Methanol and hydrogen production

This section reviews the thermochemical production of methanol and hydrogen from biomass, and briefly discusses process economics. Gasification of biomass to produce a synthesis gas followed by catalytic methanol or hydrogen synthesis is the technology selected for study and development, based on the current possibilities, the potential for short-term impact, and the significant level of industrial interest. A detailed discussion of the catalytic chemistry of steam reforming, water gas shift, and methanol synthesis will not be given. For information on the catalytic chemistry of such processes, refer to Twigg (1989). For an overall discussion of synthesis gas conversion to liquid fuels refer to Mills (1993).

### 9.2.5.1 Methanol production technology description

The thermochemical production of methanol from biomass involves the production of a synthesis gas rich in $H_2$ and CO which is then catalytically converted into methanol. Synthesis gas is produced by thermal gasification. The unit operations involved in methanol production from biomass are divided into the following major areas: 1) feed preparation; 2) gasification; 3) synthesis gas modification; and 4) methanol synthesis and purification.

The equipment downstream of the gasifier is the same as that used in producing methanol from natural gas. The syngas from the gasifier, after suitable cleaning, reforming and upgrading, is used for methanol synthesis. Finally, the methanol product goes through the methanol purification stage, which removes water and organic products present to different degrees, according to the final use of the methanol.

To summarise technology status, a number of gasifiers that have the potential to produce a synthesis gas suitable for methanol synthesis are under development. These gasifiers are operating in the 4–200 Mg/d scale. All systems under development are designed to produce a low- to medium-calorific-value fuel gas. None of the systems has been operated on an integrated process basis to determine operating parameters necessary for maximum methanol production. All downstream synthesis gas operations are commercial technologies in which operating conditions and yields are known. New developments, such as single-pass liquid-phase methanol synthesis, also have potential.

### 9.2.5.2 Methanol energy balances and economics

In this section preliminary energy balances and economics for methanol production from biomass are presented. The costs of production represent values considered feasible using a commercial gasifier for coal and values potentially obtainable given successful commercialisation of gasifiers presently being developed specifically for biomass. The economics of methanol production from biomass are based on a study performed by Chem Systems (1990) for the former Solar Energy Research Institute, now the National Renewable Energy Laboratory (NREL). In this study preliminary economics were developed for two methanol production systems, a system using the Koppers-Totzek (K-T) gasifier, designated LPO, and a system using the IGT-RENUGAS® gasifier, designated HPO. In addition, preliminary economics have been developed by NREL for an indirect methanol production system, based on the Battelle Columbus Laboratory (BCL) gasifier (Bain 1991; Wan and Fraser 1989), designated BCL. These economics were also compared to economics developed by Princeton University (DeLuchi et al. 1991; Larson and Katofsky 1992). Princeton looked at the IGT process, BCL, Wright Malta, and Shell gasifiers.

Process energy efficiencies for conceptual biomass systems (LPO, HPO, BCL) are 49–68 per cent (see Table 9.2). These efficiencies compare with 63 per cent for a typical natural gas-based system and 48 per cent for a second-generation coal-based system. Carbon conversion efficiencies reflect the relative hydrogen/carbon ratios of the

*Table 9.2* Methanol production energy balances

| Process | Biomass LPO | Biomass HPO | Biomass indirect | Natural gas | Coal |
|---|---|---|---|---|---|
| Tonne MeOH/day | 790 | 920 | 1,110 | 2,500 | 5,000 |
| Tonnes feed/tonne MeOH | 2.30 | 1.97 | 1.63 | 0.64 | 1.76 |
| Feed LHV, GJ/tonne | 18.87 | 18.87 | 18.87 | 52.25 | 24.75 |
| MeOH LHV, GJ/tonne[a] | 21.14 | 21.14 | 21.14 | 21.14 | 21.14 |
| Energy efficiency, % | 48.7 | 56.9 | 68.4 | 63.3 | 48.5 |
| Carbon utilisation, % | 32.6 | 38.0 | 45.9 | 79.2 | ND[b] |

*Notes*
[a] Methanol vapour
[b] Not determined

individual feedstocks and range from a low of 33 per cent for an LPO biomass-based system to 79 per cent for a natural gas-based system.

The cost of methanol from biomass is estimated at US$8.22–19.84/GJ methanol (1998 USD). Gasification processes range from indirectly-heated gasifiers to high-pressure oxygen-blown gasifiers, and plant sizes are 900–4,600 GJ/h. Methanol from natural gas is estimated at US$5.24/GJ for a 2,075 GJ/h facility, and methanol from coal gasification at US$16.76/GJ for a 4,150 GJ/h facility. To put these production costs in perspective, methanol prices are historically based on fuel prices and are seasonal. The present market price is about US$4.45/GJ (US$0.265/gal) (*Chemical Marketing Reporter* 1999), which is substantially below the estimated production costs.

There is significant scatter in the estimated methanol costs, but some general conclusions can be reached. First, the cost of methanol from biomass is significantly higher than the cost of methanol from natural gas. Also, although there are differences between estimates performed by different organisations, the results indicate that the indirectly-heated gasifiers give lower production costs than the directly-heated gasifiers in a methanol production process. Lastly, the results show that methanol economics from biomass compete favourably with coal-based system economics.

### 9.2.5.3 *Hydrogen production*

Hydrogen production is a commercial technology in the petroleum-refining industry, where steam-reforming and purification technology is used to convert light hydrocarbons such as natural gas, refinery fuel gas, liquefied petroleum gas/butane and light naphthas to hydrogen for hydrotreating, hydrocracking, or other refinery, petrochemical, metallurgical, and food processing uses. A number of vendors, such as Foster Wheeler, Haldor-Topsøe, and Howard Baker offer units (Hydrocarbon Processing 1992 and 1994). Heavier hydrocarbons such as fuel oil can also be used to produce hydrogen by partial oxidation (Gary and Handwerke 1984; *Hydrocarbon Processing* 1994) such as that offered by Texaco. For synthesis gas conversion steam reforming is the option considered.

In older steam-reforming units (Gary and Handwerke 1984), four processing steps are involved: 1) steam reforming to convert methane to CO and $H_2$; 2) shift conversion to convert CO to $H_2$; 3) gas purification involving the adsorption of $CO_2$; and 4) methanation to remove the remaining small amounts of $CO_2$ and CO. With the advent of commercial pressure swing adsorption technology (*Hydrocarbon Processing* 1994) new steam-reforming processes are somewhat different. Biogas at elevated pressure (typically 3.5 MPa) is steam reformed and then the product passes through high- and low-temperature shift reactors to remove CO. The gas exiting the low-temperature shift reactor is then fed to a PSA unit to produce high-purity hydrogen. The offgas from the PSA is used as fuel in the steam reformer.

Hydrogen production cost estimates from biogas have been developed by Princeton University (Deluchi *et al.* 1991; Larson and Katofsky 1992) and NREL. Hydrogen production costs range from US$8.71/GJ $H_2$ for a 1,650 Mg/d feed Battelle gasifier-based process to US$17.46/GJ $H_2$ for a 1,000 Mg/d feed IGT-based process. Comparable hydrogen prices for natural gas steam-reforming process range from US$6.26/GJ $H_2$ for a 1,120 GJ $H_2$/d process to US$7.90/GJ $H_2$ for a 712 GJ $H_2$/d process (Padró and Putsche 1999).

## 9.3 Electricity production from biomass

*Arnaldo Walter, André Faaij and Ausilio Bauen*

### 9.3.1 Introduction

The current interest regarding the production of electricity from biomass is due mainly to factors relating to environmental issues, especially global warming. In developing countries particularly, electricity consumption is forecast to increase dramatically during the next 20–30 years (rates of up to 10 per cent per year are estimated). Also in these countries about two billion people are living away from the grid, while financial constraints cause another significant proportion of their populations to have a low electric consumption level. To avoid a dramatic increase in environmental impacts, on both global and regional levels, renewable sources of energy need to contribute substantially to the future expansion of electric capacity, and in this sense biomass is a good option.

Technologies of electricity production from biomass can be classified in two broad groups: those based on direct biomass combustion and those that use derivative fuels – gas or liquid fuels. Within the first group are the power plants based on steam cycles, including those in which biomass is mixed with a fossil fuel before or during burning (co-firing). Within the second group are the technologies based on biomass gasification or pyrolysis, integrated to gas turbines, internal combustion engines or fuel cells.

The use of derivative fuels provides some advantages. One important issue is that biomass to electricity efficiencies could be higher through better combustion and higher efficiency of the devices used into power generation – gas turbines, internal combustion engines and, further, fuel cells. Owing to the better combustion, lower emissions can also be expected. Moreover, scale effects on initial investment are stronger in power plants based on steam cycles (based on direct biomass combustion), affecting the feasibility of small capacity plants. This constraint could be reduced in plants based, for instance, on gas turbines.

### 9.3.2 Technologies of electricity production based on direct combustion of biomass

#### 9.3.2.1 Conventional steam cycle power plants

Electricity production based on biomass steam cycle power plants is a commercial reality, particularly in the USA and Scandinavian countries. In the USA, for instance, these units were installed during the 1970s and 1980s (until recently there were about 7 GW connected to the grid, or about 8 per cent of all non-utility generating capacity) as result of incentives provided by PURPA (Public Utility Regulatory Policies Act). Lately, however, due to increases in biomass costs and, most importantly, due to fiercer competition in the electricity industry, these power plants have gradually been decommissioned as the contracts signed have expired.

In a general sense, these power plants present low thermal efficiency (14–18 per cent range, with the best values in the 20–25 per cent range) and are not competitive *vis-à-vis* most other sources of power generation. Their low efficiency is basically due to the fact that these plants do not operate with state-of-the-art boiler parameters (steam pressure and temperature); in fact this is a strategy to reduce unit capital costs ($/kW installed). Conventional biomass plants need to be small in scale (usually less than 100 MW, and most of the plants in operation have net capacity up to 50 MW), since in a larger unit the feedstock hauling cost would be prohibitive. Conversely, steam cycles typically present a strong scale-dependence on the unit capital cost. To compensate this effect, less technology is incorporated in the plant (e.g. using lower quality material for boilers, reducing steam pressure and temperature, and using steam turbines with lower efficiency as well), affecting the biomass to electricity conversion efficiency. Additionally, in conventional biomass boilers, steam temperature is generally reduced because of concerns about formation of ash deposits and superheater corrosion from biomass ash (Farris *et al.* 1998).

Some technology improvements concerned with steam generation processes have been recently proposed or were even incorporated into existing biomass conventional steam power plants. For instance, with technologies like circulating fluidised beds and water-cooled vibrating grates boiler efficiency can reach up to 89 per cent (on LHV basis). As a matter of comparison, it should be noted that biomass boiler efficiencies are about 70–73 per cent in conventional existing plants. As a consequence of higher boiler efficiencies, overall biomass to electricity conversion efficiency can surpass 30 per cent (LHV basis) (van Den Broek *et al.* 1996).

Additionally, in the USA the Electric Power Research Institute (EPRI) has been developing the Whole Tree Energy® technology, that allows efficient burning of wood without previous cut or even any other wood preparation, except its drying to about 25 per cent moisture. EPRI estimates that a power plant in the range 50–100 MW could operate at 39 per cent overall efficiency (McGowin and Wiltsee 1996). Due to its higher combustion efficiency, lower investments and lower operational costs, electricity production costs are estimated to be 10–30 per cent lower than a conventional biomass steam power plant (Williams and Larson 1996).

### 9.3.2.2  Co-firing

The term co-firing has been applied to designated combined use of biomass and fossil fuel in power plants as well as in industrial steam boilers. The most accepted idea is burning a mix of biomass and coal in power plants that should be adapted to this purpose at the moment the useful life of the existing boilers expires. Owing to the substantial reduction in technical and economic risk, co-firing has been considered in some countries to be the first step for the enhancement of biomass use in power generation.

The amount of biomass to be burned depends on a detailed economic assessment, considering factors such as biomass availability, the feedstock hauling distance, effects on plant performance, and the investment in adapting steam boilers or, eventually, in new steam boilers.

Direct use of biomass without any pre-treatment (e.g. drying and milling) is not possible since coal is usually burned pulverised in large boilers. Thus, different technical concepts for biomass co-firing have been proposed. The simplest idea, but with some drawbacks, is to burn biomass in a separate combustion chamber and then transport the hot gases to the coal boiler. A variation of this concept involves the installation of a grate at the bottom end of the boiler hopper, adaptation that is not easy due to the obligatory modifications. The option that allows few changes regarding the existing boilers requires biomass grinding and subsequent burning of pulverised biomass in addition to coal. The drawback of this approach is that biomass should be previously dried to make the milling process easy. Another concept is based on biomass gasification followed by gas burning together with coal in boilers.

In addition, independent of the technical concept, co-firing could present some drawbacks concerned with changes of boiler behaviour, slagging of biomass ashes, and high temperature corrosion due to high chlorine content (Anderl and Mory 1998).

Two examples of the biomass gasification co-firing concept are the power plants of Lathi (Finland) and Zeltweg (Austria). The experimental project of Kymijärvi CHP (combined heat and power production – 167 $MW_e$ and 240 $MW_{th}$), in Lahti, is described by Nieminen (1998) and was mentioned earlier, in Section 9.2.3.3. Since 1998 biomass residues available in the Lahti region have been gasified and the fuel gas has been burned in the existing coal-fired boilers. The objective of this project is to prove that wet biomass can be gasified and the resulting very low-calorific gas can be co-fired successfully. The Zetweg power plant (overall capacity of 137 $MW_e$) has a gasifier for biomass residues (bark and sawdust) of a nearby forest industry. The gasifier began its operation in 1997, producing fuel gas that is burned in the existing pulverised coal-fired boiler. In the demonstration project conducted so far, 3 per cent of the total input of hard coal was replaced (Anderl and Mory 1998).

### 9.3.3 Technologies of electricity production based on derivative fuels of biomass

#### 9.3.3.1 Integrated biomass gasifier/gas turbine (BIG-GT) systems

Biomass gasification allows power production with the use of gas turbines as primer drivers. Gas turbines are power devices that have important attributes: reasonable thermal efficiency and initial capital costs that are less affected by scale effects. The integration of biomass gasifiers to gas turbine cycles has been generically designated as BIG-GT (Biomass Integrated Gasifier/Gas Turbine) technology.

Thermal efficiencies of power plants based on gas turbines can be substantially improved when the energy of high-temperature exhausted gases is recovered and used to boost power production, through a bottoming steam cycle (resulting in a combined cycle – BIG-CC). Steam injection cycles (BIG-STIG) were previously considered but constraints, due to combustion stability, can make this option infeasible. A scheme for a BIG-CC power plant based on a biomass direct-heated atmospheric gasifier, according to the technology proposed by TPS (see Section 9.2.3.4) is shown in Figure 9.2.

A gas turbine cycle with external combustion of biomass is also possible, but this alternative has not been seriously considered so far. Drawbacks concerned with lower thermal efficiency and very high costs of a special air-combustion gases heat exchanger can be anticipated. The prototype developed by the Free University of Brussels/Dinamec (500 $kW_e$), previously mentioned in Section 9.2.4.3, is one of the R&D projects on this technology. Previous feasibility studies on external combustion gas turbines have been carried out by the Swedish Royal Institute of Technology (Eindensten et al. 1996).

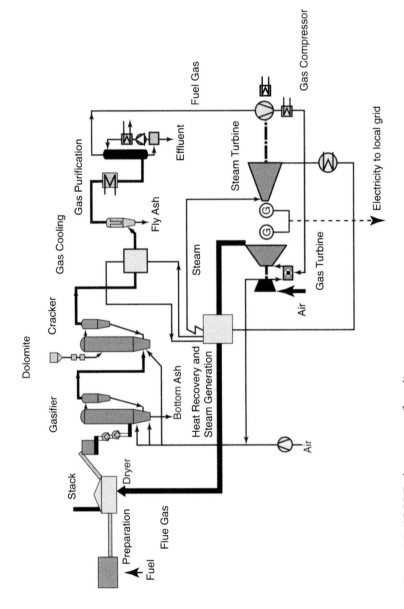

*Figure 9.2* ARBRE plant process flow diagram
*Source:* ARBRE Energy Ltd

## CURRENT STATUS OF BIG-GT TECHNOLOGY

Some biomass gasification projects which are aimed at the provision of electricity are under development at present. Table 9.3 gives an overview of the most important BIG-GT demonstration projects world-wide. The Värnamo plant, in Sweden (already mentioned in Section 9.2.3.3), is the first of its kind in the world and the only one, so far, to have full integrated operation of the biomass gasifier and gas turbine (almost 1,000 hours of integrated operation by March 1997). An extensive demonstration and development programme has been carried out since 1996, when the unit was finally commissioned, aimed at improving the project and providing more knowledge about the technology (Ståhl and Neergaard 1998).

The main technological issues in the demonstration of BIG-GT technology are concerned with scaling-up of gasifier technologies, gas cleaning and gas turbines adaptation to low-heating-content fuel. To date, the major biomass gasification processes have a capacity of biomass conversion of around 10 tonnes per day (tpd). To achieve a commercial stage, units with a capacity of at least 100 tpd need to operate efficiently and reliably (Overend *et al.* 1996).

To prevent environmental impacts and damages to the components downstream of the gasifier, fuel gas should be cleaned before its combustion in the gas turbine (see also Section 9.2.2.2). Gas quality has to meet strict standards (see Table 9.4) with respect to alkali, tars, and particulate contents to prevent erosion, fouling and corrosion of the gas turbine hot section. Low temperature gas cleaning (including scrubbing) involves commercially available components but still needs to be demonstrated for fuel gas from gasifiers. More challenging is the high temperature gas cleaning, in particular the removal of contaminants like nitrogen (present as ammonia in the fuel gas) and, when needed, sulphur, heavy metals and chlorine. Research focuses on ceramic filters, e.g. using catalytic material such as nickel for conversion of ammonia to molecular nitrogen. Nitrogen removal is an issue since ammonia will combust to form $NO_x$.

The low heating value gas (4–6 $MJ/Nm^3$) produced by direct gasification is not suited for unadapted gas turbines. Minimal modifications include adaptations of the fuel nozzles and the fuel manifold. Flame stability is crucial and should be sufficient to guarantee reliable operation of the gas turbine. This can only be demonstrated at full scale. The gas turbine industry has reasonable experience with blast furnace gas and limited experience with gas from coal gasification, but so far there is only the one real experience with biomass syngas at the Värnamo unit.

The main problems with the current demonstration projects seem to be the costs of initial first generation plants (see Table 9.3), and the difficulties with arranging a reliable biomass fuel supply for the lifetime of the project at a reasonable cost. A clear demonstration of those problems is provided by the Noord-Holland gasification project (30 $MW_e$), which was cancelled after it became clear that the total project costs would rise to levels of over 4,000 US$/$kW_e$ installed. Combined with disappointingly high biomass prices (based on limited supply) the costs became unacceptably high for the utilities involved.

Finding niches is crucial. These may especially be found for waste treatment (low fuel costs) and by installing BIG-GT technology at existing facilities, such as sugar mills, paper and pulp plants or existing biomass combustion facilities (as is done in Vermont, USA). Another interesting way to partly circumvent fuel supply risks and the difficulties

Table 9.3 Most important BIG-GT demonstration projects

| Project and location | Gasification process | Gasification technology | Biomass | Power cycle | Plant capacity | Estimated efficiency (%) (HHV) | Tar removal process | Final gas cleaning system | Investment cost – 1st plant | Proposal and status |
|---|---|---|---|---|---|---|---|---|---|---|
| Burlington, VT, USA | Indirect heating, low pressure, steam injection | FERCO/Batelle | Wood | Gas turbine | 12–15 MW$_e$ 42 MW (gasifier) | 30–35 | Catalytic, Dolomite | Cyclone and water quench | | Commercial, co-firing now, GT after |
| PICHTR – Pacific International Center for High Technology Research, Hawai, EUA | Direct heating, air or oxygen injection, pressurised, bubbling, fluidised bed | IGT/Renugas | Sugar-cane bagasse | Gas turbine | 3–5 MW$_e$, 10 t/d | 30–35 | NA | Ceramic filters | | Pilot plant, Gasifier tests |
| BDP – Brazilian Demonstration Project, Murici, BA, Brazil | Direct heating, air injection, atmospheric, circulating fluidised-bed | TPS | Wood chips (eucalyptus) | Combined cycle based on GE LM2500 | 32 MW$_e$ | 37 | Catalytic, Dolomite | Filter and wet scrubbing | 2.500 US$/kW (1998) | Commercial, to be constructed |
| ARBRE Energy (Arable Biomass Renewable Energy), Yorkshire, UK | Direct heating, air injection, atmospheric, circulating fluidised-bed | TPS | Short rotation coppice (willow and poplar) and forestry waste | Combined cycle based on EGT Typhoon | 8 MW$_e$ | 31 | Catalytic, Dolomite | Fabric filter and water scrubbing | 2100 ECU/kW (1998) | Demo, in construction |
| Energy Farm, Bioelletrica S.p.A., Cascina, Italy | Direct heating, air injection, atmospheric, circulating fluidised-bed | Lurgi | Wood chips (poplar, robinia) and agricultural waste | Combined cycle based on Nuovo Pignone PGT10B | 12.1 MW$_e$ | 32 | NA | Fabric filter and water scrubbing | 2650 ECU/kW (1998) | Demo, planned 2000/ Commercial operation in 2001 |

Table 9.3 continued

| Project and location | Gasification process | Gasification technology | Biomass | Power cycle | Plant capacity | Estimated efficiency (%) (HHV) | Tar removal process | Final gas cleaning system | Investment cost – 1st plant | Proposal and status |
|---|---|---|---|---|---|---|---|---|---|---|
| Biocycle, Finland | Direct heating, air injection, pressurised, fluidised bed | U-GAS Renugas | | District heating based on EGT Typhoon | 7.2 MW$_e$ 6.8 MW$_{th}$ | 40 (electricity) 77 (CHP) | NA | Ceramic filters | 4750 ECU/kW (1998) | |
| MVAP 'Alfafagas', Granite Falls, Minnesota, USA | Direct heating, air injection, pressurised, fluidised bed | Kvaerner/ Carbona based on IGT Renugas | Stems of alfafa | Combined cycle based on Westinghouse gas turbine | 75 MW$_e$ | 40.2 | NA | Ceramic filters | | Commercial, under design and planned 2001 |
| Sydkraft AB, Värnamo, Sweden | Direct heating, air injection, pressurised, fluidised bed | Bioflow/Foster Wheller Energy International | Wood residues and chips | Combined cycle/District heating based on EGT | 6 MW$_e$ 9 MW$_{th}$ | 32 (electricity) 83 (CHP) | Thermal cracking | Ceramic filters | 5350 ECU/kW (1998) | Demo, commissioned 1996 |

Sources: Babu (1995); Bridgwater (1995); Kaltschmitt et al. (1998); Lundqvist (1993); Naccarati and de Lange (1998)
Notes
NA: Information not available
FERCO = Future Energy Resources Corporation
MVAP = Minnesota Valley Alfafa Producers

*Table 9.4* Required gas quality for syngas use in gas turbines

| Contaminants | Tolerance level |
|---|---|
| Particulate (< 5 μm) | < 30 mg/Nm$^3$ |
| Tars | < 50 to 100 μg/Nm$^3$ |
| Alkali metals | 0.24 mg/Nm$^3$ |
| Alkali (K, Na) | 0.03 ppm (w) |
| Calcium (Ca) | 1 ppm (w) |
| Ash (2–20 μm: 7.5%; 0–2 μm: 92.5%) | 2 ppm (w) |
| Heavy metals (V, Pb) | 0.05 ppm (w) |
| Sulphur compounds (H$_2$S, COS, CS$_2$) | 20 ppm (w) |
| Halogen (HCl, HF) | 1 ppm (w) |

*Source*: Kaltschmitt *et al.* (1998) from Kloster *et al.* (1996) (TH Karlsruhe)

of adapting gas turbines with LCV gas is to combine fuel gas produced from biomass gasification with natural gas. The mixing of fuels can not only solve those problems, but also allows for gaining economies of scale (see e.g. Walter *et al.* 1998). This option has received relatively limited attention so far.

The full potential of BIG-GT technology as co-generation units in sugar-cane mills all over the world is estimated at about 40 GW. In Brazil alone, the biggest producer of sugar-cane (25 per cent of total production), the full potential was estimated as 10–12 GW for systems based on atmospheric and pressurised gasification technologies. This evaluation was made taking into account the sugar-cane production in the 1995–96 harvest season (240 million tonnes); the production in 1998 was 25 per cent higher (300 million tonnes) and the electricity generation potential would be enlarged in the same proportion (Walter and Overend 1999).

In the Kraft pulp and paper industry BIG-GT technology could be associated either with black-liquor gasification and bark or other mill residues gasification. Black-liquor gasifier/gas turbine co-generation has been considered a promising option as higher electrical efficiency with prospective environmental, safety and capital cost benefits is identified in comparison with conventional Tomlinson boiler-based co-generation systems. Depending on the gasification technology (low or high temperature, indirectly-heated or air- or oxygen-blown) and on the cycle design, net electricity production could be 2–2.5 times higher than current co-generation systems (Larson and Consonni 1997).

## THE LONGER TERM PROSPECTS FOR BIG-GT TECHNOLOGY

Despite the fact that BIG-GT technology still has to go through a commercialisation trajectory, with related high costs, it may be of far more interest to have insight into the potential performance of this technology in a somewhat longer term. This issue has been dealt with in various studies and analyses. Results of some of those evaluations are presented in Table 9.5.

The analysis of Faaij *et al.* (1998) detailed the role of technical improvement options, the economies of scale of BIG-GT systems and learning by doing, or cost reduction over time. Those three factor are the three major ways to increase the performance and reduce costs over time. Potential improvement options were inventorised and the implications

*Table 9.5* Summary of main results of studies dealing with potential cost levels and performance of BIG-CC systems. Clean wood is the assumed fuel for all studies

| Source | U$/kW$_e$ | Efficiency (LHV) (%) | Capacity (MW$_e$) | Remarks |
|---|---|---|---|---|
| Eliott and Booth (1993) | 1,300 | not given | 30 | Projection for 10th plant |
| DOE (1998) | 1,100 | 53 | 110 | Outlook for 2030 (turn-key) |
| | 2,100 | 44 | 75 | Base case |
| Williams and Larson | 1,000 | 51 | 97 | Concepts considered include |
| (1996) | 1,200 | 44 | 39 | ISTIG's (turn-key) |
| TPS (1997) | 2,700 | | 30 | Base level 1st plant |
| | 1,400 | | 55 | Obtainable after the 'Nth plant' |
| | (turn-key) | 42–52 | 21–122 | Outlook 2005 |
| | | 56–58 | >100 | Outlook 2020 |
| Faaij *et al.* (1998) | 4,000 | 39 | 29 | Base case; full project costs |
| | 2,000 | 54 | 51 | Advanced cases, outlook |
| | 1,600 | 55 | 110 | 2015 for full |
| | 1,100 | 59 | 215 | project cases |

for the overall performance analysed with help of modelling tools. Although modifications in drying and other pre-treatment, gas cleaning and gasification itself are possible, the impact on the overall performance is modest. The gas turbine is, however, the component that is expected to be developed further over time. Increased combustion temperature (due to further material development), reheating, intercooling and recuperation can all lead to increased output and higher electrical efficiencies for BIG-GT systems.

Increasing the scale leads, besides increasing efficiency due to the viability of higher pressure steam systems, to lower specific investment costs. It is expected that the scale effect would be higher for BIG-GT systems based on pressurised gasification than for plants based on direct atmospheric gasifiers. Pressurised gasifiers can more easily be enlarged in volume, while for a large amount of gas production based on atmospheric gasification more units are required. The definition of the best range of capacities requires the solution of a simple optimisation problem: for higher capacities, feedstock input increases and the costs of biomass hauling increase. Previous analyses were presented by Marrison and Larson (1995), showing that BIG-GT plants as large as 150–200 MW could be feasible.

Regarding capital costs, learning effects are indeed expected considering the high cost levels which are reported for first generation BIG-GT plants. There are no strong arguments against such cost reductions when BIG-GT technology is compared to similar equipment and installations such as power plants in general, chemical process technology, and the like.

### 9.3.3.2 Biomass gasification integrated to internal combustion engines

Electricity production by means of internal combustion engines integrated to a biomass gasifier is a commercial technology for very small capacity plants (most of them lower than 150 kW$_e$ capacity). It should be noted that commercial stage in this case does not mean that the integration of all system parts is sufficiently reliable. In fact, there is no accurate information about the long-term reliability and performance of these systems (Kaltschmitt *et al.* 1998). Larger capacity units are not predicted to be feasible because of the effects of economies of scale.

There is good potential for this option in developing countries, especially in isolated areas where grid connection is not available or expensive, using a cold gas-cleaning system and a gas-fired diesel engine, or even a dual fuel system (e.g. gas + diesel oil). The use of complementary liquid fuels is to avoid engine de-rating to a large extent. Additionally, a potential market for CHP use is also identified in Europe for systems up to 5 MW$_{th}$, using wood residues and agrobiofuels.

A serious constraint for this option is the gas-cleaning process (in particular tar and small particulate removal), as the technology is not yet well developed and costs are predicted to be high. Experiences with biomass-gas-fueled engines show that gas contaminants, such as tars, dust, soot and ash, are responsible for engine wear and high maintenance costs. Considering usual contaminant concentration levels from small-scale biomass gasifiers (fixed-bed gasifiers), the desired levels for internal combustion engine operation are down to 60–160 times for particulate matter and down to 60–1,500 times for tars (Stassen and Koele 1997; Bühler 1997).

Another important aspect is concerned with the costs of biomass pretreatment and the system's attendance. For systems up to 1 MW$_e$, biomass drying and cutting to a very specific chip size are not feasible. Systems should also be run with a minimum of attendance.

### 9.3.3.3 *Electricity generation based on biomass pyrolysis*

Electricity production based on biomass pyrolysis is one of the technological paths carefully considered through R&D projects. One of the potential advantages of this technology is that electricity production could be physically de-coupled from biomass production: pyrolysis oil could be transported to a thermal power plant and constraints concerned with plant size and environment impacts could be overcome.

Biomass fast pyrolysis and the subsequent use of the liquid fuel in internal combustion engines is a new and as yet unproven technology. As discussed in more detail in Section 9.5, R&D efforts in pyrolysis are still involving tests of small-scale units. Moreover, there are still doubts about this route, especially because of problems with alkali contamination and chemical instability of the oil due to the effects of high temperature (Bridgwater 1995).

Continuous operation of internal combustion engines integrated to pyrolysers has not been proved so far, even in small units. Obviously, long-term operational experience is required to obtain sufficient data for warranties. Results of extended engine tests report that crude-oil without any pretreatment has been successfully burnt in a modified 250 kW$_e$ dual fuel diesel engine (95 per cent of bio-oil and 5 per cent of diesel), for about 200 hours running experience (Leech 1997). Kurkela (1998) reports tests performed in Finland using bio-oil in a 80 kW Valmet engine and in a 1.5 MW Wärtsilä engine as well. Other organisations world-wide have also gained experience with the use of bio-oil in internal combustion engines.

Despite the incipient stage of this technology, previous feasibility studies (Mitchell *et al.* 1995; Solantausta *et al.* 1995) show that there could be a potential niche market for the production of electricity based on pyrolysis for small-scale units (5 to 25 MW), especially to meet peak-load demand. The analysis presented by Toft and Bridgwater (1997) explores more the advantage of de-coupling bio-oil production and its use to produce electricity, allowing exploitation of scale economies where there is a substantial biomass resource. Additionally, this alternative could maintain the supply of fuel through minor unscheduled shut-downs and improve overall reliability.

### 9.3.3.4 *Biomass gasification and Stirling engines*

Stirling engines are external combustion devices that were used in many industries during the nineteenth and early twentieth centuries. Recently, Stirling prototypes have been built with efficiencies of 30–45 per cent – much higher than the efficiencies of conventional internal combustion engines that are in the range of 20–30 per cent. However, Stirling engines are less suited for automobile use because a large heat exchanger is required to provide heat rejection, enlarging its dimensions a lot.

As an external combustion engine, solid biomass could be used as fuel. Nevertheless, in most of the R&D projects conducted so far, gasified biomass has been considered as a way to use compact Stirling engines originally developed for diesel or natural gas. The heat resulting from the external fuel combustion is transferred to the working fluid through a compact heat exchanger (at temperatures typically in the 950–1,000 K range). The use of gasified biomass has not been successful so far, mainly because of operational problems regarding clogging of the narrow passages between heat exchanger tubes. The problem could be solved with the design of a larger combustion chamber or with gas cleaning, but its cost could cancel out the advantage of Stirling engines *vis-à-vis* conventional internal combustion engines (Kaltschmitt *et al.* 1998).

The cycle itself is closed and the working fluid is a gas. To reduce engine size for a given capacity, the gas should have a low molecular weight and high heating capacity, leading to the selection of He and $H_2$ as the best candidates. However, for biomass the use of air or nitrogen as a working fluid is recommended, especially air because it is easily supplied at relatively low cost, with the additional advantage that seals against air loss are cheaper. The higher weight of Stirling engines operating with air makes no difference in stationary use (Podesser 1997).

In Europe, research projects are under way in Austria and Denmark. Research has also been conducted in the USA. Very small units (in the 3–10 $kW_e$ range) have been developed, and some projects are also dealing with units in the range 30–150 $kW_e$ (Kaltschmitt *et al.* 1998). Tests of a Stirling engine heated by the flue gas from a biomass furnace without hot gas cleaning are reported by Podesser (1997). During the test the engine produced 3.2 $kW_e$ at 25 per cent efficiency, and based on these results a model scaled-up to 30–150 $kW_e$ was considered. A unit of 35 $kW_e$ (nominal capacity) integrated with a biomass furnace has been tested in Denmark, operating in a CHP mode for more than 700 hours with only minor problems. The electric efficiency reported by Calrsen (1998) is 19 per cent (when wood chips are burned with 40 per cent moisture), with 89 per cent overall efficiency. In the USA, Stirling Thermal Motors has constructed a power plant rated at 20 $kW_e$, integrating a Stirling engine with a commercial updraft gasifier. Some tests were conducted and in the most successful one the system produced 24 $kW_e$ gross electric power (Johansson and Ziph 1997).

### 9.3.3.5  Biomass gasification and fuel cells

Biomass gasification integrated to fuel cells has also been considered for electricity production. Fuel cells allow direct electricity generation through electrochemical oxidation of the fuel. They are not limited by Carnot's law and efficiencies of 45–70 per cent have been achieved in units of small capacity (just a few hundred kilowatts). Hart and Bauen (1998) provide an overview of fuel cell technology and prospects. Hydrogen production from biomass gasification has been specially considered for this purpose. Projected efficiencies for gasifier/fuel cell systems are as high as 60 per cent (Kinoshita *et al.* 1997).

Besides its very high efficiency, even with small-scale application and under part load, other important advantages are the lower emissions of $NO_x$, CO and HC compared to the gas use in gas turbines or engines, and less wastewater than competing technologies. Fuel cells are inherently small, but a high capacity unit would be possible due to their modular characteristic. Modularity represents a great advantage for this technology as long as a better match of capacity with demand could be achieved.

High-temperature fuel cells (molten carbonate fuel cell (MCFC) and solid oxide fuel cell (SOFC)) have been considered the most promising options, because: 1) of their high overall efficiency; 2) a CO-shift reaction is not required to adjust gas composition (commercial low-temperature fuel cells do not tolerate high levels of CO); and 3) residual heat can be used (Kaltschmitt *et al.* 1998). However, these fuel cell technologies are not yet commercially available.

A fuel cell requires a gas free of impurities (especially particulates, tars and alkalis) and with a medium heating value, imposing challenges to the biomass gasification and to the gas-cleaning processes, also because these systems should be reliable and economic on a small scale.

The main emphasis of R&D projects in the US is on steam reforming using catalysts, a process similar to the one utilised to produce hydrogen from methane (Overend *et al.* 1996). Among the short-run goals of the US Biomass Power Programme, a demonstration unit with a molten carbonate fuel cell was scheduled to be built during 1998 (Craig and Mann 1996). However, more research and development is still required prior to considering this option more seriously.

## 9.4  Ethanol from cellulosic materials

### *José Roberto Moreira*

### 9.4.1  Biomass composition

Carbohydrates, including sugars, are among the most abundant constituents of plants and animals. Carbohydrates are classified as mono-, di-, tri-, tetra-, and poly-saccharides, depending on the number of sugar molecules they comprise. Practically all natural monosaccharides contain five or six carbon atoms, known as pentoses and hexoses, respectively.

The disaccharide known as cane sugar, or sucrose, can be broken down by hydrolysis into the six-carbon sugars glucose and fructose. Sugar-cane and other plants contain about 10–15 percent sucrose. On the other hand, about 70 percent of corn seed is the polysaccharide known as starch, which is a mixture of straight-chain and branched polymers of glucose. Hydrolysis of starch by acids or enzymes known as amylases forms glucose sugar.

Cellulosic biomass is actually a complex mixture of carbohydrate polymers known as cellulose and hemicellulose, plus lignin and a small amount of other compounds known as extractives. Cellulose is generally the largest portion, representing about 40–50 percent of the material by weight; the hemicellulose portion represents 20–40 percent of the material. The remaining parts are predominantly lignin with a lesser amount of extractives. The cellulose fraction is composed of glucose molecules bonded together in long chains that form a crystalline structure. The hemicellulose portion is made of long chains of different sugars and lacks a crystalline structure. For hardwoods, the predominant component of hemicellulose is xylose. Softwoods are generally not considered a viable feedstock for the dedicated production of energy in the near term because competition from the paper industry and other markets makes softwoods too expensive. In addition, hardwoods are more amenable to short-rotation production methods that are vital for large-scale energy applications.

For ethanol production by hydrolysis, the cellulosic feedstock is first pre-treated to reduce its size and open up the structure to facilitate conversion. The cellulose component is hydrolysed by acids or enzymes to produce glucose, which is subsequently fermented to ethanol. The soluble xylose sugars derived from hemicellulose are also fermented to ethanol, while the lignin fraction, which cannot be fermented into ethanol, can be used for the rest of the process, converted into octane boosters, or used as a feedstock for the production of chemicals.

### 9.4.2 Feedstocks

Cellulosic materials for ethanol production can originate either as waste materials (e.g. agriculture, forest product industries) or as energy crops grown expressly for this purpose. It is agreed that waste feedstocks have the greatest potential due to their lower costs (Wyman and Goodman 1993). Wastes costing more than US$45 per dry tonne might never be used for ethanol production, because dedicated energy crops are expected to be available at lower cost. In fact, only a fraction of the material costing less than US$45 per dry tonne is likely to be utilised in plants exclusively processing wastes, because of scale considerations. Thus, large-scale displacement of conventional transportation fuels with cellulose ethanol will require significant production from dedicated energy crops. Examples of cellulosic energy crops are short-rotation woody crops (e.g. poplar, willow) and perennial herbaceous crops (e.g. witchgrass).

Conversion of waste materials into ethanol in general raises fewer environmental issues than does conversion of dedicated feedstocks because no land is required, e.g. land may even be saved by decreasing material flows to landfills. Depending on the wastes, collection may either be easier than it is for energy crops (as in the case of waste produced at a centralised processing facility, such as a paper mill) or more difficult (as in the case of agricultural residues that would not otherwise be collected). Additionally, for some waste it is necessary to consider disposal issues, e.g. that methods such as open-field burning of straw or sugar-cane leaves is becoming environmentally unacceptable.

Most of those who have analysed energy crop production (e.g. Ranney and Mann 1994) stress that a significant number of issues related to its environmental impacts are incompletely understood, and advise a cautious approach and further research. A second common observation is that good management practices are likely to result in environmental benefits, whereas poor management represents a significant potential liability. Finally, most land-use-related impacts depend greatly on what form of land-use energy crops are replacing. These remarks may be particularly relevant to the issue of biodiversity.

Increasingly, analyses of energy crop production focus on croplands. In general, most metrics of environmental quality improve when short-rotation woody crops or herbaceous perennial crops replace conventional row crops. Compared to row crop production, energy crops involve far less erosion, slightly less fertiliser application, and much reduced pesticide application. These trends are consistent with the possibility of very positive water quality benefits resulting from energy crop planting (Ranney and Mann 1994).

### 9.4.3 Conversion technology

#### 9.4.3.1 Processes description

Steps in the conversion of cellulosic materials to ethanol in processes featuring enzymatic hydrolysis include pre-treatment, biological conversion, product recovery, and utilities and waste treatment. An overview of conversion technology is provided by Wyman (1994). Lynd *et al.* (1996) have recently addressed future improvements to conversion technology, and some detailed process design studies are available, e.g. Grethlein and Dill (1993).

Cellulose ethanol processes are differentiated primarily on the basis of the methods used to achieve hydrolysis and fermentation, steps that are the least technologically mature and the most specific to ethanol production. Hydrolysis processes can be categorised into those that use mineral acids (e.g. sulphuric acid) and those that use cellulase enzymes. Although acid-reliant processes are more technologically mature, enzymatic processes have roughly equal projected costs today. In addition, processes relying on acids generally have greater environmental liabilities than enzymatic processes. Because of these considerations, enzymatic hydrolysis is the primary focus here. However, acid-reliant processes do have one significant advantage over enzymatically-based processes: they are equally effective with softwoods and hardwoods. Current enzymatic processes are much more readily applied to hardwoods and herbaceous materials than to softwoods.

#### 9.4.3.2 Acid-catalysed processes

Several dilute acid hydrolysis pilot plants were constructed in Germany and the United States during World War II to produce ethanol as a petroleum substitute (Wenzl 1983), but the economics were too unfavourable for continued post-War operation. Dilute acid-catalysed processes are currently operated in the former Soviet Union for converting cellulosic biomass into ethanol and single-cell protein. Thus, acid-catalysed processes provide a near-term technology for production of fuel-grade ethanol from cellulosic biomass, but the low yields of sugars from cellulose and hemicellulose (about 50–60 percent of the theoretical maximum) make them unable to compete with existing fuel operations in a free market economy (Wright 1988a). Concentrated sulphuric or halogen acids achieve high yields (essentially 100 percent of theoretical). However, recycling of acid by efficient, low-cost recovery operations is essential to achieve economic feasibility and this is a difficult requirement (Wright *et al.* 1985).

Two case studies are worth discussing:

1   a real commercial facility that operated in Brazil about two decades ago; and
2   a recent feasibility study in California.

PRODUCTION OF ETHANOL FROM WOOD IN BRAZIL

In the early 1970s some Brazilian institutions became interested in the conversion of cellulosic material to ethanol. In 1980 COALBRA (Coque e Álcool de Madeira) was set up as a federal-owned company to install and operate a commercial facility for the conversion of wood into liquid fuels and other products. After evaluation of technologies available world-wide, COALBRA decided in favour of the technology based on low concentration of sulphuric acid (0.5 per cent), and in use in the former Soviet Union for many decades. A factory was assembled in the state of Minas Gerais for the production of ethanol and by-products under the technical specifications (see Table 9.6).

*Table 9.6* Technical specifications of COALBRA's low-concentration sulphuric acid unit

| Products | Units | Daily production | Annual production |
|---|---|---|---|
| Ethanol | Litre | 30,000 | 10,200,000 |
| Furfural | Tonne | 1.42 | 485 |
| Protein for cattle feeding | Tonne | 9.35 | 3,180 |
| $CO_2$ | Tonne | 14.96 | 5,090 |
| Lignin (69% wet) | Tonne | 199 | 67,660 |
| *Main inputs* | | | |
| Wood eucalyptus (without bark) | Stere | 600 | 204,000 |
| Wood residues from eucalyptus | Stere | 295 | 100,000 |
| Sulphuric acid (98%) | Tonne | 13.4 | 4,556 |
| Limestone (85% CaO) | Tonne | 13.1 | 4,454 |
| Ammonia sulphate | Tonne | 5.2 | 1,770 |
| Superphosphate | Tonne | 5.0 | 1,700 |

*Source*: COALBRA (1983)

COALBRA's plant operated from 1983 to 1987, when the facility was closed down. Major reasons for the lack of commercial success were:

1   High cost of ethanol produced compared with ethanol from sugar-cane.
2   Difficulties in finding market for furfural. Furfural had a reasonable market price but negotiations for furfural exports failed to consolidate.
3   High cost of wood. Wood suppliers immediately asked prices higher than previously estimated.

4   Lack of use for lignin. Drying lignin was difficult and very energy intensive, and its use as fuel for the industrial processing of wood to ethanol was not possible.

A study has recently been carried out in California with the purpose of reviewing the feasibility of using wood wastes and timber harvest as a source of ethanol produced by acid hydrolysis (CEC 1996). NREL prepared design and cost estimates for six sites and three different biomass-to-ethanol conversion technologies that are representative of the near-term opportunities. The technologies considered in this report are:

1   concentrated sulphuric acid (includes technology patented by Arkenol, Inc.);
2   dilute sulphuric acid (contains no patented technology);
3   dilute nitric acid (includes technology with licensing available).

The size of the ethanol facility at each site is based on the amount of feedstock available within a 40 km radius of the site. The resulting ethanol plant sizes range from 53.6 million litres per year (with dilute acid technology) to 128 million litres/year (with concentrated acid technology). Internal rate of return (IRR) was calculated assuming a 20-year project life, 100 per cent owner equity, feedstock cost of US$20 per BDT, and an ethanol selling price of US$0.26 per litre. According to the study, only the technology based on dilute nitric acid shows a rate of return above 15 per cent, which is a minimum level to attract private investments in a risky project. Even so, IRR values were obtained under conditions which are quite favourable and difficult to identify in the market.

### 9.4.3.3 Enzymatic hydrolysis technologies

Enzymatic hydrolysis emerged from US Army research during World War II aimed at finding ways to overcome microbial attack on the canvas (cellulosic) webbing and tents of soldiers stationed in the tropics. These studies in turn led to research on the possibility of promoting the decomposition of cellulose by a fungus responsible for the breakdown of cotton, now named *Trichoderma reesei*, in the hope of generating glucose syrups for application to food and then fuel ethanol production (Reese 1976).

Enzyme-catalysed processes offer several key advantages. They achieve high yields under mild conditions with relatively low amounts of catalyst. Moreover, enzymes are biodegradable and thus environmentally benign. The cost of ethanol produced from enzyme- and acid-catalysed processes may be comparable at present, but enzyme-catalysed process have tremendous potential for technology improvements that could bring the cost of ethanol down to levels competitive with those for petroleum-based fuels (Lynd *et al.* 1991a). Nonetheless, considerable improvement is required to achieve economic application of this technology.

### 9.4.4 Pre-treatment

All naturally occurring, and most refined, cellulosic materials require pre-treatment by which cellulosic biomass is made amenable to the action of hydrolytic enzymes.

Typically, hydrolysis yields in the absence of pre-treatment are less than 20 per cent of theoretical yields, whereas yields after pre-treatment often exceed 90 per cent of theoretical. The limited effectiveness of current enzymatic processes on softwoods is thought to be due to the relative difficulty of pre-treating these materials. A review of pre-treatment has been done by McMillan (1994).

The ideal pre-treatment process would:

1  produce reactive fibre;
2  yield pentoses in non-degraded form;
3  exhibit no significant inhibition of fermentation;
4  require little or no biomass size reduction;
5  entail reactors of moderate cost;
6  not produce solid residues;
7  have a high degree of simplicity; and
8  be effective at sufficiently low moisture contents to avoid significant economic penalty.

Even well-known pre-treatment processes (dilute acid hydrolysis and steam explosion) fail to meet these criteria. Ammonia Fibre Explosion (FEX) and liquid hot water pre-treatment appear promising, but additional study is required before a more definitive evaluation can be made (van Walsum *et al.* 1996).

### 9.4.5 Biological conversion

Individual steps for converting cellulosic biomass into liquid fuels can be combined in various ways to minimise the overall conversion cost. The front-running integrated microbiology-based process configurations are described below. Among all alternatives described, the SSF process has emerged as an especially promising route to low-cost fuel ethanol production within a reasonable time frame (Wright 1988b).

#### 9.4.5.1 Separate hydrolysis and fermentation (SHF)

The SHF process uses distinct process steps for enzyme production, cellulose hydrolysis, and glucose fermentation (Mandels *et al.* 1974; Wilke *et al.* 1976). The primary advantage of this configuration is that all three processes can be treated separately, thus minimising the interactions between them. However, cellulase enzymes are inhibited by the accumulation of sugars, and considerable effort is still needed to overcome this end-product inhibition, which impedes attainment of reasonable ethanol concentrations at high rates and with high yields even at high enzyme loadings.

#### 9.4.5.2 Simultaneous saccharification and fermentation (SSF)

The sequence of steps for the SSF process is virtually the same as for SHF except that hydrolysis and fermentation are combined in one vessel (Gauss *et al.* 1976; Takagi *et al.* 1977). The presence of yeast along with the enzymes minimises accumulation of sugar in the vessel, and because the sugar produced during breakdown of the cellulose slows down the action of the cellulase enzymes, higher rates, yields, and concentrations of ethanol are possible for SSF than SHF at lower enzyme loadings. Additional benefits are that the number of fermentation vessels is halved, and that the presence of ethanol makes the mixture less vulnerable to invasion by unwanted micro-organisms.

### 9.4.5.3 *Direct microbial conversion (DMC)*

The DMC process combines enzyme production, cellulose hydrolysis, and sugar fermentation in one vessel (Veldhuis *et al.* 1936; Cooney *et al.* 1978). In the most extensively tested configuration, two bacteria are employed to produce cellulase enzymes and ferment the sugars formed by the breakdown of cellulose and hemicellulose into ethanol. Unfortunately, the bacteria also produce a number of products in addition to ethanol, and yields are lower than for the SHF or SSF processes.

### 9.4.5.4 *Cellulase production*

Several organisms, including bacteria and fungi, produce cellulase enzymes that can be used to hydrolyse cellulose into glucose (Ng *et al.* 1977; Cooney *et al.* 1978). Currently, genetically altered strains of the fungus *Trichoderma reesei* are favoured because of the relatively high yields and activities of cellulase that are realised. The best performance is generally achieved in the fed-batch mode of operation. Simple batch production of cellulase with addition of all ingredients at the beginning of the enzyme production cycle may be used with good results.

### 9.4.5.5 *Hemicellulose conversion*

The hemicellulose polymers in cellulosic biomass can be readily broken down during the pre-treatment step to form xylose and other sugars. Several options have been considered for utilisation of the sugars formed from hemicellulose.

In the presence of acid, the xylose can be reacted to form furfural, either during acid hydrolysis or after xylose recovery (Wright 1988b). Furfural can be sold for use in foundry and other applications, but the market can be quickly saturated by the volume of furfural that would accompany large-scale applications of fuel ethanol. Anaerobic bacteria can convert xylose to methane gas, but methane is less valuable than ethanol.

An alternative is to ferment xylose into ethanol, using strains of some yeasts. However, these strains require small amounts of oxygen in the fermentation broth to ferment xylose and, typically, cannot achieve high ethanol yields or rates or tolerate high ethanol concentrations (Hinman *et al.* 1989). Other micro-organisms can anaerobically ferment xylose into ethanol, but ethanol tolerance, yields, and selectivity are historically low for these options. The common bacterium *Escherichia coli* has been genetically engineered to produce large quantities of xylose isomerase enzyme. This enzyme can convert xylose into an isomer called xylulose, which many yeasts can ferment into ethanol under anaerobic conditions (Jeffries 1981; Tewari *et al.* 1985). By employing the enzyme and yeast together in one vessel, ethanol yields from xylose of 70 percent of theoretical have been achieved but with some technological difficulties. In another approach, the genes from *Zymomonas mobilis* have been spliced into *E. coli* enabling it to ferment xylose directly into ethanol with high yields (Ingram *et al.* 1987). However, further evaluation of the procedure is needed. The latter two options are favoured for ethanol production at this time.

### 9.4.6 Micro-organisms and fermentation

An element common to essentially all proposed processes for producing ethanol from cellulosic biomass is microbial fermentation. A variety of micro-organisms (either bacteria, yeast, or fungi) ferment carbohydrates to ethanol under oxygen-free conditions. The maximum possible yield of ethanol corresponds to 51 per cent of the carbohydrate converted to ethanol on a mass basis, but in most proposed ethanol production processes the yield is no more than 47 per cent.

The ideal organism, or system of organisms, for producing ethanol from cellulosic biomass in a process featuring enzymatically-mediated hydrolysis would simultaneously exhibit the following properties: 1) synthesis of an active cellulase enzyme system at high levels; 2) fermentation and growth on sugars arising from both cellulose and hemicellulose; and 3) production of ethanol at high selectivity and high concentration.

Development of pentose-fermenting ethanol-producing micro-organisms is equally important for processes based on acid hydrolysis and for processes based on enzymatic hydrolysis. The matter of fermentation inhibitors is equally or more important for acid hydrolysis with respect to corrosion products. In comparison to enzymatically-based processes, inhibition by sugar and lignin degradation products is less important for processes using concentrated acid and of roughly equal importance for processes using dilute acid. Cellulase production is irrelevant for a process reliant on acid hydrolysis.

Hydrolysis of cellulose is generally rate-limiting in enzymatically-based biomass ethanol processes (South and Lynd 1994), and it occupies a similarly central position in process economics. The high cost of cellulase (which catalyses the hydrolysis of cellulose) production is associated with the cost of hydrolysis rather than the cost of cellulase production *per se*. When cellulase production is carried out separately from hydrolysis and fermentation, as in the SSF process, there is a trade-off between the cost of cellulase production and the cost of hydrolysis/fermentation. Short hydrolysis reaction times involve higher cellulase and lower hydrolysis/fermentation costs than longer reaction times.

The difference between SHF and SSF is one of arrangement of the same components. In both of these alternatives, one micro-organism is used to produce cellulase, and a second is used to carry out fermentation. The SSF approach appears to enjoy a substantial economic advantage over SHF. Processing (CBP) is the logical endpoint in the evolution of biomass conversion technology. CBP is less well developed than SSF, but it is expected to offer the lowest costs if limitations of current systems can be overcome. The key difference between CBP and other biomass-processing strategies is that a single microbial community is employed for both cellulase production and fermentation. This difference has several significant ramifications, including no capital or operating costs for dedicated enzyme production, greatly reduced diversion of substrate for enzyme production, and compatible enzyme and fermentation systems.

Regardless of the fermentative organism used, ethanol production from cellulosic biomass is likely to involve operation at lower product concentrations (e.g. <5 wt per cent) than are typical of ethanol production from corn. This situation arises as the result of both biological and processing constraints.

### 9.4.7 *Current process economics*

No commercial facility currently produces ethanol from cellulosic biomass via enzymatic hydrolysis and, consequently, cost-estimating of such processes is necessarily based on projections developed from laboratory and, in some cases, pilot plant data. As NREL's research is the largest and most comprehensive effort dedicated to developing cellulosic ethanol conversion technology, their publicly available designs are used as a benchmark in this review and elsewhere. The most recent results are based on a poplar energy crop assumed to cost US$42 per delivered dry tonne and on an SSF-based process involving fermentation via yeast and cellulase production by *T. reesei*. Table 9.7 presents a cost and selling price summary, developed by Lynd *et al.* (1996), based on an NREL design.

*Table 9.7* Biomass ethanol cost summary-base-case technology

| | Selling price breakdown | | | | |
| --- | --- | --- | --- | --- | --- |
| | Cents/litre | | | % of total | |
| | Capital and labour | Energy | Total | Processing | Overall |
| Raw materials | | | | | |
| Feedstock | | | 10.11 | | 39.00 |
| Other | | | 2.15 | | 8.30 |
| Subtotal | | | 12.26 | | 47.30 |
| Processing | | | | | |
| Pre-treatment | 3.02 | 1.44 | 4.47 | 32.70 | 17.20 |
| Cellulase production | 0.34 | 0.37 | 0.71 | 5.20 | 2.70 |
| SSF | 3.04 | 0.73 | 3.78 | 27.70 | 14.60 |
| Pentose conversion | 0.71 | 0.22 | 0.93 | 6.80 | 3.60 |
| Distillation | 0.60 | 1.12 | 1.72 | 12.60 | 6.70 |
| Power cycle | 6.29 | −5.93 | 0.36 | 2.70 | 1.40 |
| Other | 1.61 | 0.08 | 1.69 | 12.40 | 6.50 |
| Subtotal | 15.63 | −1.97 | 13.66 | 100.00 | 52.70 |
| Grand total | | | 25.92 | | 100.0 |

*Source*: Lynd *et al.* (1996)
*Notes*
Feedstock: 658,000 dry tonnes/year; plant capacity: 273 million litres/year; installed capital cost: US$150.3 million (first quarter, 1994)

Of the overall projected selling price of 25.9 cents per litre (117.8 cents per gallon) in the NREL design, capital recovery accounts for 11 cents per litre (50 cents per gallon), indicating that the cost of capital has a strong impact on economic viability. The data presented in Table 9.7 are consistent with a return on investment of 14.2 per cent, based on construction time and capacity build-up assumptions anticipated for mature technology, and of 10 per cent based on assumptions likely to be more representative of a first-of-kind process (Lynd *et al.* 1996).

## 9.4.8 Energy balance

The relative magnitude of energy outputs and inputs associated with ethanol production is a key index for evaluating the efficacy of large-scale implementation. Relevant measures of energy output include the energy yielded as ethanol and electricity on either an unadjusted (e.g. kJ) or a displaced fossil fuel basis. These measures differ because of fossil energy required to produce electricity; a representative efficiency of 35 per cent is used here.

In addition, if ethanol is used in neat form in optimised engines, then more than one unit of fossil fuel is displaced per unit of ethanol used; a ratio of 1.175 units of displaced fossil fuel per unit of ethanol is used here, indicative of the long-term potential of this fuel, assuming optimised engines like the ones already available in Brazil. Relevant measures of energy input include the energy content of the biomass (e.g. high heating value) and the energy inputs for feedstock production, feedstock transportation, chemical inputs, product distribution, and plant amortisation.

Table 9.8 shows a comparison of energy–output/energy–input ratios based on the base-case, advanced, and best-parameter scenarios which can be expected in the future. An additional scenario is also defined, based on the combining of the ethanol production technology in the advanced case with electricity co-production via a first-generation biomass gasification/combined cycle (BIG-CC), rather than a conventional Rankine cycle, as in the other three cases. No cost analyses have been done for the BIG-CC case. Ratios of energy output, relative to the biomass energy content, show an efficiency between 0.5 and 0.97, depending on the level of maturity of the technology and whether energy outputs are valued on an unadjusted or a displaced fossil fuel basis. Ratios of

*Table 9.8* Output–input energy ratios for current and projected cellulosic ethanol technology

| | Energy output (% HHV)[a] | | EtOH + electricity | Displaced fossil fuel[b] | EtOH + electricity | Displaced fossil fuel |
|---|---|---|---|---|---|---|
| | Ethanol | Electricity | Biomass (HHV) | Biomass (HHV) | Feedstock and process[c] | Feedstock and process |
| Base case | 46.1 | 4.2 | 0.503 | 0.662 | 4.36 | 5.74 |
| Advanced | 54.4 | 6.8 | 0.612 | 0.834 | 6.58 | 8.96 |
| Best parameter | 61.4 | 7.9 | 0.693 | 0.947 | 7.40 | 10.10 |
| Advanced/BIG-CC[d] | 54.4 | 11.4 | 0.659 | 0.965 | 7.10 | 10.40 |

*Source*: Lynd (1996)

*Notes*

[a] HHV: high heating value of the biomass. All values are from Lynd *et al.* (1996) except the electricity value for BIG-CC (see footnote d).

[b] 2.86 × electricity + 1.175 × ethanol, see text.

[c] Energy values for energy crop production, raw material transport, chemical inputs, fuel distribution, and plant amortisation from Lynd *et al.* (1991). Fuel distribution inputs are assumed constant per unit of ethanol produced, with other inputs assumed constant per unit of feedstock.

[d] BIG-CC: values based on electricity output increasing to 11.4% of the biomass heating value. The calculation is as follows: gross power output = (3.77 kWh/gallon produced (Rankine cycle) × [(40% efficiency)(combined cycle)/(25.8% efficiency)(Rankine cycle)] = 5.84. Net power output = 5.84 gross − 0.71 required for the process = 5.13/gallon exported. Electricity output goes up by 67.6% to 11.4% of output.

energy output relative to energy input for feedstock production and processing range from 4.4 to 10.4, again depending on the level of maturity assumed and the way in which the energy outputs are valued.

Electricity is expected to be an increasingly significant co-product of ethanol manufacture from woody materials as the technology matures. For example, the amount of energy exported as electricity in the advanced/BIG-CC turbine scenario is one-fifth the amount exported as ethanol.

To reach feasibility on ethanol production from cellulosic biomass, it is important to derive value from the lignin. Because lignin has a high energy content, it can be used as a process fuel (Hinman *et al.* 1989). The amount of lignin in most feedstocks is more than enough to supply all the heat and electricity required for the entire ethanol production process, and even surplus electricity or heat could be produced. Alternatively, the phenolic fraction from lignin can be reacted with alcohols to form methyl or ethyl aryl ethers, which are oxygenated octane boosters (Johnson *et al.* 1990), although high product yields must be realized at low costs to provide a net income gain for the ethanol plant.

### 9.4.9 *Near-term commercialisation prospects*

A straightforward analysis of a cost breakdown such as that shown in Table 9.7 indicates that the following items are key determinants of the price of cellulosic ethanol: feedstock, power cycle and other infrastructure components, pre-treatment, and capital. If the cost of any one of these is unusually low because of advantageous local circumstances, the overall cost may be significantly lower than for a stand-alone plant with a dedicated energy crop, such as that assumed in Table 9.7. Many waste feedstocks cost less than dedicated feedstocks, and waste feedstocks that require active disposal (e.g. MSW, paper sludge, certain agricultural residues) may have negative costs. Large cost savings may be realised by locating at a site that provides various infrastructure functions required at an ethanol plant (e.g. existing ethanol plants, paper mills, refineries, and power plants). Certain feedstocks are particularly susceptible to enzymatic attack, e.g. many paper sledges and corn fibber, which can result in substantial savings in biological conversion and/or pre-treatment.

## 9.5 Pyrolysis

### *Guilherme Bezzon and José Dilcio Rocha*

### 9.5.1 *Introduction*

Biomass pyrolysis can be defined as the thermal decomposition of this organic material under partial or total lack of an oxidising agent, or even in an environment with such oxygen concentration to avoid extensive gasification. Pyrolysis usually takes place in the temperature range 400–800°C. Gases, liquids, and solids are generated in different proportions, depending on pyrolysis parameters such as final temperature, residence time, heating rate, gas environment, and initial biomass properties (Beenackers and Bridgwater 1989). The main objective of pyrolysis is to obtain products with higher energy density and better properties compared to the initial biomass.

The most used pyrolysis variation is the carbonisation to produce wood charcoal for energy purposes. In Brazil, the world's largest charcoal producer, this product is mostly used as a reducing agent for pig-iron and steel-making. Carbonisation is a very old concept, but one which is still extensively utilised today, especially for domestic cooking.

In the last two decades, the uses of biomass as a potential source of transportation fuels, chemicals and materials have given a new impulse to pyrolysis. The concept of fast pyrolysis for organic liquid production has found increasing interest, and the research and commercial applications are growing quickly, mainly in North America and Europe.

The control of some pyrolysis parameters, such as heating rate, final temperature, residence time, and maximum pressure, results in diversified products for special applications. For instance, increasing the heating rate under controlled conditions of temperature, pressure, and gas environment results in high yields of liquids (bio-oil) with elevated calorific value. Both high heating and cooling velocities minimise solids formation, resulting in low charcoal production. The pyrolysis under these circumstances is called *fast*, *flash*, or even *ultra-flash pyrolysis* (Bridgwater and Bridge 1991). On the other hand, elevated pressures with relatively low temperatures increase solid formation (charcoal production) (Antal *et al.* 1992a). In addition, for process temperatures higher than 650°C, gases are the main products. To increase overall efficiency and allow product diversification, inert (nitrogen and vacuum) or reactive (hydrogen and methane) environments can be used. In this last case, fuels with lower oxygen content and better properties can be obtained.

For vacuum pyrolysis, heating rates are between those of carbonisation and fast pyrolysis process, but with a shorter residence time for the contact between solid product and organic vapours generated in the process. The operation of the vacuum reactor minimises secondary reactions of biomass decomposition, increasing liquid yields, which are typically 65 per cent (dry basis) (Roy *et al.* 1992).

### 9.5.2  Fast pyrolysis

The main characteristics of the fast pyrolysis process are the short residence time for vapours inside the reactor, the high heating rates, the high heat transfer coefficients, and the relatively low temperature of the heating source. Residence times shorter than one minute are currently used. All new pyrolysis technologies apply this principle to maximise the bio-oil yields. The production of a liquid derivative that could be stored and transported is exactly the main potential advantage of fast pyrolysis regarding other thermochemical conversion processes of biomass. The biomass pyrolysis liquid produced in this way is a primary tar; secondary reactions are avoided due to the short residence time. Secondary tar is a result of primary tar cracking (Antal 1983).

Fast pyrolysis has achieved some commercial success for production of chemicals and is being actively developed for liquid fuels production (Bridgwater 1997). In the last twenty years a number of research, pilot and commercial reactors have been tested using different technologies. Some of these reactors are still in operation, producing charcoal and bio-oil. The process itself and the bio-oil composition are better known now than before, but many fundamental, industrial, and economic aspects of the pyrolysis are still unclear. Pyrolysis reactions are very complex, and the chemical composition and yields are not completely understood. The scale-up of fast pyrolysis reactors with cost reduction is the next step to be reached towards industrial applications (Bridgwater 1998).

One important development concerned with fast pyrolysis is the ablative principle. The technology has been developed at the NREL (Golden, USA) in order to allow large-scale production of liquids. A pilot plant of the so-called vortex reactor technology has been operating since 1996. The facility, named Thermochemical Users Facility (TCUF), is able to reach very high heat transfer rates. The plant has a feed capacity of 20 kg/h and is able to provide recalculation of unreacted material. The TCUF is also able to test reactors, filters, catalysts, and many other unit operations.

According to an economic assessment conducted in the early 1990s, an ablative pyrolysis plant for 907 metric tonnes of biomass could produce 680 tonnes of crude bio-oil per day, with a cost of US$100/tonne. This estimate corresponds to an interest rate of 20 per cent per year and biomass charged at US$44 per tonne. The biomass share of the total cost would be US$58.7/tonne (53 per cent). The estimate equipment cost is US$11 million and the total investment US$44.5 million. For a smaller plant (227 tonnes of biomass per day), the bio-oil cost would be U$158/tonne, indicating an important scale effect (Gregoire 1992).

A commercial example of fast pyrolysis is the plant development in the PYROVAC Institute, in Quebec, Canada. This plant is using a vacuum pyrolysis technology in a horizontal moving bed. The plant presents capacity of 1.8 tonnes per day of biomass (Roy *et al.* 1997). Other plants and companies in Canada and the USA have commercial and pilot units. Ensyn Technologies Inc., another Canadian company, has installed some pyrolysis units in the USA, Canada and Europe. Ensyn reactor technology is based on a recirculating fluidised bed operating in the range of 500–550°C and applying 2.5 seconds as biomass residence time (Graham *et al.* 1988).

### 9.5.3 *Obtention of biofuels*

The liquids obtained from biomass pyrolysis, traditionally referred as pyroligneous tar, have recently been called *bio-fuel*, *bio-crude oil* or even *bio-oil*. Despite this reference to petroleum they are very different. Bio-oil is sulphur-free, very low in polyaromatic hydrocarbons (PAHs), high in oxygen content (ca 50 per cent wt), partially water soluble, and it has higher acidity (pH 2) and lower molecular weight and heating content than petroleum. It is a clear reddish-brown coloured substance when free of char particles, with a distinctive odour, and it is an eye irritant. Bio-oil is a very complex organic mixture formed by hundreds of different compounds belonging to many chemical groups. The heating value for bio-oil is in the range of 16–18 MJ/kg or, in other words, about half of the conventional value for fuel oil. The water content is from 15 per cent wt to an upper limit of about 40 per cent wt. The bio-oil density is also high, around 1.2 kg/l (PyNE 1997).

Bio-oil is also very unstable at temperatures higher than 100°C. The high reactivity of the oxygenated groups (carboxylic, alcohol, metoxi, etc.) makes distillation a non-recommended process for bio-oil fractionation. Raising the temperature can make bio-oil react rapidly and produce a solid char residue. Solvent fractionation is possible but usually very expensive; this process has been used on a laboratory scale to isolate the phenolic fraction. In this case, ethyl acetate is the recommended solvent (Chum and Black 1990).

Catalysis has been widely applied for bio-oil upgrading. A number of catalysts have been tested in bio-oil hydrocracking, hydrodeoxygenate, and heteroatom elimination. Most of these catalysts are already working in petroleum refining. A combination of pyrolysis in a reactive atmosphere, like hydrogen (hydropyrolysis), with a catalyst is another possible route to submit the bio-oil. The HZSM-5 zeolite catalyst is probably the most studied catalyst for bio-oil upgrading. The process was proposed to convert bio-oil into a kind of aromatic hydrocarbon to substitute gasoline. As the final product yield was low, about 23 per cent wt, the process was evaluated during the 1980s as non-economic (Diebold and Scahill 1987).

Deoxygenation of bio-oil to produce hydrocarbons using catalysts is based on carbon oxides elimination. Since the bio-oil is not rich in hydrogen, the yield in organics is not expected to be very high. Although expensive, the use of an external source of hydrogen to boost the product is very desirable. Biomass hydropyrolysis followed by bio-oil conversion over a catalytic bed in the vapour phase will be able to hydrodeoxygenate bio-oil via water formation instead of carbon oxides. This route saves carbon in the final products and avoids carbon oxide emissions. The catalysis in the vapour phase reduces the catalyst contamination by coke formation (Rocha *et al.* 1996).

Another problem of bio-oil upgrading through a catalyst is its water content. The commercial catalysts are usually active metals supported in alumina and a potentially good catalyst for bio-oil conversion is a petroleum hydrocracking catalyst of nickel and molybdenum in alumina. The water content in bio-oil can attack the alumina structure and damage the catalyst. One possible solution is to use carbon as a catalyst support to avoid water attack and coke contamination. Preliminary results have shown lower yields with carbon as a catalyst support when compared with the same catalyst in a different support. Despite the problems with bio-oil catalysis, substantial advances are expected in the field of biomass liquid pyrolysis catalytic upgrading (Elliott and Maggi 1997).

Some applications are suggested for bio-oil. Liquid premium fuels, such as light hydrocarbons and aromatic mixture gasoline or diesel-like substances, could be produced via catalysis. The use in off-road engines as a substitute for fuel oil is possible, but before that it is necessary to solve problems such as corrosion, low heating value and ageing (polymerisation reactions) during storage.

Bio-oil is also a source of fine chemicals with a high retail price. For instance, with bio-oils it is possible to produce hydroxyacetaldehyde ($CH_2OH$-CHO), an agent to reduce thermal and light-induced brightness reversion in lignin-containing pulps, that costs about US$5 per kg. Many compounds for food additives and aroma like allylsyringol (charged at US$1,000 per kg), syringaldehyde and syringol (both charged at US$400 per kg) are among the most valuable components in bio-oil (Pakdel *et al.* 1997).

The development of materials using bio-oil fractions has also received considerable attention in the past few years. The phenolic derivatives presented in the insoluble bio-oil, mainly derived from lignin depolymerisation, replace very successfully petrochemical phenol in fenolic resins (PF-resins). These kinds of resins are the binder in plywood, particle board, wafer board, oriented stand board composites, and they are also the basic material in the friction and abrasive industries. A 50 per cent replacement was found feasible in keeping the resin performance as with pure phenol (Chum and Kreibich 1992). Activated short carbon fibres, suitable for making a filter for water treatment, were produced using a residual material in slow carbonisation tar distillation (Otani *et al.* 1990). The residual pitch recovered during bio-oil distillation is also a binder in pre-baked electrodes. Bio-pitch is more reactive than coal tar pitch and results in a hard cross-linked network with coke (Luengo *et al.* 1995).

Increasing bio-oil use and the scale-up of the new pyrolysis technologies will also impact on the environment and expose a large number of workers to potential health hazards. Bio-oil production and utilisation can cause impacts such as atmospheric emissions, contamination due to water process effluent and human contact with acids, phenolics and aldehydes. All these topics have been under investigation. Carbon oxides and volatile organics are the most likely gas emissions in biomass pyrolysis plants. Despite the lower potential toxicity of bio-oil compared with petroleum and coal tar, water, air and soil contamination have to be avoided (Diebold 1997; Elliott *et al.* 1995).

### 9.5.4 *Slow pyrolysis for charcoal production*

Charcoal produced from either native or planted forests is used around the world for domestic cooking, and refining of metals (copper, bronze, steel, nickel, and aluminium, among others), for the production of chemicals (carbon disulphide, calcium carbide, silicon carbide, sodium cyanide, activated carbon, etc.) and also as an energy source for industrial processes (see also Chapter 8).

#### 9.5.4.1 *Pyrolysis technologies for charcoal production*

More efficient pyrolysis technologies and equipment for charcoal production were developed to increase charcoal yield and/or to recover liquids and gases for energy or other purposes. Some technologies and equipment developed for improving charcoal production are described below:

RECTANGULAR KILN WITH TAR RECOVERY

A rectangular kiln with large wood capacity, adapted with a condenser for tar recovery, was developed by Mannesmann Florestal (MAFLA) in Brazil. This kiln operates to supply charcoal for pig-iron companies, substituting the traditional round-shaped kiln at MAFLA facilities. The kiln improves carbonisation yield, utilisation of by-products, productivity, charcoal quality, work and environment conditions. The recycling of products (gases) is used as an energy source during carbonisation and between kilns for starting a new carbonisation cycle. Tar is recovered and stored for further energy use or to obtain more valuable products by distillation (see Table 8.4).

HIGH-YIELD BIOMASS PYROLYSIS AT ELEVATED PRESSURES FOR CHARCOAL PRODUCTION

Charcoal yields ranging from 38 to 48 per cent were obtained in the Hawaii Natural Energy Institute (HNEI), University of Hawaii (Antal 1992a, b). The charcoal presented similar calorific value, fixed carbon, and volatile contents, compared to the commercial charcoals. The process is based on the pyrolysis at elevated pressures with controlled heating rates and final temperatures in a fixed-bed reactor. The biomass moisture acts as a catalyst for the pyrolysis reaction, increasing charcoal yield. Due to the high pressures, the pyrolysis vapour phases ($H_2O$ and pyrolysis liquids) are in effective contact with the solid phase, maximising charcoal formation (Antal *et al.* 1996).

The experiments on biomass pyrolysis at elevated pressures indicated that the best operation conditions are pressure of 1 MPa, final temperature of 450°C, and total time of two hours. The pyrolysis of different kinds of biomass was conducted in a pressurised batch reactor, configured to convert up to 2 kg of biomass to high-yield charcoal. The pyrolysis occurs under constant pressure in a self-generated gaseous atmosphere. The pyrolysis gases and liquids are burnt in an atmospheric flare and a dry charcoal is obtained. Table 9.9 shows the results for four different kinds of biomass.

*Table 9.9* Results of biomass pyrolysis at elevated pressures

| Biomass | Yield (%)[a] | Fixed carbon content (%)[a] | Volatile content (%)[a] | Ash content (%)[a] | Heating value (MJ/kg)[a] |
|---|---|---|---|---|---|
| Macadamia shell | 48 | 74.1 | 25.6 | 0.3 | 32.1 |
| Macadamia tree | 43 | 74.0 | 25.4 | 0.6 | 30.4 |
| Eucalyptus wood | 46 | 78.6 | 20.2 | 1.2 | 32.6 |
| Sugar-cane bagasse | 44 | 64.7 | 32.6 | 2.7 | 28.3 |

*Source*: Antal (1992b)
*Note*
[a] Dry basis

The results from Table 9.9 show that the charcoal produced at elevated pressures presents a similar fixed carbon content and heating value compared to commercial charcoals, but with a higher mass yield. This charcoal can be used for energy as well as metal-reducing purposes.

CONTINUOUS CHARCOAL PRODUCTION (see also Chapter 8)

A continuous process of pyrolysis for charcoal production and liquids recovery was developed by a Brazilian steel company, ACESITA, and was operated in the late 1980s and early 1990s. Downdraft kilns were used, loading biomass on the top and unloading charcoal from the bottom. No external energy source was required for carbonisation as pyrolysis gases were recirculated and burnt, supplying enough energy for the process. The condensable liquids were recovered and used as fuel or processed to generate chemical products. In the latter case, a further treatment was necessary, due to the concentration of water, tar, and pyrolysis acids in the liquids. The charcoal and tar yields were, on average, 33 per cent and 11 per cent respectively. The plant presented a productivity of 0.5 t/hour and charcoal production capacity of 1800 t/year (Rezende *et al.* 1992). However, after the privatisation of ACESITA this plant was dismantled.

*9.5.4.2 Economic aspects of charcoal production*

Charcoal is generally used as a thermal energy supplier or as a carbon source for metal reducing. The final use depends on the consumer and its price can be in the range

US$0.09–1.00/kg. Usually, for domestic and commercial markets, the prices are higher and market rules are very flexible. For the industrial sector (e.g. cement, chemical, beverages, paper and pulp industries), the prices are lower because charcoal is sold in large amounts.

Brazil is the word's largest producer, representing 6 million tonnes per year. The main consumer is the steel and pig-iron sector, which uses charcoal in the blast furnaces for iron reducing. In this case, charcoal price range is US$0.09–0.12/kg and the market is quite stable, dependent on whether it is originated from native or planted forests and on the distance from the producer to the final consumer (increasing, on average, US$3/t for each 100 km). The price of charcoal produced from planted forests decreased slightly in the last ten years in Brazil, which indicates a reduction in the forestation costs. Also, eucalyptus charcoal is more homogeneous compared to native wood charcoal, presenting better characteristics for use in blast furnaces (Abracave 1997).

### 9.5.5 *Biomass torrefaction*

Charcoal production is frequently an inefficient and predatory activity, causing serious environmental problems. An alternative to reduce these impacts consists of using new technologies to produce charcoal, or other solid fuel, which can be used for energy purposes. The efficient use of agricultural and forest residues can be a rational solution to decrease wood use and consequently forest exploitation. These residues are generated from agriculture, industrial activities and wood management. In general, they present large polymorphism, heterogeneous particle size, high moisture content, low density, and low calorific value. Thus, to increase their energy efficiency, the application of conversion processes which improve their energy properties is necessary. Most of the time this procedure is economically feasible, due to the low cost of these materials. The main industrial activities which generate biomass residues in Latin America are the sugar-cane industry, wood exploitation and some agricultural activities. The most important residues, considering their large production are: sugar-cane bagasse, trash, wood residues, and agricultural residues.

Torrefaction of biomass residues is a feasible alternative to improving their energy properties. The torrefaction process consists basically of heating biomass in an inert atmosphere, with low heating rates up to the maximum temperature of 300°C. Thus, it can be considered as slow-rate pyrolysis. For powdered residues, such as bagasse and sawdust, the densification of the torrefied materials results in solid fuels with high density, low moisture, and a desired shape. The torrefied biomass presents the following properties (Battacharya 1990):

1  hydrophobic nature, as it does not absorb water;
2  higher calorific value and less smoke when burnt;
3  it can be used in the steel industry, and also in gasification and combustion processes;
4  similar density and mechanical strength compared to the initial biomass.

A bench unit for biomass torrefaction was designed by the Group for Alternative Fuels (GCA), University of Campinas, Brazil. In some applications, torrefied residues may substitute firewood and charcoal, reducing costs and environmental impacts during forest exploitation (Felfly *et al.* 1998). The unit is basically composed of two chambers, one for combustion and the other for thermal treatment. The combustion chamber supplies thermal energy for the process, by burning biomass residues. Hot vapours and gases produced in the torrefaction chamber are recirculated and burnt in the combustion chamber, improving overall energy efficiency and avoiding atmospheric pollution. Torrefied products of several experiments have presented low moisture and a hydrophobic nature. According to the process conditions and properties of the starting material, the fixed carbon content of the products ranged from 25 to 40 per cent and the overall yield from 70 to 90 per cent. The average calorific value was near 23 MJ/kg, which is an intermediate value between biomass and charcoal. Torrefied biomass presents good quality for combustion and gasification purposes, with favourable characteristics for storage and transportation, mainly because of its low moisture, high density and hydrophobic nature.

This torrefaction system is easy and cheap to build and can be used in isolated regions without qualified labour. Any solid biomass, like wood and briquettes, can be used as feedstock. With proper control of the amount of air and biomass for combustion, it is possible to change process conditions and get torrefied materials with different properties. This system is operated using wood briquettes (Felfly *et al.* 1998).

In the torrefaction of biomass materials, only water and some volatile components with low heating value are driven out. The heavier components with larger calorific value stay in the products, resulting in high yields and energy content. The hydrophobic behaviour of torrefied materials facilitates their storage and preservation for long periods. These materials are also good feedstock for combustion, gasification, and steel-making processes. The application of torrefaction units for improving energy properties of biomass residues can contribute to reducing forest exploitation and pollution from burning residues in the field.

### 9.5.6 *Charcoal activation*

The controlled oxidation of charcoal or other carbon can give rise to activated carbon, a special material characterised by a large specific surface area of 300–2500 m$^2$/g, which allows the physical absorption of pollutants from liquid and gas flows (Kirk 1992). The activation of carbon is accomplished by two basic processes – chemical and physical activation, depending upon the starting material and whether a low- or high-density, powdered or granular-activated carbon is desired.

### 9.5.6.1 *Chemical activation*

The material is treated with an activating agent in solution. A chemical reaction takes place between the activating agent and the internal carbon atoms, increasing the volume and creating new internal pores. The most used chemical activating agents are phosphoric acid, zinc chloride, sulphuric acid, potassium sulphide, and calcium chloride (Jagtoyen and Derbyshire 1993).

### 9.5.6.2 *Physical activation*

An activating gas ($O_2$, $CO_2$ or steam) at elevated temperatures reacts with the internal carbon atoms, increasing internal pore volume and surface area. The activation is usually preceded by a primary carbonisation of the raw material (Jankowska *et al.* 1991).

Activated carbons are basically used for purification of water and products from the chemical, pharmaceutical, and food industries, prevention of environmental pollution and for meeting the ever-increasing demands for purity of natural and synthetic products (Bezzon *et al.* 1997a, b). The consumption of activated carbon has, in recent years, developed a consistent upward tendency, mainly due to global and local environmental regulations concerning air and water pollution. The USA is the largest producer and consumer of activated carbon, followed by Europe and Asia (Ullman 1991). The main industrial activation technologies are based on physical activation, using oxidising gases at high temperatures. The most common equipment comprises shaft furnaces, rotary kilns, and fluidised bed furnaces.

The price of most products is in the range US$0.70–6.00/kg, but it can be as high as US$12.00/kg. Powdered activated carbon, a less expensive form, usually used for liquid phase applications, is generally utilised once and then disposed of. In some cases, however, granular and shaped products are regenerated and reused. The production capacity for granular and shaped products is split, with about two-thirds for the liquid phase and one-third for the gas phase applications.

### 9.5.7 *Briquetting*

Due to the manipulation of charcoal during production and transportation, around 10–20 per cent of the total mass is converted into powdered fines, which cannot be used for reducing metal or for some energy applications. Charcoal fines present low density and economic value, with an undesirable shape for some applications. The densification of fines reduces volume, increasing density, and in some cases, moisture content decreases. The result is a charcoal with high density and the proper shape and size for a desired application.

Briquetting is the densification of fines using high-pressure presses, with or without heating. Small particles are pressed, usually with a suitable binder, to form solids with higher commercial value (briquettes). The most common presses for briquetting are the extrusion and roll presses. The selection of the binder depends on the final use for the briquette. In general, binders are more expensive than charcoal and are very important for the briquette's final cost. Pitch, molasses, starch, bitumin, and calcium carbonate are the most widely used binders for briquetting. The binder must be highly agglomerating, cheap and easy to use, providing high mechanical strength to the briquette. Depending on the binder characteristics, densification can occur at room or elevated temperatures. For pitch or bitumin, for example, the briquette must be heated in order to improve the binder agglomerating properties. The elevated temperature briquetting can be applied to biomass residues without the use of a binder, due to the decomposition of the lignin in the biomass at elevated temperatures, which acts as a natural binder for the biomass particles (Bezzon and Luengo 1993).

The production and consumption of briquettes from charcoal fines is consolidated in the USA and some European countries, since these countries have a specific market for these fuels, especially for domestic cooking. In Brazil, use of charcoal in the steel industry means that there is a great opportunity to increase the production of briquettes of fines generated by coal manipulation. The main problem is the high cost of conventional binders, which increases the final briquette price. Despite the higher prices, the briquette presents some advantages, like uniform properties (moisture, heating value, and carbon content), desirable shape and size, easy storage, and regular supply. Its production also decreases potential environmental problems by disposing of or storing charcoal fines.

## 9.6 Bibliography

Abracave (1997) *Statistical Yearbook – 1997*, Belo Horizonte, Brazil.

Anderl, H. and Mory, A. (1998) 'Operation experiences in the CFB gasification project BioCoComb for biomass with co-combustion of the gas in a PF boiler at Zeltweg power plant, Austria', *Proceedings of VTT Seminar, 'Power Production from Biomass III'*, Espoo, Finland, September 14–15.

Antal, Jr., M.J. (1983) 'Biomass pyrolysis: a review of the literature, part 1 – carbohydrate pyrolysis', in Boer, K.W. and Duffie, J.A. (eds) *Advances in Solar Energy*, American Solar Energy Society, 1, 61–111.

Antal, Jr., M.J., Mok, W.S.L., Szabó, P., Várhegyi, G. and Zelei, B. (1992a) 'Formation of charcoal from biomass in a sealed reactor', *Industrial and Engineering Chemistry Research*, 31, 1162–6.

Antal, Jr., M.J., Mok, W.S.L., Várhegyi, G. and Szekely, T. (1992b) 'Review of methods for improving the yield of charcoal from biomass', *Energy and Fuels*, 4, 221–5.

Antal, Jr., M.J., Croiset, E., Dai, X., de Almeida, C., Mok, W.S.L. and Norberg, N. (1996) 'High yield biomass charcoal', *Energy and Fuels*, 10, 652–8.

Babcock and Wilcox (1992) 'Atmospheric pressure fluidized-bed boilers' (Chapter 16), in *Steam*, 40th edn, Barberton, USA: Babcock and Wilcox.

Bain, R.L. (1991) 'Methanol from biomass: assessment of production costs', paper presented at the Hawaii Natural Energy Institute Renewable Transportation Alternatives Workshop, Honolulu, USA, January 9–11.

Barducci, G., Ulivieri, P., Plzinetti, G.C., Donati, A. and Repetto, F. (1997) 'New developments in biomass utilization for electricity and low-energy gas production in the gasification plant of Greve in Chianti', in Bridgwater, A.V. and Boocock, D.G.B. (eds) *Developments in Thermochemical Biomass Conversion*, London: Blackie.

Battacharya, S.C. (1990) 'Carbonized and uncarbonized briquettes from residues', *Proceedings of Workshop on Biomass Thermal Processing*, London: Shell.

Beenackers, A.A.C.M. and Bridgwater, A.V. (1989) 'Gasification and pyrolysis of biomass in Europe', in *Pyrolysis and Gasification*, 1, 129–55, Elsevier Applied Sciences.

Bezzon, G. and Luengo, C.A. (1993) 'Briquetting of agricultural residues at elevated pressures and temperatures', *Proceedings of XII Congresso Brasileiro de Engenharia Mecânica*, Brasilia, 2, 1101–4.

Bezzon, G., Luengo, C.A., Dai, X. and Antal, Jr., M.J. (1997a) 'High yield carbons from eucalyptus wood at elevated pressures', *Proceedings of XXIII Biennial Conference on Carbon*, Pennsylvania, 2, 258–9.

Bezzon, G., Luengo, C.A., Capobianco, G., Dai, X. and Antal, Jr., M.J. (1997b) 'Experimental and numerical analysis of the eucalyptus wood activation', *10th Jornadas Argentinas de Catalisis*, Buenos Aires.

Bridgwater, A.V. (1995) 'The technical and economic feasibility of biomass gasification for power generation', *Fuel*, 74, 631–53.

Bridgwater, A.V. (1997) 'Biomass fast pyrolysis and applications in Europe', *Proceedings of Third Biomass Conference of the Americas*, Montreal, Canada, 797–809.

Bridgwater, A.V. (1998) 'The status of fast pyrolysis of biomass in Europe', *Proceedings of 10th European Biomass Conference and Technology Exhibition*, Würzburg, Germany, 268–71.

Bridgwater, A.V. and Bridge, S.A. (1991) 'A review of biomass pyrolysis and pyrolysis technologies' in Bridgwater, A.V. and Grassi, G. (eds) *Biomass Pyrolysis Liquids Upgrading and Utilization*, 11–92.

Bühler, R. (1997) 'Fixed bed gasification for electricity generation application in Europe', in Kaltschmitt, M. and Bridgwater, A.V. (eds) *Biomass Gasification and Pyrolysis – State of the Art and Future Prospects*, Newbury: CPL Press, 117–28.

Calrsen, H. (1998) 'Status and prospects of small-scale power production based on Stirling engines – Danish experiences', *Proceedings of VTT Seminar, 'Power Production from Biomass III'*, Espoo, Finland, September 14–15.

CEC (California Energy Commission) (1996) *Northeastern California Ethanol Manufacturing Feasibility Study*, California.

Chem Systems (1990) *Assessment of Cost of Production of Methanol From Biomass*, USDOE Report No. DOE/PE-0079P, Washington DC, US Department of Energy.

*Chemical Marketing Reporter* (1999) New York: Schnell Publishing Company, **255**, (1).

Christianson, D.P., Niemi, G.J., Hanowiski, J.M., and Collins, P. (1994) 'Perspectives on biomass energy tree plantations and changes in habitat for biological organisms', *Biomass Bioenergy*, **6** (1/2), 31.

Chum, H.L. and Black, S.K. (1990) *Process for fractionating fast-pyrolysis oils, and products derived therefrom*, US patent number 4,942,269.

Chum, H.L. and Kreibich, R.E. (1992) *Process for preparing phenolic formaldehyderesole resin products derived from fractionated fast-pyrolysis oils*, US patent number 5,091,499.

COALBRA (Coque e Álcool de Madeira) (1983) *Produção de Etanol da Madeira*, Brasília: COALBRA.

Cooney, C.L., Wang, D.I.C., Wang, S.D., Gordon, J. and Jiminez, M. (1978) 'Simultaneous cellulose hydrolysis and ethanol production by cellulolytic anaerobic bacterium', *Biotechnology Bioengineering Symposium Series*, 8, 103.

Craig, K.R. and Mann, M.K. (1996) *Cost and Performance Analysis of Biomass-Based Integrated Gasification Combined Cycle (BIGCC) Power Systems*, NREL Report, Golden, CO: NREL.

DeLuchi, M.A., Larson, E.D. and Williams, R.H. (1991) *Hydrogen and Methanol Production from Biomass and Use in Fuel Cell and Internal Combustion Engine Vehicles: A Preliminary Assessment*, Report No. PU/CES 263, Princeton: Princeton University Press.

Diebold, J.P. (1997) *A Review of the Toxicity of Biomass Pyrolysis Liquids Formed at Low Temperature*, NREL Report TP-430-22739, Golden, CO: NREL.

Diebold, J.P. and Scahill, J. (1987) 'Upgrading pyrolysis vapors to aromatic gasoline with zeolite catalysis at atmospheric pressure', *Biomass Thermochemical Conversion Contractor's Technical Review Meeting*, Atlanta.

Eindensten, L., Yan, J. and Svedberg, G. (1996) 'Biomass externally fired gas turbine cogeneration', *Journal of Engineering for Gas Turbine and Power*, 118, 604–9.

Elliot, D.C. and Maggi, R.R. (1997) 'Workshop report: catalytic process', in Bridgwater, A.V. and Boocock, D.G.B. (eds) *Developments in Thermochemical Biomass Conversion*, 2, 1626–30.

Elliot, D.C., Hart, T.R., Neuensschawander, G.G., McKinney, M.D., Norton, M.V. and Abrams, C.W. (1995) *Environmental Impacts of Thermochemical Biomass Conversion*, NREL Report TP-433-7867, Golden, CO: NREL.

Elliott, P. and Booth, R. (1993) *Brazilian Biomass Power Demonstration Project*, special project brief, London: Shell International.

Faaij, A., van Ree, R., Waldheim, L., Olsson, E., Oudhuis, A., van Wijk, A., Daey-Ouwens, C. and Turkenburg, W. (1998) 'Gasification of biomass wastes and residues for electricity production', *Biomass and Bioenergy*, 12, 387–407.

Farris, M., Paisley, M.A., Irving, J. and Overend, R.P. (1998) 'The Batelle/FERCO biomass gasification process – design, engineering, construction and startup phase', *Proceedings of VTT Seminar 'Power Production from Biomass III'*, Espoo, Finland, September 14–15.

Felfly, F.F., Luengo, C.A. and Bezzon, G. (1998) 'Bench unit for biomass torrefaction', *Proceedings of 10th European Biomass Conference and Technology Exhibition*, Würzburg, Germany, 1593–5.

Gary, H.H. and Handwerke, G.E. (1984) *Petroleum Refining Technology and Economics*, 2nd edn, New York: Marcel Dekker.

Gaur, S., Anuradha, G., Rao, T.R., Iyer, P.V.R. and Grover, P.D. (1985) 'Development and operation of a biomass gasification system for 5HP irrigation pumps', *3rd Annual Workshop of Biomass and Coal Conversion Programs of the USAID/GOI Alternative Energy Resources Development*.

Gauss, W.E., Suzuki, S. and Takagi, M. (1976) *Manufacture of alcohol from cellulosic materials using plural ferments*, US patent 3,990,944.

Graham, R.G., Freel, B.A. and Bergougnou, M.A. (1988) 'The production of pyrolytic liquids, gas and char from wood and cellulose by fast pyrolysis', in Bridgwater, A.V. and Kuester, J.L. (eds) *Research in Thermochemical Biomass Conversion*, Phoenix, 269.

Gregoire, C.E. (1992) *Technoeconomic Analysis of the Production of Biocrude from Wood*, NREL Report TP-430-5435, Golden, CO: NREL.

Grethlein, H.E. and Dill, T. (1993) *The Cost of Ethanol Production from Lignocellulosic Biomass – a Comparison of Selected Alternative Process*, Final Report No. 58-1935-2-050, US Department of Agriculture, Washington DC.

Hansen, J.L. (1992) 'Fluidized bed combustion of biomass: an overview', *Proceedings of Biomass Combustion Conference*, Reno: US Department of Energy, Washington DC.

Hart, D. and Bauen, A. (1998) *Fuel Cells: Clean Power, Clean Transport, Clean Future*, London: Financial Times Energy Publishing.

Hinman, N.D., Wright, J.D., Hoagland, W. and Wyman, C.E. (1989) 'Xylose fermentation: an economic analysis', *Applied Biochemical Biotechnology*, **20/1**, 391.

*Hydrocarbon Processing* (1992) Gas Processing Handbook '92: A Special Report, Houston: Gulf, **71**, 83–140.

*Hydrocarbon Processing* (1994) *Gas processes '94: A Special Report*, Houston: Gulf, **73**, 67–116.

Ingram, L.O., Conway, T., Clark, D.P., Sewell, G.W. and Preston, J.F. (1987) 'Genetic engineering of ethanol production in Escherichia coli', *Applied Biochemical Biotechnology*, **53**, 2420.

Jagtoyen, M. and Derbyshire, F. (1993) 'Some considerations of the origins of porosity in carbons from chemically activated wood', *Carbon*, **31**, 1185–92.

Jakobsen, H.H., Pedersen, L.T. and Houmoller, S. (1998) *Technologies for Small-scale Wood-Fueled Combined Heat and Power*, DK-Technik, Soeborg, Denmark, available through NTIS, Springfield, DE98771786.

Jankowska, H., Swiatkowski, A. and Choma, J. (1991) *Active Carbon*, Ellis Horwood Limited.

Jeffries, T.W. (1981) 'Fermentation of xylulose to ethanol using xylose isomerase and yeast', *Biotechnology Bioengineering Symposium Series*, **11**, 315.

Johansson, L. and Ziph, B. (1997) 'Co-generation from wood chip combustion using a Stirling power conversion system and gasifier', *Proceedings of Third Biomass Conference of the Americas*, Montreal, Canada, 577.

Johnson, D.K., Chum, H.L., Anzick, P. and Baldwin, R.M. (1990) 'Preparation of a lignin-derived pasting oil', *Applied Biochemical Biotechnology*, **24/5**, 31.

Kaltschmitt, M., Rösch, C. and Dinkelbach, L. (eds) (1998) *Biomass Gasification in Europe*, European Commission Report AIR3-CT-94-2284, Brussels: EC.

Kinoshita, C.M., Turn, S.Q., Overend, R.P. and Bain, R.L. (1997) 'Power generation potential of biomass gasification systems', *Journal of Energy Engineering*, **123**, 88–99.

Kirk, R.E. (1992) *Encyclopedia of Chemical Technology – Activated Carbon*, 4, 1015–36.

Kloster, R., Oeljekaus, G. and Pruschek, R. (1996) 'Dry high temperature coal gas cleaning for gasification combined cycles system integration and process optimisation', in Schmidt, E. (ed.) *High Temperature Gas Cleaning*, TH Karlsruhe, 743–56.

Knoef, H.A.M. and Stassen, H.E.M. (1997) *Small-scale Biomass Gasification Monitoring; Volume II: Gasifier Performance Final Report*, Enschede, The Netherlands: BTG University of Twente, prepared for the Danish Energy Agency.

Kurkela, E. (1998) 'Progas – gasification and pyrolysis R&D programme 1997–1998', *Proceedings of VTT Seminar 'Power Production from Biomass III'*, Espoo, Finland, September 14–15.

Larson, E.D. and Consonni, S. (1997) 'Performance of black-liquor gasifier/gas turbine combined cycle co-generation in the kraft pulp and paper industry gasifier', in Overend, R.P. and Chornet, E. (eds) *Proceedings of Third Biomass Conference of the Americas*, Montreal, Canada, August 24–29, 1495–512.

Larson, E.D. and Katofsky, R.E. (1992) *Production of Hydrogen and Methanol from Biomass*, Report No. PU/CES 271, Princeton: Princeton University Press.

Lau, F.S., Carty, R.H., Onishak, M. and Bain, R.L. (1993) 'Development of the IGT RENU-GAS® process', *Proceedings of EPRI Strategic Benefits of Biomass and Wastes Conference*, Washington DC.

Leech, J. (1997) 'Running a dual fuel diesel engine on crude pyrolysis oil', in Kaltschmitt, M. and Bridgwater, A.V. (eds) *Biomass Gasification and Pyrolysis – State of the Art and Future Prospects*, Newbury: CPL Press, 495–7.

Luengo, C.A., Rocha, J.D., Julião, J.T., Martin, Y., Garcia, R. and Moinelo, S.R. (1995) 'Electrode grade carbon prepared with different pitch binders', *Proceedings of 8th International Conference on Carbon Science*, II, Oviedo, Spain, 1173–6.

Lundqvist, R.G. (1993) 'The IGCC demonstration plant at Värnamo', *Bioresource Technology*, 46, 49–53.

Lynd, L.R. (1996) 'Overview and evaluation of fuel ethanol from cellulosic biomass: technology, economics, the environment, and policy', *Annual Energy Environment Review*, 21, 403–65.

Lynd, L.R., Ahn, H.J., Anderson, G., Hill, P., Kersey, D.S. and Klapatch, T. (1991a) 'Thermophilic ethanol production: investigation of ethanol yield and tolerance in continuous culture', *Applied Biochemical Biotechnology*, 28/9, 549.

Lynd, L.R., Cushman, J.H., Nichols, R.J. and Wyman, C.E. (1991b) 'Fuel ethanol from cellulosic biomass', *Science*, 251, 1318–23.

Lynd, L.R., Elander, R.T. and Wyman. C.E. (1996) 'Likely features and costs of mature biomass ethanol technology', *Applied Biochemical Biotechnology*, 57/8, 741–61.

MAFLA (Mannesmann Florestal) (1997) *Internal Report*, Belo Horizonte, Brazil.

Mandels, M.L., Hontz, L. and Nystrom, J. (1974) 'Enzymatic hydrolysis of waste cellulose', *Biotechnology Bioengineering*, 16, 1471.

Marrison, C.I. and Larson, E.D. (1995) 'Cost versus scale for advanced plantation-based biomass energy systems in the USA and Brazil', *Proceedings of Second Biomass Conference of the Americas*, NREL/CP-200-8098.

McGowin, C.R. and Wiltsee, G.A. (1996) 'Strategic analysis of biomass and waste fuels for electric power generation', *Biomass and Bioenergy*, 10, 167–75.

McMillan, J.D. (1994) 'Conversion of hemicellulose hydrolyzates to ethanol', in Himmel *et al.* (eds) *Enzymatic Conversion of Biomass for Fuel Production*, ACS Symposium Series 566, 411–37, Washington DC.

Merhling, P. and Vierrath, H. (1989) *Gasification of Lignite and Wood in the Lurgi Circulating Fluidized-Bed Gasifier*, Lurgi GmbH, Frankfurt, Germany, EPRI Report GS-6436, Palo Alto: Electric Power Research Institute.

Mills, G. (1993) *Status and Future Opportunities for Conversion of Synthesis Gas to Liquid*

*Energy Fuels: Final Report*, Center for Catalytic Science and Technology, Department of Chemical Engineering, University of Delaware, for the National Renewable Energy Laboratory, Golden, CO, NREL Report TP-421-5150.

Mitchell, C.P., Bridgwater, A.V., Stevens, D.J., Toft, A.J. and Watters, M.P. (1995) 'Techno-economic assessment of biomass to energy', *Biomass and Bioenergy*, 9, 205–26.

Mukunda, H.S., Dasappa, S. and Shrinvasa, U. (1993) 'Open-top wood gasifiers' (Chapter 16), in Johansson, T.B. *et al.* (eds) *Renewable Energy: Sources for Fuels and Electricity*, Washington DC: Island Press, 699–728.

Naccarati, R. and de Lange, H.J. (1998) 'The use of biomass-derived fuel in a gas turbine for the Energy Farm project', *Proceedings of VTT Seminar 'Power Production from Biomass III'*, Espoo, Finland, September 14–15.

Nieminen, J. (1998) 'Biomass CFB gasifier connected to a 350MW$_{th}$ steam boiler fired with coal and natural gas – THERMIE demonstration project in Lahti in Finland, *Proceedings of VTT Seminar Power Production from Biomass III'*, Espoo, Finland, September 14–15.

Ng, T.K., Weimer, P.J. and Zeikus, J.G. (1977) 'Cellulolytic and physiological properties of clostridum thermocellum', *Archives of Microbiology*, 114.

Novem (1996) *State of Small-scale Biomass Gasification Technology*, study performed by Enschede, The Netherlands: BTG University of Twente.

Otani, C., Pasa, V.M.D. and Carazza, F. (1990) 'The structure and chemical characteristics variation of wood tar pitch during its carbonization', *Proceedings of International Symposium on Carbon*, Tsucuba, 1, 546–9.

Overend, R.P. (1998) *Biomass and Bioenergy in China in 1998*, The Department of Environmental Protection and Energy, People's Republic of China, for the National Renewable Energy Laboratory, Report No. NREL/SR-570-24860, Golden, CO: NREL.

Overend, R.P., Kinoshita, C.M. and Antal, Jr., M.J. (1996) 'Bioenergy in transition', *Journal of Energy Engineering*, 122 (December).

Padró, C.E.G. and Putsche, V. (1999) *Survey of the Economics of Hydrogen Technologies*, NREL Report, Golden, CO: NREL (forthcoming).

Paisley, M.A. and Overend, R.P. (1994) 'Biomass power for power generation', EPRI Coal Gasification Conference, San Francisco, October.

Paisley, M.A., Litt, R.D., Taylor, D.R., Tewksbury, T.L., Hupp, D.E. and Wood, R.D. (1992) *Operation and Evaluation of an Indirectly Heated Biomass Gasifier*, Battelle Columbus Laboratory, for the National Renewable Energy Laboratory, Golden, CO: NREL.

Paisley, M.A., Farris, G., Slack, W. and Irving, J. (1997) 'Commercial operation of the Battelle/FERCO biomass gasification process – initial operation of the McNeil gasifier', in Overend, R.P. and Chornet, E. (eds) *Proceedings of Third Biomass Conference of the Americas*, Montreal, Canada, 579–88.

Pakdel, H., Piskorz, J., Himmelblau, A. and Clements, D. (1997) 'Workshop on Chemical from Biomass', in Bridgwater, A.V. and Boocock, D.G.B. (eds) *Developments in Thermochemical Biomass Conversion*, I, 1621–5.

Parikh, P.P., Bhave, A.G., Kapse, D.V. and Shashikantha (1989) 'Study of thermal and emission performance of small gasifier-dual fuel engine systems', *Biomass Journal*, 19, 75–97.

Parikh, P.P., Banerjee, P.K. and Veekar, S.S. (1994) *Biomass Gasifiers and Producer-Gas Engine Systems – Overview of Indian R&D Activities*, Bombay: Biomass Gasification Research Group, Department of Mechanical Engineering, Indian Institute of Technology.

Perry, R.H. and Chilton, C.H. (1973) *Chemical Engineers' Handbook*, 5th edn, New York: McGraw-Hill.

Podesser, E. (1997) 'Small-scale co-generation in biomass furnaces with Stirling engines', *Proceedings of Third Biomass Conference of the Americas*, Montreal, Canada, 747–55.

PyNE (1997) 'What is pyrolysis liquid?', *Pyrolysis Network for Europe*, Birmingham: Aston University, 4, 11.

Ranney, J.W. and Mann, L.K. (1994) 'Environmental considerations in energy crop production', *Biomass and Bioenergy*, 6, 211.

Reed, T.R. and Gaur, S. (1998), *A Survey of Biomass Gasification 1998; Volume 1, Gasifier Projects and Manufacturers Around the World*, Golden, CO: Biomass Energy Foundation Press.

Reese, E.T. (1976) 'History of the cellulose program at the US Army Natick Development Center', *Biotechnology Bioengineering Symposium Series*, 6, 9–20.

Rezende, M.E.A., Pasa, V.M.D. and Lessa, A. (1992) 'Continuous charcoal production: a Brazilian experience', *Proceedings of Advances in Thermochemical Biomass Conversion*, Interlaken, Switzerland, 2, 1289–98.

Rocha, J.D., Luengo, C.A. and Snape, C.E. (1996) 'Hydrodeoxygenation of oils from cellulose in single- and two-stage hydropyrolysis', *Renewable Energy*, 9, 950–3.

Roy, C., Blanchette, D., Caumia, B. and Labrecque, B. (1992) *Conceptual Design and Evaluation of a Biomass Vacuum Pyrolysis Plant*, Quebec: Laval Universty.

Roy, C., Yang, J., Blanchette, D., Korving, L. and de Caumia, B. (1997) 'Development of a novel vacuum pyrolysis reactor with improved heat transfer potential' in Bridgwater, A.V. and Boocock, D.G.B. (eds) *Developments in Thermochemical Biomass Conversion*, I, 351–67.

Solantausta, Y., Bridgwater, A.T. and Deckman, D. (1995) 'Feasibility of power production with pyrolysis and gasification systems', *Biomass and Bioenergy*, 9, 257–69.

South, C.R. and Lynd, L.R. (1994) 'Analysis of conversion of particulate biomass to ethanol in continuous solids-retaining cascade reactors', *Applied Biochemical Biotechnology*, 45/46, 467–81.

Ståhl, K. and Neergard, M. (1998) 'Experiences from the Värnamo IGCC demonstration plant', *Proceedings of VTT Seminar on 'Power Production from Biomass III'*, Espoo, Finland, September 14–15.

Stassen, H.E. and Koele, H.J. (1997) 'The use of LCV-gas from biomass gasifiers in internal combustion engines', in Kaltschmitt, M. and Bridgwater, A.V. (eds) *Biomass Gasification and Pyrolysis – State of the Art and Future Prospects*, Newbury: CPL Press, 269–81.

Takagi, M., Abe, S., Suzuki, S., Evert, G.H. and Yata, N. (1977) 'A method of production of alcohol directly from yeast', *Proceedings of Bioconversion Symposium*.

Tewari, Y.B., Steckler, D.K. and Goldberg, R.N. (1985) 'Thermodynamics of the conversion of aqueous xylose to xylulose', *Biophysical Chemistry*, 22, 1815.

Toft, A.J. and Bridgwater, A.V. (1997) 'How fast pyrolysis competes in the electricity generation market', in Kaltschmitt, M. and Bridgwater, A.V. (eds) *Biomass Gasification and Pyrolysis – State of the Art and Future Prospects*, Newbury: CPL Press, 504–15.

Twigg, M.V. (1989) *Catalyst Handbook*, 2nd edn, London: Wolfe Publishing.

Ullman (1991) *Ullman's Encyclopedia of Industrial Chemistry – Activated Carbon*, 5, 124–40.

Van Den Broek, R., Faalj, A. and van Wuk, A. (1996) 'Biomass combustion for power generation', *Biomass and Bioenergy*, 11, 271–81.

van Walsum, P., Allen, S.G., Spencer, M.J., Laser, M. and Antal, M. (1996) 'Conversion of lignocellulosics pretreated with hot compressed liquid water to ethanol', *Advanced Biochemical Biotechnology*, 57/58, 157–70.

Veldhuis, M.K., Christensen, L.M. and Fulmer, E.I. (1936) 'Production of ethanol by thermophilic fermentation of cellulose', *Ind. Eng. Chem.*, 28, 430.

Walter, A. and Overend, R.P. (1999) 'Financial and environmental incentives: impact on the potential of BIG-CC technology in the sugar-cane industry', *Renewable Energy*, 16, 1045–8.

Walter, A.C.S., Souza, M.C. and Overend, R.P. (1998) 'Preliminary evaluation of co-firing natural gas and biomass in Brazil', *Proceedings of Brazilian National Meeting on Thermal Sciences*, Rio de Janeiro, 1432–7.

Wan, E.I. and Fraser, M.D. (1989) 'Economic assessment of advanced biomass gasification systems', paper presented at the *Proceedings of IGT Biomass Conference*, New Orleans, USA, February.

Wenzl, H.F. (1983) *The Chemical Technology of Wood*, New York: Academic Press.

Wilen, C. and Kurkela, E. (1997) 'Country report, Finland', in *State of the Art Report on Biomass Gasification*, European Union, AIR3-CT94-2284 and IEA Bioenergy Biomass Utilization, Task XIII.

Wilke, C.R., Yang, R.D. and Stockar, U.V. (1976) 'Preliminary cost analysis for enzymatic hydrolysis of newsprint', *Biotechnology Bioengineering Symposium Series*, 6, 155.

Williams, R.H. and Larson, E.D. (1996) 'Biomass gasifier gas turbine power generating technology', *Biomass and Bioenergy*, 10, 149–66.

Wright, J.D. (1988a) 'Ethanol from biomass by enzymatic hydrolysis', *Chemical Engineering Progress*, 84, 6211.

Wright, J.D. (1988b) 'Ethanol from lignocellulose: an overview', *Energy Progress*, 8, 81.

Wright, J.D., Power, A.J. and Bergeron, P.W. (1985) *Evaluation of Concentrated Halogen Acid Hydrolysis Processes for Alcohol Fuel Production*, SERI/TR-232-2386, Golden, CO: NREL.

Wyman, C.E. (1994) 'Ethanol from cellulosic biomass: technology, economics, and opportunities', *Bioresource Technology*, 50, 3–16.

Wyman, C.E. and Goodman, B.J. (1993) 'Near term applications of biotechnology for fuel ethanol production from cellulosic biomass', in Busche, R.M. (ed.) *Opportunities for Innovation: Biotechnology*, Gaithersburg: National Institute Stand. Technology, 151–90.

Wyman, C.E., Bain, R.L., Hinman, N.D. and Stevens, D.J. (1992) 'Ethanol and methanol from cellulosic biomass' (Chapter 21), in Johansson, T.B. *et al.* (eds) *Renewable Energy: Sources for Fuels and Electricity*, Washington DC: Island Press, 865–924.

# 10

# CONCLUDING REMARKS AND FUTURE DIRECTIONS

*Frank Rosillo-Calle*

In this concluding chapter we bring together the main issues and findings examined throughout the book and indicate the key policy and technical tasks facing those who would seek to enhance the contribution that will be made by biomass to our future energy needs.

## 10.1 Chapter 1 (Hall, House and Scrase)

Biomass has been the main source of energy for humanity since time immemorial. It was only with the age of oil and coal in the twentieth century, and particularly since World War II, that biomass was gradually replaced. But as the authors of Chapter 1 clearly show, biomass not only remains the main source of energy in many developing countries, but it is also growing in importance in many industrial countries. Biomass is used both in its primitive traditional forms and in advanced modern industrial applications.

Biomass currently supplies one-third of the primary energy in developing countries as a whole, but for many such countries, e.g. Ethiopia, Uganda, Rwanda, Nepal, and Tanzania, biomass supplies over 90 per cent of their primary energy requirements. Even in large countries, such as China, India and Brazil, biomass supplies over 30 per cent of their primary energy. In some industrialised countries biomass is also a major source of energy, e.g. 19 per cent in Finland, and 18 per cent in Sweden.

In this chapter the authors ask what the consequences will be for the future of the two billion people who currently depend on biomass. A major problem is that in many poor parts of the world, biomass continues to be used in its traditional forms, with very low efficiency and considerable potential for environmental damage. In this chapter Hall *et al.* make a strong case for the need to modernise biomass to increase energy efficiency and environmental sustainability. The numerous advantages and potential entrepreneurial opportunities of modern and efficient uses of biomass energy are also discussed.

The authors argue that nearly 90 per cent of the world's population will reside in developing countries by 2050, thus implying that biomass will not go away but stay with us for ever. Also, as Table 1.1 shows, all major energy scenarios already include biomass as a significant source of energy in the near future. A major problem of biomass energy has been (and still is) grossly inadequate data which has seriously affected sound policy decision-making. These problems and the need for time series data are discussed, together with problems of image and land uses.

Section 1.2 examines the implications of potential future use of biomass energy according to various scenarios, e.g. RIGES, ECES, FFES, WEC, SHELL, IPCC, IIASA/WEC and the IEA, all of which include a significant role for biomass in the future energy supply mix. Land availability, yields and large-scale energy plantations are also addressed.

Land availability is perceived by many as a major constraint of large-scale modern uses of biomass energy, but as the authors show, and as has been demonstrated by a large number of studies, there is plenty of land available. For example, a recent study by FAO indicates that Latin America presently uses only 15 per cent of its potential cropland and predicts about 23 per cent use by 2025, and this would be capable of producing nearly eight times its present biomass energy consumption. In the EU and USA large areas of surplus land are increasingly being taking out of food production, which could become available for biomass energy. The food versus fuel issue is a complex one that needs to be examined within a wider context of under-capitalisation of agriculture in poor countries, land-use problems, wars, excessive use of land for grazing and animal feed, etc. After examining all the pros and cons, Hall *et al.* come to the conclusion that 'on a global scale there is enough land available to allow biomass to make a significant impact on atmospheric $CO_2$ levels and energy production'.

In Section 1.3, the authors assess the costs and potential markets for modern bioenergy. The most important conclusions are: 1) there is considerable potential for marketing bioenergy; 2) costs of bioenergy are already competitive in certain instances; 3) as technologies become more developed, costs will decline further. Currently renewable energies (REs) are systematically put at a disadvantage when compared to conventional sources because the commercialisation of RE technologies ignores the social and environmental costs associated with fossil fuel use. In addition, conventional energy sources tend to receive large subsidies which are often ignored. The authors argue that 'what is needed is to pursue all policy options that stimulate RE and put them into a fair "playing field" to compete with fossil fuels'.

Section 1.4 provides some examples of modern biomass energy uses in industrial countries, e.g. EU (Austria, Denmark, Finland, Sweden) and USA. Two main conclusions can be derived: 1) biomass in its modern forms is increasing rapidly in many industrialised countries for a combination of environmental, socio-economic and energy reasons; 2) this increase has been largely possible thanks to political and financial support at local, regional and national levels.

The merits of $CO_2$ sequestration versus fossil fuel substitution strategies are discussed in Section 1.5. Since they were first proposed in the late 1970s, these strategies have received considerable attention. The authors, after discussing the different merits of each strategy, conclude that direct substitution of fossil fuels by biomass is the most environmentally and economically feasible option. However, Hall *et al.* quote Marland and Marland to indicate that this 'depends on the current status of the land, the productivity that can be expected, the efficiency with which the forest harvest is used to substitute fossil fuels, and the time perspective of the analysis'.

A key point for policy-makers is that trees and other forms of biomass can act as carbon sinks, but at maturity or at their optimum growth rate there must be plans to use the biomass as a source of fuel to offset fossil fuel energies or as long-lived timber products. To this end the Kyoto Protocol offers good opportunities for the Joint Implementation programme and community development mechanism (CDM) for biomass fuel projects in both industrial and developing countries.

In Section 1.6 Hall *et al.* deal with the environmental concerns of modern biomass energy carriers. The authors affirm that 'ensuring environmental sustainability will be the single most important determining factor in future biomass for energy development'. In addition to energy balances, the authors assess possible environmental impacts of energy forestry/crop plantations, management practices to ensure sustainability, site establishment, species selection, erosion control, etc., ending with an excellent analysis of 'good practice guidelines'. The key message is that site and crop selection must be managed sustainably, and people must be involved from an early stage. In the end there is no single best way to use biomass for energy, and environmental acceptability will depend on sensitive and well-informed approaches to new developments in each location. The authors conclude that 'perhaps the single greater environmental benefit of biomass is that it can help to prevent the build up of greenhouse gases in the atmosphere'.

## 10.2 Chapter 2 (Bajay, Carvalho and Ferreira)

In this chapter, the authors present a general overview of biomass energy in Brazil, with particular reference to the sugar-cane and alcohol, pulp and paper, and charcoal-based industrial sectors. The first section is a general view of the Brazilian energy situation, which has experienced a rapid growth in the past decade. For example, in 1983–97 the contribution of RE increased by about 46 per cent and conventional energy by about 61 per cent. Despite the fact that the share of RE decreased as a percentage of the total primary energy consumption, total consumption increased by 10 per cent in absolute terms. The main fall in the use of biomass energy is attributed to firewood consumption, which decreased from 15 per cent of final energy in 1983 to 6 per cent in 1997.

The main industrial sectors that use substantial amount of biomass energy are those described above. The charcoal-based pig-iron and steel sector has seen a decline in the use of biomass energy (charcoal) from a peak of 45 M m³ in 1989 to 29 M m³ in 1997; the production of ethanol fuel (both anhydrous and hydrated) has significantly increased over the same period, while the pulp and paper sector is also modernising and increasing biomass use.

Currently Brazil is the world's largest producer of sugar-cane with over 300 M tonnes in the 1997–98 harvest, or 15 per cent of the total world production. The authors describe various aspects of the National Alcohol Programme (PNA), policy implications and possible options. The programme is at a crossroads and is facing serious difficulties, e.g. the share of neat ethanol-fuelled vehicles has declined from 96 per cent of total sales in 1985 to less than 1 per cent in 1997. The authors argue that the compulsory addition of anhydrous ethanol to gasoline up to 24 per cent and the ageing neat alcohol vehicles fleet, estimated in 1998 at about 3.8 million, are the only reasons the PNA continues to operate.

After describing the earlier years of the programme, the authors discuss the new policies under consideration by the government to revitalise the PNA. Some of these policies include the compulsory addition of ethanol to diesel engines in lorries and buses, and converting the government vehicle fleet to use neat ethanol. One of the problems, the authors argue, is lack of consensus among the different interested parties about what kind of programme is economically and politically feasible.

The authors recognise that there are various justifications for the existence of such a programme beyond economic considerations, e.g. environmental, social, technological and strategic, but say that important changes are needed to ensure the greater competitiveness of ethanol fuel with gasoline, e.g. increasing productivity, eliminating inefficiency, and significantly cutting production costs. The authors propose a 'Third Phase' of the PNA that should include the following measures: 1) phase out the old practice of keeping artificially low fuel prices; 2) create a price mechanism that rewards the most efficient alcohol producers and penalises the most inefficient ones, even if that means forcing them out the alcohol business; 3) provide tax incentives to modernise existing distilleries and to build new and more efficient ones; and 4) negotiate with all parties concerned for a balanced share of the expected benefits of the programme.

In Section 2.4 the authors analyse the charcoal-based pig-iron and steel industry and its main environmental problems. The main conclusion is that for this industry to have a future it is necessary to use highly productive forest plantations and large-scale and efficient furnaces to produce cheap and good quality charcoal with minimum environmental costs. The authors point out the lessons being learned from the pulp and paper sector, which derives considerable benefits from owning its own plantations.

A discussion follows of the pulp and paper industry, which is one of the most successful in Brazil. This sector owns about 1.4 M ha of plantations and directly employs about 45,000 people. This success is due to a combination of factors, including: 1) high forest productivity, mainly due to favourable natural conditions; 2) availability of cheap land and labour; and 3) tax incentives introduced in the early 1960s.

This industry has gone through a process of modernisation, including greater energy efficiency, with significant results if one excludes the small mills which are often run inefficiently from an energy point of view. The authors conclude that to increase the use of biomass energy in this industry, efforts and new investment should be channelled to improve forest management and energy efficiency in the mills, together with the modernisation of old co-generation units. Installations of new and efficient units capable of burning firewood/forestry residues and black liquor should be encouraged.

New opportunities for electricity generation in the alcohol distilleries and the pulp and paper industry are discussed in Section 2.5. Government policy towards the privatisation of energy supply in general, and electricity generation in particular, from the mid-1990s, will have profound implications, with major challenges and opportunities ahead. This new situation is explained by the authors, together with the institutional implications and the creation of new bodies to deal with the post-privatisation era, e.g. the creation of ANEEL. Their conclusion is that privatisation will open up new opportunities in the energy sector, including greater freedom for building new plants for energy self-sufficient units and IPPs. This is particularly the case of the sugar-cane and ethanol and pulp and paper industries, where a better use of residues provides a good opportunity for energy self-sufficient units and sales of surplus power to the public grid. The authors state that 'the new rules of the game tend to encourage a larger use of such industrial residues for process steam and power generation in co-generation units'.

Regulatory barriers continue to be obstacles, especially for energy self-sufficient and IPP units, and the attempts that are being made to overcome them are discussed, together with the main incentives to co-generation and power generation from RE.

In Section 2.6 there is a discussion of the environmental implications of biomass energy and the potential for encouraging its further expansion in Brazil, given the favourable natural conditions for growing biomass. Biomass energy potentially offers considerable energy and environmental benefits for Brazil if proper policies are put in place to ensure that it is produced and used in an environmentally sustainable manner, especially in the three main sectors under discussion.

Sugar-cane residues are among the most important and promising sources of biomass energy around the world, and they have been widely studied in Brazil. For example, it has been estimated by Macedo (1998: chap. 3) that the reduction of $CO_2$ emissions arising from the direct use of ethanol fuel in 1996 alone was over 9 Mt carbon. The sugar-cane industry fixed 12.74 M tonnes of carbon in 1998. There is a growing debate about the future of the PNA and the currently under-utilised energy potential of the sugar-cane industry, which could generate a significant amount of energy and income for this sector, making it more competitive, and at the same time significantly contributing to mitigating greenhouse gases.

A similar argument applies to charcoal production, where for each tonne of charcoal consumed 0.4 to 1.2 tonnes of $CO_2$ are fixed, compared to 1.86 tonnes released when coke is used instead. However, the authors associate the production of charcoal more often than not with poor working conditions and environmental degradation, particularly when native forests are involved. In the case of pulp and paper, the authors describe the efforts of the industry to innovate and become more environmentally friendly and conclude that the main long-term strategic goal of this industry must be a radical reduction in pollution levels. This could be achieved through the recycling of all by-products, in a closed system, without any effluent and through the sustainable management of plantations.

There is discussion of some of the policy tools required to create an adequate regulatory environment body through direct and indirect tools and market forces which can benefit RE in general and biomass energy in particular. One of the main difficulties is the Government's inability to enforce environmental legislation, albeit well intended. For example, despite the existence of tough environmental legislation against charcoal production from native forests, many illegal practices still continue. With better law enforcement, such practices could be prevented and, at the same time, the taxes avoided on such activities could be collected, making charcoal from plantations more competitive. Another example is the coke-based pig-iron plants that still require considerable investment to fully meet the environmental requirements of Brazilian legislation. The authors end by discussing various options that could further the implementation of biomass energy schemes, e.g. environmental certificates and CDM.

## 10.3  Chapter 3 (Rothman and Furtado)

Technology Assessment (TA) was first institutionalised by the USA's Office of Technology Assessment (OTA) in 1972, and was the culmination of political struggle and various trends in technology policy studies. TA has variously been defined as the 'purposeful, timely and alternative search for unanticipated consequences of an innovation derived from applied science or empirical development, identifying affected parties, evaluating the social, environmental and cultural impacts, . . . revealing constructive opportunities, with the intent of managing more effectively to achieve societal goals' (Medford 1973). It has also been said to be 'neutral and objective-seeking, to enrich the information for management decisions' (Hetman 1973).

In this chapter the authors argue that we are entering the new millennium with a big question mark against our current unsustainable technological paradigms. They pose the question of how we change to sustainable technologies and sustainable economies able to serve our current wants while ensuring an ecological basis for the future. The authors point out that we already possess knowledge of energy technologies that are more benign, e.g. bioenergy and other renewables. They suggest that these may form the basis for an alternative energy paradigm, in which TA could have an important role to play by helping to create a sounder policy decision-making process.

The authors explain well the pros and cons of TA and some of the problems posed by the advance of science and technology in the twentieth century. The OTA expressed the TA objective well, when it stated, in the TA Establishment Act, it was that '. . . to the fullest extent possible, the consequence of technological applications be anticipated, understood and considered in determination of public policy on existing and emerging national problems'.

Rothman and Furtado discuss some of the underlying assumptions of TA, e.g. its holistic approach to technological programmes, and the problems caused by technological change. Despite the democratic nature of TA (or simply because of it), during its early years it was characterised by methodological debates, which are well described here. The authors go on to try to apply TA to bioenergy by examining various case studies related to biomass energy technology.

In the first case study, the authors examine alternative fuels for transportation to illustrate the role of TA, and the main advantages and disadvantages of such an approach. It is admitted that in the light of many uncertainties surrounding TA, projections of the costs and benefits rely on a series of assumptions about technology success, capital charges, feedstock costs, vehicle efficiency, etc. Changing assumptions to other plausible values could alter cost–benefit results quite dramatically.

In the second case study, the authors deal with bioenergy crop production to show how environmental impact analysis was incorporated into the OTA study in 1993 with the aim of generating a debate about the potential benefits and impacts of energy crop programmes in the USA, particularly with regard to short rotation woody crops (SRWC).

Case study three examines eucalyptus plantations in Brazil and the social and economic impacts associated with the development of such large plantations for energy and for pulp and paper. Brazil has about 6 Mha of plantations, mostly originating during the 1960s when the Brazilian government provided generous fiscal incentives to afforestation. These large plantations have caused considerable environmental concerns which are not always necessarily justified (see Rosillo-Calle *et al.* 1996: chap. 8). The authors argue that the continued scientific and political struggle over such potential impacts illustrates the difficulties in coming up with any final assessment about the value of technological programmes. This is particularly so where scientific knowledge is insufficiently developed, and political and economic power is extremely unequally distributed. The authors cite two reports (Couto and Betters 1996 and Carrere and Lohmann 1996) to make their case, while pointing out that in all TA we need to examine the reasons for the existence of the technology, e.g. monoculture eucalyptus plantations, and try to identify the actors and their potential roles. The authors of the reports cited above, for example, both sought to marshal the scientific literature to support their case and yet they come to very different conclusions. The differing interpretations, Rothman and Furtado argue, seem to illustrate a fundamental problem of TA, and that we have to live with and accept such seeming contradictions. The experts do not always agree, and the school of Constructive TA therefore argues that part of the function of the answers should be to bring such expert conflict to the attention of interested parties rather than to brush it aside.

In Section 4.4 the authors explore some of Brazil's biomass policy experiences, and remind us that biomass technology is a *set* of technologies rather than a single one, forming a technological system in which they are linked together as interrelated minor and major innovations. Policy-makers, therefore, need to consider this systematic complexity of biomass in their policies. The authors state that the Brazilian experience in promoting biomass energy is very illuminating. It is dependent upon a specific set of circumstances, but for most of the time policies are not carried out consistently within the technological system, e.g. by fostering at the right time complementary technologies or introducing the necessary institutional changes.

Two specific examples are discussed to prove their point, e.g. FINEP and the PROALCOOL or PNA. FINEP created an energy department to promote alternative renewable energies (AREs), which was initially quite successful in carrying out research to create a technological base for future developments. However, at a later stage, the agency's emphasis on industrial projects linked to commercial returns affected its ability to promote industrial innovation because the technological base was still too weak, combined with poor articulation among different ARE technologies and the market. According to the authors, a linear approach to technology policy and the distribution of a limited volume of resources in a heterogeneous group of technologies, without real demand from industry and market, compounded by low oil prices and well-established conventional energy industries (e.g. oil and hydro), frustrated FINEP's attempts to promote ARE.

PROALCOOL, on the contrary, is a good example of successful government policy to promote an alternative liquid bio-ethanol fuel. This is because the programme was large enough to embrace many sectors (e.g. financing, creation of a captive ethanol fuel market, guaranteed ethanol prices, etc.), and was combined with the influence of the sugar-cane lobby and institutional historical experience with ethanol fuel, and with government pressure to reduce dependence on imported oil. The fall in oil prices from the mid-1980s, which also forced a cut in ethanol prices, led to a sharp increase in cane yields and to substantial improvements in the industrial sector, as it responded directly in order to survive.

In TA, the authors argue, it is necessary to consider the social and economic context of technological generation and diffusion. The success or failure of biomass energy compared to fossil fuels is connected with the strength and weakness of technological opportunities and to social support for each one of these sources of energy. Biomass technologies will only succeed if they are able, during the development phase, to acquire enough social support to be technically feasible, economically viable and have the adapted institutions. This success relies on effective public policy to encourage the constitution of a convergent network of actors.

TA can be a good instrument for improving the level of technical decision-making and thus has an important part to play. Despite the fact that TA can be partisan and unrepresentative, the authors argue that 'contemporary CTA thinking does contain elements of a critical emancipatory approach to the problems of identifying solutions in society's management of technical change, [e.g.] the multi-disciplinary holistic approach . . . , the search for affected parties, alternative technologies, and environmental and social impacts.'

The authors make some comparisons between TA and CTA (Constructive TA), which has emerged in the EU, principally in The Netherlands and Denmark, which represents a revision of the original TA concepts, reflecting certain European trends. CTA places greater emphasis on the initial design steps of technological innovation, so as to increase the opportunities of identifying impacts prior to a technology becoming entrenched. In this sense CTA offers greater openness than traditional TA concepts.

There is a need to ensure that the question of vested interest is faced and TA can be a good instrument with which to do so. The authors argue that if, for example, Brazil at national and state levels is honestly committed to goals of sustainable development, full employment, social equity and education, it is possible to see TA playing a key role in improving policy decisions and democratic participation. This is a utopian vision, given current asymmetries of economic and political power; in practice TA could be quite partisan and unrepresentative.

## 10.4  Chapter 4 (Bauen)

Evaluation of externalities has received considerable attention over the past few years, particularly in the USA and the EU. In Brazil, however, the idea of internalising externalities in the energy sector is quite novel and hardly any serious work has been carried out to date. In this chapter the author presents an excellent overview of the latest assessment evaluation of the energy externalities evaluation, both environmental and non-environmental, and its role in sustainable development, with particular emphasis on the industrial uses of biomass energy in Brazil. The most important neoclassical concepts surrounding externalities issues, together with the main steps towards the internalisation of externalities, are reviewed.

A discussion follows on the externalities of energy. Externalities occur at all stages of a fuel cycle, but energy externalities can be reduced by improving fuel cycles, switching between fuel cycles, the more efficient end-use of energy, and reductions in energy consumption. The author states that 'the ultimate goal of externalities valuation is to achieve an economically efficient allocation of resources through the integration of externalities in energy prices'. Despite the difficulties and uncertainties surrounding 'externality evaluation' – the monetary evaluation of externalities, and in particular environmental externalities – it now seems to be the dominant paradigm in the comparative environmental appraisal of energy options.

The author goes on to discuss various approaches to externalities assessment and the various evaluation techniques. The ExternE methodology – the most exhaustive study to date on external costs of energy – is discussed in some detail, followed by a review of the main approaches to the externalities of energy.

Externalities are difficult to measure in monetary terms, almost impossible in some instances, because of our value judgements, a lack of clear scientific knowledge in the valuation process, etc., and this is recognised by all studies on externalities. There are considerable differences in the values obtained by different studies, stemming mainly from methodological differences, differences in impacts, assumptions, etc., which the author explains well.

Undoubtedly methodological difficulties with externalities evaluation have hindered considerably their application in policy decision-making. However, as our scientific understanding of externalities increases, they are bound to play an increasing role in determining energy choices.

Bauen then discusses the externalities specifically related to biomass energy fuel cycles compared to those of fossil fuels, together with other recent studies in this area. Of particular relevance is the BioCosts study in Europe, which has identified four main categories of priority impact: 1) impacts of atmospheric emissions on human health; 2) impacts of greenhouse gases on climate change; 3) impacts of the fuel production stage on the local biosphere; and 4) impacts on rural amenity. These impacts, and the potential relevance to Brazil, are then considered by the author.

The variety of possible biomass energy sources, variations in practices used for the procurement of feedstock, and the site-specificity of biomass means that careful examination of any additional potential effects on soil, water, biodiversity, rural amenity, etc. is required.

The BioCosts study has shown that, in most cases, the environmental externalities (excluding the impacts of greenhouse emissions) represent a small fraction of the private costs and are lower than three of the reference fossil fuel cycles at the original sites.

Biomass energy can play a major role in reducing greenhouse gases compared to fossil fuels, although the external costs of these emissions and the benefits of their avoidance compared to the reference fossil fuel cycles remain uncertain and difficult to estimate.

The main problem is the uncertainties and value judgements surrounding the valuation of climate change damages, which make it extremely difficult to apply a methodology based on conventional economic theory. For these reasons it seems that the best way to deal with externalities and climate change may be by applying principles of ecological economics (advocating strong sustainability), rather than those of classical economics.

Brazil is a world leader in industrial uses of biomass energy, and trying to internalise the energy externalities could have important implications for the energy sector. The author makes a serious attempt to evaluate the energy externalities in ethanol production and electricity generation in the sugar-cane sector. Sugar-cane is a major crop in Brazil, with many social, economic and environmental ramifications. Certainly, externalities evaluation of sugar-cane-based ethanol fuel and electricity generation from sugar-cane bagasse could have a significant impact in this sector, e.g. when comparing costs of ethanol versus gasoline. External costs have been ignored, partly because the concept of 'externalities' is little understood by those involved in the policy decision-making process.

One of the greatest problems in Brazil is the lack of any serious study on externalities evaluation. Despite these difficulties, the author identified significant environmental benefits from the use of ethanol fuel in vehicles, when compared with fossil fuels. However, when all the environmental effects of ethanol fuel are considered, it is more difficult to assess because all stages of sugar-cane production have to be included. This can also cause negative effects, e.g. water, soil impacts, burning of pre-harvest sugar-cane, which the author acknowledges. For example, the impact of the pre-harvest burning is most likely to lead to significant externalities related to human health and urban amenity, compared to end-use of ethanol which it is generally accepted has helped to abate pollution in Brazil, particularly in urban areas. However, the internalisation of externalities of sugar-cane ethanol may not necessarily imply a rise in the internal costs of sugar-cane production, and it even could lead to a win–win situation, when environmental and economic gains occur simultaneously. Little is known about the use of stillage and its potential external impacts, and this is currently a major barrier for externalities evaluation.

The priority impacts and benefits of electricity from bagasse led to the conclusion that to reduce the externalities per unit of energy generated, more efficient energy systems (for which there is considerable scope) should be introduced. Bauen provides an indication of the magnitude of the externalities of electricity from sugar-cane based on the externality values specific to the EU experience, simply because there are no pollutant and specific externality values for Brazil.

One conclusion is that the internalisation of externalities may significantly reduce any private cost based on competitive advantage of power generation from more conventional technologies. Despite the many uncertainties of externalities evaluation, it seems that the internalisation of external costs can be a significant factor in the ranking of energy options based on social costs. Bauen states that ' biomass energy appears generally to possess lower external costs compared to conventional energy, in particular if the externalities of climate change are considered'.

There is a considerable need for research on the externalities of energy and sustainable energy options in Brazil. Evidence indicates that the internalisation of the external costs of biomass energy will put biomass on a more equal footing with fossil fuels, and make this energy source more competitive with them.

## 10.5  Chapter 5 (Cortez and Braunbeck)

In this chapter Cortez and Braunbeck present an excellent review of sugar-cane practices and of sugar-cane residues in Brazil. The chapter falls into four main sections. In Section 5.1 the authors explain the historical ups and downs of sugar-cane in Brazil and its importance to the national economy, agro-technical developments and the use of residues.

Brazil is currently the world's largest sugar-cane producer, an industry that represents about US$5 billion annually. Cultural practices, R&D policy, introduction of new sugar-cane varieties, breeding programmes, technical changes, etc., are discussed. The authors conclude that, despite the considerable technical advances of the past forty or so years, considerable efforts are still needed to reduce production costs. For example, harvesting methods remain largely unchanged and systematic burning of sugar-cane fields still continues to facilitate traditional manual harvesting methods. The authors also assess various techniques for increasing the productivity of sugar-cane.

The main residues of sugar-cane from sugar and ethanol production, e.g. bagasse, trash (or barbojo) and stillage, which represent the main challenges of the industry, are discussed in detail. In Section 5.2 the authors analyse mechanical harvesting and identify three major implications: 1) employment and the benefits of lower costs; 2) elimination of sugar-cane burning with its significant environmental advantages; and 3) technical difficulties in developing a satisfactory whole sugar-cane harvester. Currently there is no harvesting system that satisfies the handling of the wide range of field conditions anywhere in the world.

The authors examine various harvesting systems around the world and conclude that none meets the technical specifications of the Brazilian conditions, and that specific technical adaptations are needed to meet topographic conditions and agricultural practices in Brazil. Various constraints to mechanisation are discussed, e.g shortage of skilled labour, lack of capital, unsuitable feeding systems for whole cane at the mills, etc. They argue that many of the present difficulties can be overcome relatively easy, concluding that a successful implementation of mechanical harvesting in Brazil needs to address a series of technical issues based on topographic, agronomic and sugar-cane processing conditions typical of that country.

Section 5.3 concentrates on the use of bagasse as fuel, e.g. co-generation of electricity for industrial uses, and non-energy alternatives. It is clear that if bagasse and trash are used efficiently, they can make a substantial contribution to the economics of the sugar-cane and ethanol fuel industry, with additional social and environmental benefits. Despite this potential and increased used of these residues in co-generation, bagasse is still used very inefficiently in most sugar-cane mills, and trash is not used at all. Modern electricity generation systems, e.g. CEST and BIG-GT, could generate substantial surpluses of electricity which could be sold to the national grid. However, there are important institutional, technical, and political obstacles that the authors address well.

The sugar-cane sector is conservative by nature and it has been shy to explore new alternatives to surplus bagasse. Thus bagasse is hardly used outside the sugar and alcohol industry, e.g. some use in the orange juice industry, in the production of pellets and briquettes and as animal feed after hydrolysis.

The final section deals with stillage (or vinasse). Stillage has been a major pre-occupation, especially since the creation of the PROALCOOL, because of the large quantities produced and the potential negative environmental impacts if not properly treated. Cortez and Braumbech provide a detailed analysis of the production and use of stillage in Brazil and, in particular, its use as fertiliser, for biogas production, animal feeds, fungi production, and direct disposal by incineration, together with the economic and environmental aspects. Only two approaches are currently being used in Brazil, the use as fertiliser and biodigestion to produce biogas for some mill vehicles.

The use of vinasse as fertiliser is an old practice in Brazil, which has witnessed significant technical improvements in the past twenty-five years. Evidence indicates that the utilisation of stillage as a fertiliser for such ends increases sugar-cane productivity between 5 and 10 per cent. Biodigestion also looked quite promising, but in practice failed to live up to expectations and currently only two large distilleries are committed to biogas production for use as fuel, and even this will not be for long. This failure is due to a combination of economic and technical factors, e.g. low prices of diesel and difficulties in obtaining spare parts for the modified diesel engines used by the mill's vehicles. Despite these drawbacks the authors indicate that there is considerable potential for increasing utilisation of these by-products.

## 10.6 Chapter 6 (Macedo and Cortez)

As we have already seen, the sugar-cane and ethanol industry is of considerable socio-economic and energy importance for Brazil. Over 300 million tonnes of cane were crushed in 1998, representing about 14 per cent and 55 per cent respectively of the world's total sugar and ethanol production. Sugar and ethanol production is intertwined in Brazil, partly determined by the sugar prices in the international market. Ethanol is not only a major source of transportation fuel, but it also acts as a kind of level in controlling sugar prices world-wide.

Ethanol production has remained more or less stable for the past fifteen years, e.g. from 9.2 billion litres in 1984–85 to about 14.5 billion litres in the 1997–98 harvest, with some fluctuations. However, there have been important changes with regard to the use of ethanol as fuel. For example, in its peak days of 1985, 96 per cent of all new passenger cars ran on neat ethanol, but since then there has been a sharp decline to reach almost zero in 1997, and the fleet now comprises only about 3.8 million cars. The reasons have been clearly stated in this book. Currently only hydrated ethanol fuel is responsible for maintaining PROALCOOL and the industry is desperately trying to find other long-term alternatives. What this clearly illustrates is the strong relationship between the sugar industry and the energy sector in Brazil.

The authors have comprehensively explained the major processes involved in the production of sugar and ethanol, the advances of the past fifteen or so years, together with the energy and environmental implications.

The first section gives a good summary overview of the processes regarding raw material (cane), transport, weighing, stocking, juice extraction, cane preparation, cane feeding and milling; inhibition, energy generation in the mill, juice treatment and weighing and various other juice processes; heating, filtration, evaporation, etc. The authors explain the process of sugar crystallisation, centrifugation and drying; and the process of ethanol production from juice treatment to fermentation, distillation, storage, etc.

In Section 6.3 the authors examine the main technological advances of the past fifteen years, e.g. milling, fermentation, automation, etc., and the use of energy in the industry. The final section assesses the energy and environmental implications of the sugar-cane and ethanol industry. PROALCOOL has important national implications and here the authors pay particular attention to the environmental advantages of adding ethanol to gasoline.

Their main conclusion is that sugar-cane technology in Brazil does not differ very much from that in any other major sugar-cane producing country, except for ethanol production, where Brazil is a world leader. Another important difference is energy self-sufficiency – Brazilian mills are mostly self-sufficient and are increasingly able to sell surplus energy to the grid. Further improvements are expected in the near future, e.g. whole cane harvesting and better use of bagasse, which could result in greater energy production and savings.

However, the major challenge facing the industry is in the agricultural sector, which still represents 60 per cent of the total costs. Here, for example, better management practices, introduction of new and improved varieties, mechanisation, better and new uses of bagasse, etc., must all combine to increase productivity and cut costs.

From an environmental point of view the most important change envisaged by the authors is dry sugar-cane cleaning, which could save large amounts of water. The final point made by the authors is the need for long-term and clear government policy towards this industry.

## 10.7 Chapter 7 (Bajay, Berni and de Lima)

The pulp and paper industry is a major industrial and economic activity in Brazil. It is among the world's largest, with over 100,000 jobs and 1.4 Mha of plantations. In this chapter the authors present a detailed analysis of all the industry's main activities, ranging from forestry activities, pulp and paper making, to paper recycling.

Large-scale forest plantations have caused serious environmental concerns and thus ensuring sustainability is an important objective of this industry. Economic sustainability, the driving concern of the pulp and paper industry, is a complex issue which depends on many varying factors, ranging from establishment costs and tax incentives to subsidies. For years, the Brazilian Association of Pulp and Paper Producers (BRACELPA) has been complaining that contrary to many other pulp and paper producing countries, Brazilian companies do not receive any fiscal incentives, which puts them at a disadvantage to their international competitors. All these issues are well examined by the authors.

Section 7.3 contains a detailed explanation of the processes and economics of pulp and paper making. The authors also thoroughly assess energy intensity and energy conservation in this industry. They explain the various ways of improving energy efficiency in the pulp and paper industry world-wide. In the case of Brazil, the conclusion is that there is considerable scope for energy savings with only minor improvements. Despite the modernisation drive of the Brazilian pulp and paper industry in recent years, the corresponding energy savings have been relatively small because the main driving force was an increase in production.

The electricity conservation potential in the pulp and paper industry is very large, e.g. the authors identified potential ranging from 33 per cent in pulp making to 47 per cent for various other types of paper. The authors state that ' the Brazilian pulp and paper industry decision-makers should realise that increasing production is not the only way to cut costs and increase revenue and, eventually, market shares, and that well planned and managed energy conservation programmes can lead to substantial cost reductions . . . and environmental gains'.

Paper recycling is clearly a good thing, resulting in better use of natural resources and helping the environment. Paper recycling is an activity now encouraged in many parts of the globe. In Brazil it is also an old activity but because of economics not environmental interests. Despite the fact that about 36 per cent of the paper there is recycled (about 2.2 Mt in 1997), this figure is far below the technical capacity, mainly because it is determined by the price paid for such paper.

Forestry wastes have not been used as fuel in the co-generation plants of the Brazilian pulp and paper mills, despite their high potential to increase the mills' electricity self-generation capacity. The increasing mechanisation of forestry activities, allied with better prospects for surplus electricity sales to the public grid, will very likely change this picture. Assuming an efficiency of 40 per cent for a BIG-GT plant with fluidised bed gasifiers, a national average productivity of approximately 30 m³/ha of eucalyptus, a specific mass of the wood feedstock of 390 kg/m³ and a conversion factor of 1.2 kWh/kg of forest waste, 24 per cent of the mills electricity consumption could be met by burning only forest residues. If black liquor is added, using the same sort of equipment (BIG-GT system), 50–100 per cent self-sufficiency on electricity consumption can be achieved, depending on the underlying assumptions regarding black liquor and forestry residues availability.

## 10.8 Chapter 8 (Rosillo-Calle and Bezzon)

This chapter deals with industrial uses of charcoal in Brazil. Charcoal-making is a growing activity in many developing nations. For example, the International Energy Agency (IEA) has estimated that charcoal production and consumption in developing countries is expected to increase from 22.3 Mtoe in 1995 to over 58 Mtoe in 2020. However, these are very conservative figures since in most cases global charcoal production is an integral part of the informal economy and hardly finds it way into official statistics and hence estimates can vary quite considerably. Brazil is the world's largest producer and consumer of industrial charcoal.

Charcoal-making is among man's oldest activities, lost in prehistory, and in Brazil it goes back to the late sixteenth century when it was first reported in the state of Minas Gerais. The historical events can be broadly divided into five main periods, which are well described by the authors. But it was not until the late 1960s that one could speak of an industry coming of age.

For centuries native forests were the sole source of charcoal. This has led to many misconceptions about the role that charcoal-making activities played in deforestation and environmental degradation. The charcoal industry suffers from a poor image and is associated with social and economic backwardness. These complex issues are discussed at length by the authors.

Brazil is unique in the world in the sense that the country turned to charcoal as a thermal and reductor agent at a time when other countries were turning to coal, coke, etc. This was largely because of the availability of natural forests near large iron-ore deposits. The industrial uses of charcoal are described by Rosillo-Calle and Bezzon, with particular reference to pig-iron and steel plants. The charcoal-based industry consumed about 45 M m$^3$ of charcoal in 1989 compared to 26 M m$^3$ in 1996.

A major difficulty facing charcoal consumers until recently was high transportation costs due to the long transportation distances as native forest are progressively cut down around the main consuming centres, together with the negative environmental impacts which, rightly or wrongly, are associated with this type of activity. As the authors explain, however, the main source of deforestation has been the expansion of agricultural and pasture lands rather than charcoal production. This is a complex and often misunderstood issue.

Greater environmental concerns are having considerable impact on the charcoal-based industrial sector. Tougher environmental laws are forcing a radical rethink in the way charcoal is produced from native forests (trees and forestry residues) in favour of plantations. For example, in 1987 over 80 per cent of charcoal was produced from native forests compared to about 30 per cent in 1996. In comparison, charcoal from plantations increased from 19 per cent to 70 per cent over the same period. Thus such environmental pressures are creating a rapid shift towards plantations. There are, of course, other advantages such as lower transportation costs, better quality of charcoal, increased professionalisation, etc. The chapter describes various afforestation programmes being implemented with support from the major charcoal consumers with the aim of achieving energy self-sufficiency.

In Section 8.6 the authors deal with the technical aspects of charcoal production in Brazil, one of the few countries in the world that still carries out R&D in some meaningful way. What is remarkable is that the technology for making charcoal has changed so little with time. Brazil is among the world's most efficient producers of charcoal, with efficiencies of around 35 per cent in the most professionally run production sites.

The most common processes still used in Brazil are the internal heating systems. The three more common kilns are the *rabo quente* (hot tail), *superfície* (beehive), and more recently the rectangular kiln, which is, perhaps, the greatest innovation in this area. The process of technical change is being facilitated by the rapid increase of charcoal production from plantations, including better use of by-products.

The authors also discuss the main implications of the use of charcoal versus coke in the pig-iron and steel industry. Despite the attractiveness of coke, charcoal presents particular advantages over coke, e.g. thermal, environmental, social and, increasingly, economic. The environmental advantages are briefly discussed in Section 8.8. The industry has been identified with deforestation, backwardness and socio-economic exploitation. These issues must be analysed within the wider socio-economic and political context of the Brazilian society. There are positive and negative costs and benefits of charcoal production, but, as a whole, the benefits far outweigh the negative effects if charcoal is produced sustainably.

The most important socio-economic and environmental changes are: 1) the gradual phasing out of charcoal production from native forests in favour of plantations, with its consequent environmental benefits; 2) a significant reduction in employment, that is directly affecting many small farmers and landless labourers who depend on charcoal as their main or major source of income. Although charcoal produced more professionally from plantations is improving the living standards of the labour force, this is translating into less employment elsewhere.

The authors' concluding remarks are that the charcoal-based industrial sector has a future but charcoal production methods, together with that of pig-iron, will have to be improved significantly. Large-scale industrial charcoal production from plantations rather than from native forests presents many challenges and opportunities. The opportunity is to produce charcoal in a more environmentally friendly and socially acceptable manner. The challenge will be to modernise and to cut costs to make it more competitive with coke, and also to maintain it as the important source of employment that it is in rural areas. It is also a challenge to combine modernisation with socio-economic and environmental responsibility.

## 10.9 Chapter 9 (Walter *et al.*)

This chapter is a detailed overview of modern technologies for converting biomass to modern energy carriers. Broadly speaking, the chapter considers four main technological categories: 1) gasification for heat and power; 2) generation of electricity; 3) ethanol from cellulosic material; and 4) pyrolysis. Each section has been written by experts, who bring together a considerable amount of information about these technologies.

Section 9.2 assesses all major types of gasifier, and a describes process applications. Both large- and small-scale gasifiers are described, together with their manufacturers. A large number of gasification systems have been developed since World War II with varying degrees of success.

Methane and hydrogen from biomass have received increasing attention in the past few years because of their considerable potential, thanks to new technological developments. Methanol (currently produced from both natural gas and coal) is particularly attractive because advances in gasifier development allow production of a synthesis gas suitable for methanol production. Costs have been a major barrier for producing biomass-derived methanol, but new developments are reducing such costs significantly and it can currently compete favourably with coal-based methanol.

Hydrogen is another area of considerable interest. Petroleum-based hydrogen production is a commercially mature and cost-competitive technology. However, hydrogen production costs from biogas are becoming quite attractive and may be able to compete with hydrogen obtained from natural gas in the near future.

Various factors have combined to increase interest in the generation of electricity from biomass, e.g. environmental considerations, perceived increased demand for electricity and changes in the electricity sector, availability of residues particularly in the sugar-cane and pulp and paper industries, etc. These, together with new technological developments, have increased significantly the scope for co-generation from biomass. For example, co-firing of biomass with coal is becoming particularly attractive in the EU and the USA. These opportunities and challenges are all addressed by the authors. There is already considerable experience in co-generation in many sugar-cane and pulp and paper industries around the world.

The production of ethanol from cellulosic material is an old dream, since it is the most abundant source material. There have been two major impediments: 1) technological difficulties; and 2) high production costs. However, recent and prospective developments in genetic engineering are radically changing the possibilities of obtaining large-scale production of ethanol from cellulose, e.g. in the USA costs have been cut by about 50 per cent in recent years to about US$0.26/litre. This theme is taken up by Moreira (Section 9.4), who explains in detail the various aspects of producing ethanol from cellulose, including feedstock production, land use and related environmental impacts, conversion technology, acid and enzymatic processes, biological conversion, product recovery, energy balance, etc. The author's final note is dedicated to near-term prospects for commercial applications for ethanol production from cellulosic material.

Bezzon and Rocha (Section 9.5) present an excellent overview of pyrolysis technology around the world. Their analysis includes fast pyrolysis and its potential for obtaining bio-oils, slow pyrolysis for the production of charcoal, economic aspects; and other technologies to produce charcoal and other forms of solid fuels.

What comes out of these reviews is that these new technological developments are opening up many new opportunities for bio-energy which were regarded only a few years ago as long-term prospects. These could be summarised as follows: 1) advanced steam cycle technology with co-generation; 2) co-firing with fossil fuels, particularly coal; 3) integrated gasification/advanced technology; 4) bio-crude-fired combustion turbine technology; 5) production of methanol and hydrogen from biomass, etc. The more established technologies, e.g. combined heat and power, and ethanol, have seen important advances and cost reductions in a number of countries. The large-scale production of ethanol from cellulosic material may become a reality in the medium term. Biogas production has also seen a considerable reduction in cost and is currently turning waste disposal problems into energy and economic opportunities.

## 10.10 Conclusion

We have tried throughout this work to present the biomass situation realistically. We have been able to show that it should not be neglected by those planning our energy futures. Indeed, most energy scenarios recognise the important role of biomass energy in the future energy supply matrix. Brazil's experiences are important because Brazil has, for reasons which should now be apparent, given biomass energy a significant role in the national energy programme. Elsewhere biomass has generally failed to obtain such a position in national planning and practice. Often this failure, along with associated low visibility and low prestige, is due to ignorance: ignorance of its potential and ignorance of the externalities of fossil fuel consumption. There is also the matter of vested interest; Brazil was one of the few places where there existed a lobby for biomass energy (from the sugar-cane industry) able to compete with the oil lobby.

Biomass energy matters for our future, not just because it is renewable and the fossil fuel alternatives are not. Yes, fossil fuels reserves will run down to low levels one day, but not before their emissions have altered our global climate to society's and nature's disadvantage. The key value of biomass, therefore, lies in its unique role as a carbon sink and as a fuel. Indeed, as Hall *et al.* put it 'perhaps the single greater environmental benefit of biomass is that it can help to prevent the build up of greenhouse gases in the atmosphere'. The lessons from Brazil's experiences which we draw for the future are: first, that biomass energy, if it is to fulfil its potential, must modernise its technical systems and become increasingly efficient so that it is seen to be more competitive with fossil fuels, and it must do so in an environmentally sustainable manner. Second, bearing in mind that that technology is embedded in our social relations, energy policy needs to become more democratic and transparent; and decisions need to be made on the best available evidence of energy costs that include social and environmental costs. At the moment these externality data are inadequate to say the least. Our blindness, social denial even, to these externalities, in the light of the global warming trends, will remain our major concern for the future. Providing we make these changes biomass energy is, in our view, a major part of the necessary energy paradigm change required to ensure our transition to a sustainable energy economy in the twenty-first century.

Predicting energy trends is notoriously difficult, but by increasing the use of biomass energy future energy supplies will become more decentralised. Fossil fuels will continue to dominate well into the next century, while bio-energy, in its various forms, will also increase its market share. Oil will still be dominating the transportation system but biomass-based liquid fuels will increase their share, creating a more sustainable and cleaner environment.

The world is presently undergoing important changes which are leading to a new political and economic order that, undoubtedly, is having a considerable impact on the way we produce and use energy e.g. a more diversified, decentralised, and privatised energy market. The growing interest in bioenergy relects a combination of factors ranging from environmental, ecological, and sustainability concerns; its potential energy contribution; its versatility and global availability; substantial local benefits; technological advances; improved economic viability, etc. Biomass energy is gradually taking a more central stage in the energy supply matrix, from about 55 EJ in 1999 to as much as 145 EJ in 2025. Its final contribution, however, will depend on a combination of political, social, cultural, economic, and environmental considerations.

# INDEX